POWER OF THE PLANTS

Nutritional, Therapeutic and Commercial Applications

POWER OF THE PLANTS

Nutritional, Therapeutic and Commercial Applications

— Authors —

Prof. Arun Arya

Dr. Upma Singh

Dr. Bina Rani

Dr. A.K. Kakodia

Dr. R.K. Maheshwari

2025

Daya Publishing House®

A Division of

Astral International Pvt. Ltd.

New Delhi – 110 002

ISBN: 978-93-5919-407-3 (HB)

Publisher's Note:

Every possible effort has been made to ensure that the information contained in this book is accurate at the time of going to press, and the publisher and author cannot accept responsibility for any errors or omissions, however caused. No responsibility for loss or damage occasioned to any person acting, or refraining from action, as a result of the material in this publication can be accepted by the editor, the publisher or the author. The Publisher is not associated with any product or vendor mentioned in the book. The contents of this work are intended to further general scientific research, understanding and discussion only. Readers should consult with a specialist where appropriate.

Every effort has been made to trace the owners of copyright material used in this book, if any. The author and the publisher will be grateful for any omission brought to their notice for acknowledgement in the future editions of the book.

Published by : **Daya Publishing House**®
A Division of
Astral International Pvt. Ltd.
– ISO 9001:2015 Certified Company –
4736/23, Ground Floor, Ansari Road,
Darya Ganj, New Delhi-110 002
Ph. 011-43549197, 8130496929
e-mail: info@astralint.com
Website: www.astralint.com

Printed at : **Replika Press Pvt Ltd**

Beauty of Banyan Tree

It's midday
Simmering summer scorching the streets
Mirages galore, mankind in despair
Shining hot sun's sweltering heat
Soaking the world with shocking brightness
Sulking world just wiping its sweat
Yet I was standing under the beautiful shade
Roof of that lovely banyan groove
Sixteen grown up banyans and their aerial roots
Like all of them arm-linked and warding off sunrays
Oh! that lovely cool breeze, and what a respite in heat
Hundreds of birds were jumping in branches
Koels were hiding in the coolest corners
Yet their bleary eyes surveyed fiery sunrays
Mynas were teaching fight techniques to kids
Peacocks were rehearsing at the brim of the shade
With half-blown feathers and peahens in pursuit
And, a question got beamed with laser like precision
Like small strand of sunrays that escaped leaves
Ye dear mankind! Possessor of sixth sense!
We bear all the heat and render you shade
Many an animal and birds including mankind
Get soul gratifying relief from this scorching heat
Yet what do you give to one another and world
Banyans' question echoes in my soul.

— Dr. A.P.J. Abdul Kalam

Devashree Vala
X-A
Tejas Vidyalaya
Challenges in ecosystem restoration

Preface

Om Sham no Bhavant Wapa Oushdhayah Shiva. *Atharva Veda-2.3.6*

Water and Medicinal herbs may be good, health giving and prosperous to us, is mentioned in Atharva Veda

The Waters be to us for drink, Goddesses, for our aid and bliss:
O waters, teem with medicine to keep my body safe from harm,
So that I may long see the Sun.
The Waters bless us, all that rise in desert lands or marshy pools!
Bless us the Waters dug from earth, bless us the Waters brought in jars,
bless us the Waters of the Rains!

The five elements – the space, air, fire, water and earth – are the foundation of an interconnected web of life.

It is said**:** O Som! come to that path of Aditi, where in twelve friendly Adityas loiter with Aditi spreading your auspicious influence. Grant us your that power which can subdue even the demoniac power. O Gods! Grants us power and make us happy.

In the Vedas, Varuna is depicted as the Lord of Rita, the universal natural order. He is sovereign god, great king, law-maker and ruler of cosmos and even of the Gods. Basically, he is regarded as the Lord of water and ocean but chiefly he controls and keeps the world in order. From his throne on high he looks down upon all that happens in the world, and into the heart of man. 'By the law of Varuna heaven and earth are held apart. He made the golden swing, the sun to shine in heaven. He has made wide path for the sun. By his ordinances the moon shining brightly moves at night, and the stars placed up on high are seen at night but disappear by day. He causes the rivers to flow. As a moral governor Varuna stands far above any other deity. Thus, the concept of Varuna represents the consciousness of Vedic seers in respect to controlling and balancing the natural forces in environment.

Green leafy vegetables in Sanskrit called Shaka, and fruits have been part of several thousands year of history now. Lord Rama was consuming fruits and vegetables particularly under ground parts, while in exile. I developed interest in fruits when my grand father Swami Shankaranand introduced me with Dushehri mango and ber (Indian jujube- *Zizyphus marutiana*). In early days mango used to be small, fibrous and it was used by sucking the juice. After domestication it was improved by Mughals and Potuguese. Chiraonji (*Buchanania lanzan*) is also indigenous to India. India is also third largest producer of coconut (*Cocos nucifera*) or shri phal. Double coconut (Coco de Mer) of Sycheles has the biggest seed in the world.

The wilderness of this country was emphasized by William Roxburgh (Tim Robinson 2008) by his visit to four villages where he found that culture of sugarcane cultivation was scattered throughout the whole and for indigo, on the lighter soils. Along the mountains he recommended, immediately on the back side of the Godawari plantation of Mulberry trees, Which he wanted to extend, if it meets with the wishes of Government.

A balanced diet contains all the essential nutrients like carbohydrates, proteins, vitamins, fats, minerals, and water in the correct proportion. A balanced diet is important for our body's normal growth and development. It can be achieved by eating a variety of food items since there is no single food item with the correct amount of all the essential nutrients. The quantity of nutrients in a balanced diet always differs with age, sex, and physical activities taken by an individual. Eating a balanced diet requires a certain amount of knowledge and planning. The basic composition of a balanced diet is highly variable as it changes from country to country depending upon the availability of food items. The nutrition group of ICMR (Indian Council of Medical Research) has recommended the composition of a balanced diet for Indians. The major components of balanced diet includes cereals (like maiz, wheat, rice and jowar or other millets), pulses, roots and tubers, fruits, milk, and dairy products, fats and oils, sugar, and groundnuts. They also recommended an intake of meat, fish, and eggs for non-vegetarians.

In a bid to foster a dialogue between modern medicine and traditional systems of healing in tribal communities, the ICMR-National Institute of Research in Tribal Health, Jabalpur, Madhya Pradesh (which is the only permanent institution of the Indian Council of Medical Research dedicated to improving tribal health and also WHO Collaborating Centre for the Health of Indigenous Population in India) organized an interactive symposium in Dindori district, Madhya Pradesh, on the occasion of International Day of the World's Indigenous Peoples on 9 August 2019. In India, Madhya Pradesh has the highest tribal population (15,316,784)1. In Dindori district, the abode of Baiga tribe known as the '*Baigachak*' is situated. Baiga are one of the three particularly vulnerable tribal groups (PVTGs) in Madhya Pradesh.

Thirty one chapters included in the book are written by five of us except chapter number 15 written by Dr. S.P. Shukla and Dr. Mehari G. Micheal from Ethiopia, Africa. Chapters are divided into two sections **A.) Plants with Nutritional value** and **B.) Plants with Medicinal value.** Along with angiospermic plants uses of **Ferns** are also described. Chapters are also included on topics like Biochemical aspects of Tropical fruits, and other on Nutraceuticals.

The book **Power of the Plants** encompasses nutritional and therapeutic potential of plants. It is dedicated to the loving memory of Father of Green Revolution Bharat Ratna Dr. M.S. Swaminathan, who introduced high yielding varieties of wheat and tried to solve the need of hungry people of India. I am also thankful to my coauthors who worked tirelessly and penned the book chapters. I pay my regards to Prof. Bihari Lal ji, Late Dr. S. N. Bhargava from the University of Allahabad and Prof. J.J. Shah and Dr. S. D. Sabnis from The Maharaja Sayajirao University of Baroda, who guided me, motivated me, trained me to work for the betterment of the Society. I take this opportunity to thank Dr, S. K. Nanda, IAS, Gandhi Nagar, Shri Shraddha Nand Tyagi IFS from Meerut (U.P.), Dr. M. Sanjappa Ex- Director BSI, now settled in Bengaluru and Dr. Kamlesh Lulla from NASA, USA, who have been source of inspiration for me. All five of us are thankful to our family members as well. This volume will be immensely useful for the students and teachers as well as experts working in the field of Economic Botany, Horticulture and human health. I thank Dr.B.B. Singh and other managers of Astral International Pvt Ltd. New Delhi for beautifully printing it in a presentable form.

Prof. Arun Arya

Dr. Upma Singh

Dr. Bina Rani

Dr. A.K. Kakodia

Dr. R.K. Maheshwari

Contents

PLANTS WITH MEDICINAL VALUE

The Authors

Prof. Arun Arya served as Head, Department of Botany and Environmental Studies in The Maharaja Sayajirao University of Baroda. He is a Botanist, Phytopathologist, Philatelist and a Popular Hindi Science writer. He is Fellow of Gujarat Science Academy, I B S, I P S, IAAPC and is a recipient of Young Scientist award by DST Govt. of India. His areas of specialization include Biodiversity, Fungal taxonomy, Aerobiology, Biodeterioration and Endo-mycorrhizal research. He is Vice President of International Council for Biodeterioration of Cultural Property, and Joint Secretary SOCLEEN, Vadodara. He had an honour to co-chair the session on Eco regions in 12th World Forestry Congress held at Canada in 2003. He was awarded World Food Organization fellowship to attend this World Conference. He has presented his scientific findings in Japan, Canada, Italy, Morocco and Singapore. Bill and Melinda Gate's Foundation selected him as an expert member for the ISO standards committee meeting held in Singapore. He has supervised 13 Ph. Ds. and has published more than 175 research papers, 2 video C.D. and 16 books. One of his book entitled Management of Fungal Plant Pathogens, is published by CABI, U.K. and another Biology, Cultivation and Application of Mushrooms by Springer-Singapore. His philatelic collection on Medicinal Plants has won prizes at National level. He has been awarded prestigious Prof. V. Puri Memorial Medal by Indian Botanical Society in 2019 and Environment Excellence award by VNM T.V.

Dr. Upma Singh is serving as Head, Department of Applied Chemistry, University School of Vocational Studies and Applied Sciences, Gautam Buddha University, Greater Noida. She did her Ph.D. in the field of metal complexes. She has published more than 50 research papers, review articles and book chapters encompassing the field of coordination compounds, natural products, bio-pesticides, environmental issues and dietary habits. She authored a book entitled, "Green Technologies for Environmental Security". She has participated and presented papers in number of Conferences/Seminars/ Workshops, and chaired the sessions also. More than 15 students did their M.Sc. projects and awarded two Ph.D. degrees under her supervision. She is associated to various scientific and academic societies. She takes pride in being into the teaching profession and has completed successfully two decades in teaching.

Dr. Bina Rani is Associate Professor, Department of Chemistry and Environmental Engineering at Poornima College of Engineering, Jaipur, Rajasthan. She has also served as Associate Professor in the Department of Humanities and Applied Sciences at IIMT, Greater Noida, Gautam Buddha Nagar, Uttar Pradesh. She pursued her PhD from ICAR-National Dairy Research Institute, Karnal, Haryana. She has diploma in Nutrition and Health Education awarded by IGNOU. She has 25 years of teaching and 5 years of research experience. Her area of research is water treatment technologies, sustainability of Environment, Green Chemistry and new engineering materials. She has published more than, 140 review and 25 research papers in journals of International repute, and 35 book chapters. Eight text books and 2 edited books are also to her credit. She is fellow of Society of Education, Agra, and Society for advancement of Sciences, Lucknow. She is also associated to a number of organizations working on social and environmental issues. She has been conferred Indira Gandhi Sadbhavna Award by International Business Council, New Delhi and Life Time Achievement Award by Venus International Foundation, Channai. She has guided 15 Inspire DST scholars at UG and PG level.

Prof. Ashok Kumar Kakodia obtained his Ph. D degree from MLS University Udaipur, Rajasthan under the guidance of Dr. B.K. Sharma in 2014. He has published more than 70 research papers and review articles in National and International Journals of repute in the area of chemistry of Photo catalytic, Photo degradation, Nanoparticles and Environmental pollution. He is actively involved organizing the various international and national conferences, seminars, workshops and faculty Development programmes. He is served as Professor of chemistry in Department of College Education, Government of Rajasthan and He also worked as OIC Academic and Director Research at Govind Guru Tribal University Banswara during 2019 to 2022. In his academic career spreading over about 22 years, he has supervised many PhD scholars. He has published twelve patents and eight edited books. He is the life member of various academic associations *e.g.* Indian Science Congress Associations, Association of Chemistry Teachers, ISST and Society for Materials Chemistry. He was awarded for the scientist of the year award by Koshambi Foundation at the three days national conference on Multidisciplinary research and practice during 12-14 November, 2021. He served as a resource person to deliver lectures in different international and national conferences. Dr. Ashok Kumar Kakodia has been teaching several courses to undergraduate and postgraduate Student dealing with Organic, Physical, Inorganic Chemistry Spectroscopy, photochemistry and Organometallic compounds, *etc.* e-mail: kakodia77@gmail com

Dr. Raaz K. Maheshwari has completed his Ph.D. from ICAR's NDRI Karnal, Haryana. He served as Associate Professor in SBRM Govt. P.G. College, Nagaur and S.K. Govt. P.G. College, Sikar. He served as Head, Department of Chemistry at MIER Lakshmanagarh and SBDT P.G. College, Lakshmangarh, Sikar. He has published more than 600 papers in different Journals and books and has presented 56 research papers in National and International conferences. He has also edited 16 books. Dr. Maheshwari has guided 5 M.Phil. and 328 Inspire Scholarship projects funded by DST at U.G. and P.G. levels. He has been associated with editorial teams of several International Journals. He has been conferred Scientist of the Year award 2016 by NESA at Punjab University, Punjab. He has been awarded Creative and Innovative Social Worker Practice of the year 2020 by DHS Foundation, New Delhi, Distinguished Scientist Award by VD Good Professional association, Environmental Protection Research Award by Academy of Environment and Life Sciences, supported by DBT, Global faculty award 2020, Shiksha Bharti Award 2021 for his outstanding teaching in Chemistry and Environment, Indian Achievers' award 2020. He was conferred with Magic book of Records presented by Outstanding Professor Service in 2021. e-mail: rkmgreenchem.jaipur@gmail.com

Part A

Plants with Nutritional Value

વધુ વૃક્ષો વાવો
પ્રદૂષણ ઘટાડે છે.
ખોરાક આપે.
પશુ-પંખીનું રહેઠાણ
લાખ, મધ અને ગુંદર આપે છે.
કાગળ બનાવવા માટે ઉપયોગી
ઇમારતી લાકડું આપે.
હવાને શુદ્ધ કરે.
વરસાદ લાવવામાં ઉપયોગી
પર્યાવરણને સંતુલિત રાખે છે.
જમીનનું ધોવાણ અટકાવે છે.
રંગબેરંગી ફૂલ અને પાન પૂરા પાડે.
પ્રાણીઓનું આશ્રયસ્થાન
ફર્નિચર બનાવવા લાકડું આપે
ઔષધિ અને દવા બનાવવા.
રણને આગળ વધતું અટકાવે.
પક્ષીઓનું આશ્રયસ્થાન
વટેમાર્ગુને છાંયો-વિસામો આપે
વાતાવરણ ઠંડું રાખે.
બળતણ માટે લાકડું આપે છે.
છાંયો આપે છે.
ઓક્સિજન આપે.
સુશોભન માટે ફૂલ-પાન આપે.
ફળ આપે છે.
વૃક્ષો આપણાં મિત્રો
એક વૃક્ષનું મહત્ત્વ

Chapter 1

Promising Edible Fruit of Garhwal: Timla (*Ficus auriculata* Lour.)

Timla fruit is scientifically known as Ficus auriculata Lour. and is also called as Roxburgh Fig. It is a wide spreading evergreen fruit tree found in moist climate of Uttarakhand. This tree is often found growing along the banks of streams, and may have more than one trunk. The fruits are formed in large clusters on the trunk, on branches and on the roots. It is not uncommon to discover a tree, where the base is covered with hundreds of fruits. They can be eaten raw or cooked, used in jams, juice and curries. Fruit consists of phyto-constituents viz. triterpenes, flavonoids, polyphenols, alkaloids, sterols, coumarins and other secondary metabolites. These chemicals may be attributed to be antimicrobial, anti-inflammatory, antidiabetic, antioxidant, and hepatoprotective. Timla can be used to maintain Blood Pressure (BP) in people having hypertension. It can also be used to treat constipation, as it acts as a laxative and helps to regulate the digestive system. The main oil components present are UFAslinolenic acid, linoleic acid stearic acid, palmitic acid, oleic acid, 3-hydroxy lauric acid and vaccenic acid which possess massive curative and health preventive properties. SFAs (i.e. palmitic acid and stearic acids) are found in the oil that has huge industrial applications. It was deemed of interest and importance to scrutinize the bioactive compounds of this plant. In the present study, phytochemical constituents and imperative biological activities of Timla are discussed.

Keywords: *Antidiabetic, Phytoconstituents, Lupeol, Flavonoids, Bergapten, Nutritive value, Alkaloid.*

1.1 Introduction

Ficus auriculata Lour. (Figures 1.1–1.3) commonly known as Timla in Garhwali language is a wide spreading evergreen short tree of 3-4 m. Its gigantic and

Figure 1.1. ***Ficus auriculata*** **a Small Tree.**

Figure 1.2A-B. Large Leaf ***Ficus auriculata*****.**

rounded leaves are used to make plates (Figure 1.2B). This tree is often found growing along banks of streams with thick large trunk. Although it is classified as an evergreen plant, it can be deciduous during extended periods of cold. Its fruit is edible and often harvested from the wild, tasting like a cross between a fig and a strawberry. The fruits form as large clusters on the trunk, on branches and even on the roots. It is not uncommon to discover a tree where the base is covered with hundreds of fruits. Bark, root, leaves, fruit and latex of this plant are frequently used for the treatment of various illnesses. Large edible, rounded figs may be produced after 3-5 years of plantation. It produces a unique fruit which develops from actually an inverted inflorescence called syconus. Flowers are pollinated by a wasp which enters from distal end. The average yield of a tree was recorded to be 32.4 kg. (Kavitha Chandran and Revikumar 2017). The species is indigenous in India (Himalaya region), Bhutan, Nepal, Sikkim, Myanmar, Thailand, Vietnam, South China (South Guangdong, Guangxi, South West Guizhou, Hainan, South West Sichuan, Yunnan). It is introduced and cultivated elsewhere. It is found in forests in moist valleys, or in midlands and mountainous areas, 100-1700 m altitudes. It thrives in moist, friable soil, rich in organic matter. The plant is in endangered species and its status is stable.

Figure 1.3: *Ficus auriculata* a Part of Main Trunk Showing Fruits Attached in Bunches.

Fruits of *Ficus* are very tasty. These are rich in flavonoids, polyphenolic compounds, which have strong antioxidant properties that help in prevention and treatment of various oxidative stress related diseases such as and hepatic and neurodegenerative diseases. They have been used as traditional remedy as stomachic, astringents, vermicides, antihelmintics, carminatives, hypotensives and anti-dysentery drugs. Phytochemical screening of *F. auriculata* leaf extract showed the presence of glycosides, resins, tannins. Alkaloids were absent. Leaf extract of *F. auriculata* also revealed the presence of carbohydrates, phenols, tannins, flavonoids and terpenoids *etc.*

1.2 Phytochemical Composition of the Leaves and Fruits

The phytochemical constituents present in leaves, fruit and stem bark of *F. auriculata* showed the presence of carbohydrates, glycosides, tannins, flavonoids, and sterols. *Ficus* fruit contains 87.1 per cent moisture and 7.5 per cent total soluble solids. The total sugars are 60.15 per cent, reducing sugars are 60.12 per cent; the non reducing sugars are 0.03 per cent and the amount of pectin is 0.48 per cent. The vitamin C content of the fruit is only 3.35mg per 100g. The protein content of the fruit is 0.59 per cent. The fully ripe fruit contains 1.068 per cent total minerals as represented by its ash. The percentage of mineral elements in the fruit such as P,K, Ca, Mg, and is, 0.039, 0.331, 0.039, 0.045 and 0.003 respectively. Compounds including betulinic acid, lupeol, stigmasterol, bergapten, scopoletin, beta sitosterol-3-O-Beta- D- glucopyranoside, myricetin and quercetin-3-O-BetaD-glucopyrawere found in leaves and fruits of *F. auriculata* (Fishawy *et al.*, 2011). Phenolic compounds showing wide bioactivity including antioxidant properties. Phytochemical analysis of stem bark extracts of *F. auriculata* showed the presence of fatty oils, alkaloids, carbohydrates, saponins, glycosides, phytosterols, resins, phenols, tannins, diterpenes, flavonoids, proteins and amino acids. Eight carbon compounds were isolated including betulinic acid (I), lupeol (II), stigmasterol (III), bergapten (IV), scopoletin (V), β-sterol-3-O-β-D-glucopyranoside (VI), myricetin (VII) and quercetin-3-O-β-D-glucopyranoside (VIII) from leaves and fruits of *F.auriculata*. Also, 5,7,42 -trihydroxy-32 -hydroxymethylisoflavone, 32 -formyl-5,42 -dihydroxy-7-methoxyisoflavon (II), ficuisoflavone, lpinumisoflavone (IV), Rutin(quercetin-3-rhamnosyl glucoside) [i], lactones (ii), ficusine D (1) (3R,4R)-4-hydroxy-de-O-methyllasiodiplodin isolated from stems along with six other known compounds *i.e.* (2) (3R,4R)-4-hydroxymellein, (3) (3S,4R)-4-hydroxymellein[II], (4) shoreaphenol, (5) heimiol A, (6) (+)-balanocarpol, (7) (+)-ampelopsin. Phenyl methyl pyridine, dibutyl phthalate (iii), phytol (iv), 3β-lup-20(29)-en-3-ol-acetate [v], 3R, 4R)-4-hydroxy-de-O-methyllasiodiplodin, 6-oxolasiodiplodin were determined in this plant. Volatile oil was extracted from the leaves of *F. auriculata* (Shao *et al.*, 2013).

1.3 Pharmacological Properties

Ethanolic fruit extract of *F. auriculata* has significant antibacterial and antifungal activity 14±1mm, 13±1mm and 12±1mm against *Shigella flexneri*, *Escherichia coli* and *Staphylococcus epidermidis* for food poisoning bacteria (Gaire *et al.*, 2011; Saklani and Chandra 2011: Salem *et al.*, 2013). Carrageenan- induced rat hind paw edema model was used for the anti inflammatory activity (Mohamed *et al.*, 2013) of the leaves and fruits extracts of *F. auriculata*. Edema was induced in the left hind paw of all rats by subcutaneous injection of 0.1 ml of 1 per cent (w/V) carrageenan in distilled water into their foot pads. The paw volume of each rat was

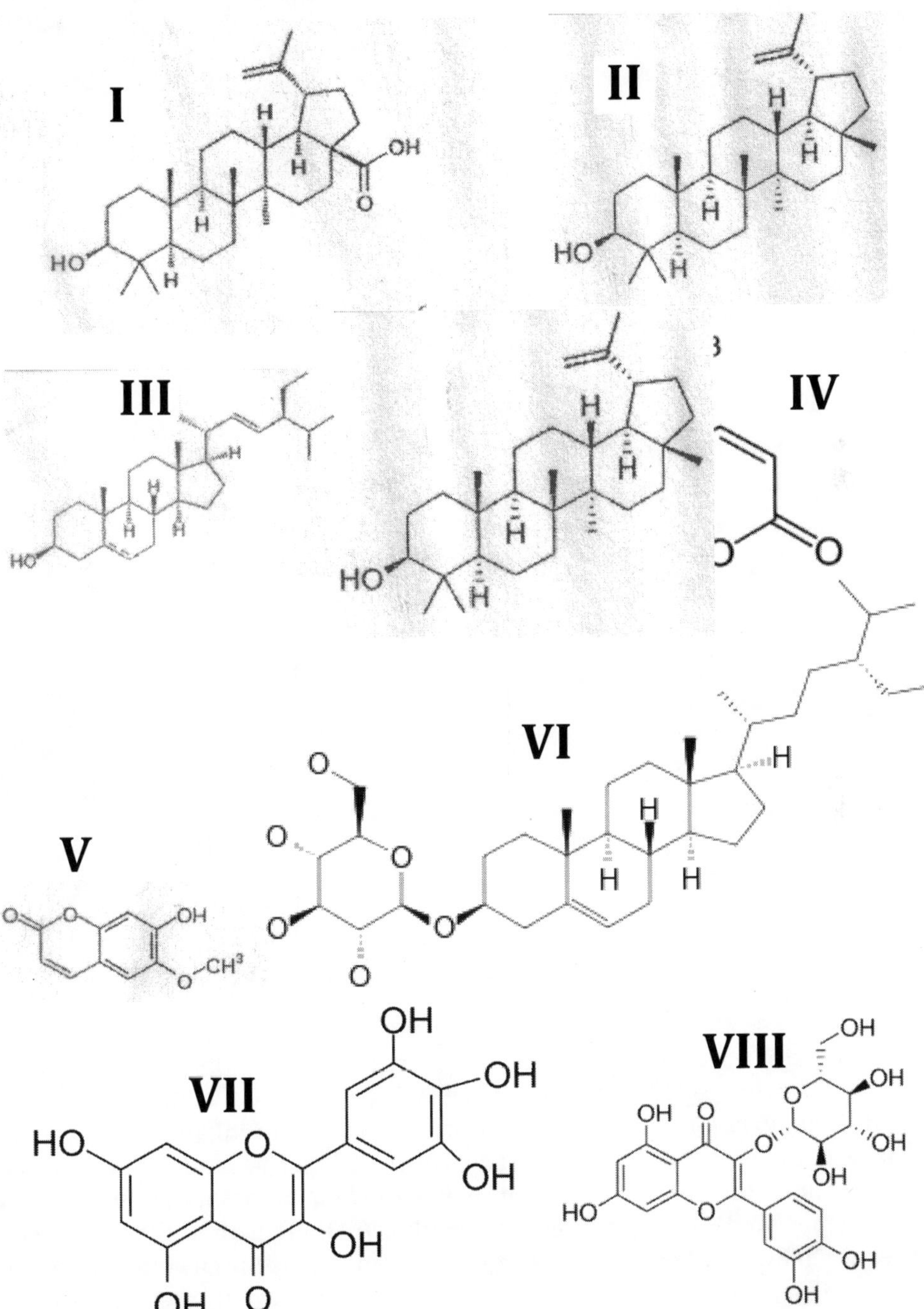

Figure 1.4. Chemical Constituents from the Leaves and Fruits of *F. auriculata*: Betulinic acid (I), lupeol (II), stigmasterol (III), bergapten (IV), scopoletin (V), β-sterol 3-O-β-D-glucopyranoside (VI), myricetin (VII), quercetin-3-O-β-D-glucopyranoside (VIII).

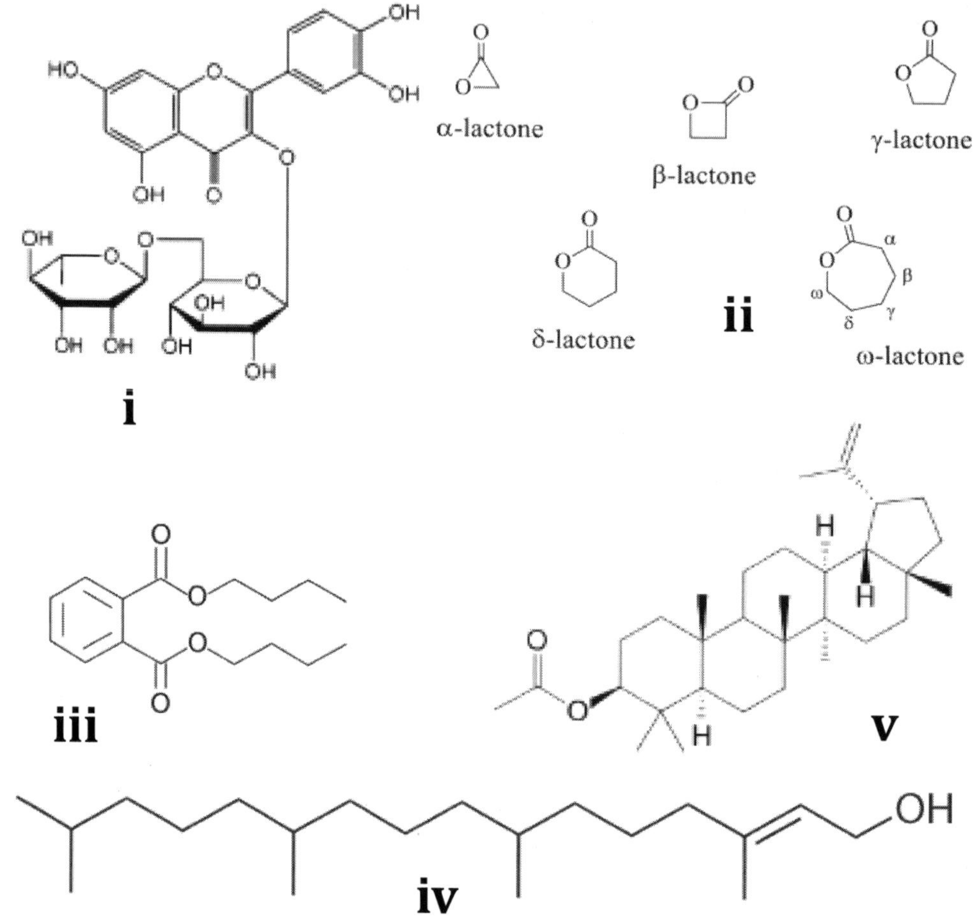

Figure 1.5. Chemical Structures of Rutin (quercetin-3-rhamnosyl glucoside) [i], lactones (ii), dibutyl phthalate (ii), phytol (iv), 3β-lup-20(29)-en-3-ol-acetate [v].

measured using Plethysmometer; before carrageenan injection. Concentrations of the extracts reduced the inflammation in a dose dependent manner. The leaves extract (500mg/kg) have found to possess significant anti-inflammatory activity on the tested experimental model (Fishawy *et al.*, 2011). The antioxidant activity of the ethanolic extract of leaves and fruits of *F. auriculata* was assayed by using DPPH radical method and reducing power assay method in DPPH method, the results were expressed in EC50 (Effective concentration to reduce the initial concentration of DPPH to 50 per cent). Lesser the value of EC50 value for an extract was considered to be associated with higher ability to donate hydrogen radical *i.e.* antioxidant activity. In reducing power assay method the amount of Fe2+ complex monitored by measuring the formation of Perl's blue at 700nm. Increasing absorbance indicates an increase in reductive ability. This proved that increasing the concentration of the

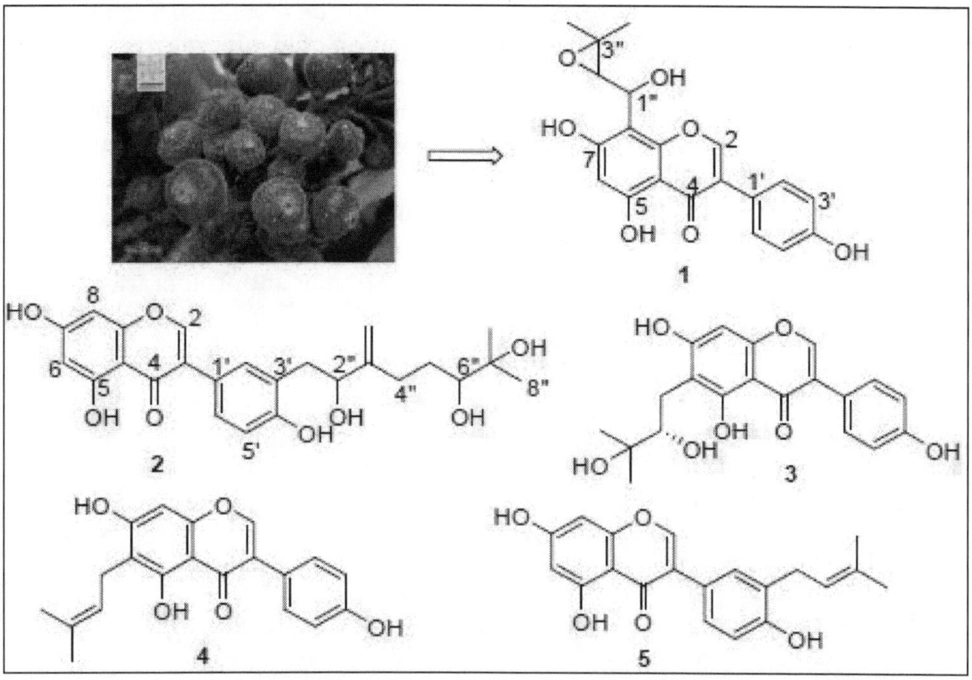

Figure 1.6: Isoflavonoids Occurring in the Root of the *Ficus auriculata*
5,7,4'-trihydroxy-8-(1-hydroxy-2,3-epoxy-3-methylbutyl) isoflavone (1), 5,7,4'-trihydroxy-3'-(2,6,7-trihydroxy-3-methylene-7-methyloctyl) isoflavone (2), 5,7,4'-trihydroxy-6-[(*S*)-2,3-dihydroxy-3-methylbutyl] isoflavone (3), 5,7,4'-trihydroxy-6-pre-nylisoflavone (4), and 5,7,4'-trihydroxy-3'-prenylisoflavone (5).

extracts increased the antioxidant effect in a dose dependent manner. Fruit extract has more activity than leaves (Rosalind *et al.,* 2012). The fruit extract is used as aphrodisiac (Kumar *et al.,* 2018). Puangpradab *et al.* (2018) reported antioxidant value of fruits in Northern Thailand. Hepatotoxicity was induced by animals treated with toxic dose of carbon tetrachloride had markedly elevated levels of the serum AST and ALT, compared to normal mice. Serum enzyme values of the crude leaves and fruits extracts of *F. auriculata* were nearly at the same levels as those of toxic control. The histopathological examination of the liver sections of positive control group showed necrosis associated with vacuolization of cytoplasm and fatty degeneration of hepatocytes. The liver sections of the mice treated with the crude extracts exhibited slight improvement with slight vacuolization of hepatocytes (Fishawy *et al.,* 2011).

The extracts of *F. auriculata* induced 8.2 per cent reduction of blood glucose level of diabetes rats as compared to the values before treatment. Diamicron as standard induced 39.8 per cent reduction of blood glucose level of diabetic rats. Methanolic leaf extracts of *F. auriculata* in Streptozotocin induced diabetic mice produced significant decrease in blood glucose level. The results were comparable with that of glibenclamide, which acts by stimulation of insulin release. The level of total protein in serum decreased in diabetic groups when compared to those of normal control group. Methanolic leaf extract may be able to improve the impairment of fatty acid metabolism in diabetes. In the histopathological studies of pancreas, normalization of beta cells in the islet of Langerhans of diabetic mice fed with the methanol leaf extract of *F. auriculata* was observed and this may explain both hypoglycemic and hypolipedemic actions of the plant (Thinbaijam *et al.*, 2012). Acetone extract of *F. auriculata* showed potent antiproliferative activity against C33A cells of cervical cell lines (22-63 per cent) (Saini *et al.*, 2012; Kumari *et al.*, 2018). Aphrodisiac properties of fruits were reported by Kumar *et al.* (2018)

Figure 1.7. Timla Jam.

Figure 1.8. Jelly Prepared from Timla Fruits.

1.4 Traditional Uses

Leaves are crushed and the paste is applied on the wounds. They are also used in diarrhea and dysentery. Leaves are lopped for fodder. Leaves show antioxidant and antimicrobial properties (Bertoletti *et al.*, 2018). Stem bark juice is effective for diarrhea, cuts and wounds. Fruits are edible and can be made into jams and curries. Roasted figs are taken for diarrhea and dysentery. Root latex is used in mumps, cholera, diarrhea and vomiting. Several tribes in Northern eastern India especially in Manipur use the leaf of *F. auriculata*, traditionally for the treatment of diabetes (Singh *et al.*, 2014). It's stem bark juice is effective for cuts and wounds and diarrhea.

1.5 Culinary Uses

Also known as Roxburgh fig tree, it is sweet and brownish or purplish in colour. They can be eaten raw or cooked, used in juice, curries, squash, jam and jelly (Figures 1.7 and 1.8). A juice procured from stem bark is effective in treatments of diarrhea, cuts and wounds. Roasted figs are taken for control of diarrhea and dysentery. Root latex is used in mumps, cholera, diarrhea and vomiting.

1.6 Conclusion

Timla fruit is consumed as dessert or cooked, and used in different forms. The fruits have great medicinal significance in diabetes and other ailments. The application of phytochemicals present in fruit, stem bark and roots is useful as anti-rheumatic, mild laxative, antihelminthic, digestive, anti-dysentery agents. Their anticoagulant potential have been also reported.It has antioxidant and anti-inflammatory properties, too. The phytochemical constituents present in leaves, fruit and stem bark of *F. auriculata* showed the presence of alkaloids, carbohydrates, glycosides, tannins, flavonoids, and sterols. Root latex is used in mumps, cholera, diarrhea and vomiting. The excellent properties of *F. auriculata* which are evident by a number of studies makes it a very helpful, easily available, low cost fruit with great commercial potential to be exploited in near future.

REFERENCES

Bertoletti LL., Skoronski E., Schittler L., Kempka AP. (2018). Extracts of leaves of *Ficus auriculata*Lour: Antioxidant, antimicrobial and phytotoxic activity. Agri. Conspec. Sci. 83(4): 321-328.

Fishawy A.E.,Zayed R., Afifi S (2011) Phytochemical and Pharmacological studies of *Ficus auriculata* Lour., Journal of Natural Products, 2011; 4: 184-195.

Gaire BP., Lamichhane R., Sunar CB., Shilpakar A., Neupane S.,Panta S (2011). Phytochemical Screening and Analysis of Antibacterial and Antioxidant activity of *Ficus auriculata*Lour. Stem bark, Pharmacog. J. 3(21): 49-55.

Kavitha Chandran C I., Revikumar KG. (2017). *Ficus auriculata* (Elephant ear fig) : A Phytochemical and pharmacological Review. World J. Pharma. Pharmacog. Sci. 6(5): 274-283.

Kumar SS., Rekka R., Thankachan J. Santhosh T. (2018). Identification of aphrodisiac properties of *Ficus auriculata*Lour. fruit extract. *Eur. J. Biomed.,*5(2): 454-46.

Kumari A., Verma R., Sharma M., Chauhan P., Kumar A. (2018). Evaluation of phytochemical, antioxidant, antibacterial and anti-cancerous activity of *Ficus auriculata* Lour. and *Osyris wightiana* Wall. ex Wight. Bull. Env. Pharmacol. Life Sci. 7(8): 64-70.

Mohamed Z.M., Salem, A.Z.M. Salem, L.M. Camacho L.M., Ali. H.M. (2013) Antimicrobial activities and phytochemical composition of extracts of *Ficus* species: An over view, African Journal of Microbiology Research. 7(33): 420-4219.

Puangpradab R., Suksathan R., Saratan N., Puangsombat P. (2018) Antioxidant properties and nutritive values of native figs consumed in northern Thailand. J. Na.t Sci., 12(8): 25-28.

Rosalind TH., Dutta BK., Paul SB. (2012). *In vitro* antioxidant capacity of estimation of total phenolic and flavonoid content of *Ficus auriculata* Lour. Int. J. Pharm. Pharmacognol. Sci., 4(4): 518-521.

Saini R., Garg V.,Dangwal K. (2012) Comparative study of three wild edible fruits of Uttrakhand for Antioxidant, Antiproliferative activities and Polyphenolic composition. Int. J. Pharm. Bio Sci.3(4): 158-167.

Saklani S., Chandra S. (2012). In *vitro* Antimicrobial activity, Nutritional Profile and Phytochemical Screening of Wild Edible Fruit of Garhwal Himalaya (*Ficus auriculata*), Int. J. Pharmaceut. Sci. Rev. Res.12(2): 61-64.

Shao T., Song X., Chen G., Li X., Yao G., Deng K., Huang K. (2013). Analysis of the chemical constituents of volatile oil from the leaves of *Ficus auriculata* Lour by GC-MS. Chem. Indus. Forest Prod. 33(3): 135-137.

Singh S R., Phurailatpam A K., Wangchu L., Ngangbam P., Chanu T M. (2014). Traditional medicinal knowledge of underutilized minor fruits as medicine in Manipur. Int. J. Agri. Sci. 4(8): 241-247.

Thinbaijam R., Dutta B K., Paul S B. (2012). *In vitro* antioxidant capacity, Estimation of Total Phenolic and Flavonoid content of *Ficus auriulata* Lour. Int. J. Pharma. Pharmaceut. Sci. 4(4): 518-521.

Chapter 2

Nutritious Plant with Super Powers: Dragon Fruit [*Selenicereus undatus* (Haworth) D.R. Hunt]

Dragon fruit is an incredible tropical cactus that produces strange looking fruit resembling a magical dragon egg. The vibrant pinkish red skinned fruit is gaining importance due to its nutritional qualities and peculiar taste. The taste is hard to pin down with some saying it's a sweet mix of mild kiwi fruit, watermelon, strawberry and pear flavours. Others describe it as only vaguely sweet or even savoury. Well growing conditions and ripeness of the fruit can impact taste but nonetheless it's visually impressive and packed full of nutrients, like Vitamin C, calcium, iron and phosphorus. Fruits are rich in pectin and betalins making it natural food. It is said to lower blood sugar in type 2 diabetes. Antioxidants present in the fruit makes it suitable to prevent cancer like diseases.

***Keywords**: Nutrition, Super food, Dragon fruit, Antioxidants, Betalins, Diabetes, Cancer.*

2.1 Introduction

Fruits are rich in nutritious elements and minerals. In mango the starch is broken down into glucose and fructose at maturity (Randhawa and Chadha 1994). Dragon fruit, also known as pitaya or strawberry pear, on the outside, it's a pink oval with green scales (hence the "dragon" name). Inside, it has white flesh with tiny black seeds. The fruit's strange appearance also gives off "psychedelic artichoke" vibes."It's a tropical fruit that comes from a cactus. It's available everywhere around the world, but it's indigenous to Mexico and South America."

If the fruit isn't enough to you tempt you then don't forget the flowers. Their stunning large flowers are easily over 20cm wide and appear in summer. They are yellowy green on the outside and open to a scented white, lily like bloom. Flowers open in the evening and only last one night. It's the perfect excuse to host an evening cocktail party and enjoy their blooms. Dragon fruits come from Mexico and Central and Southern America.

2.2 Botanical Aspects

Dragon fruit refers to fruit of the genus *Selenicereus* (formerly *Hylocereus*), of the family Cactaceae. It is a climbing cactus called *Hylocereus*, which you'll find in tropical regions around the world. The plant's name comes from the Greek word "hyle," which means "woody," and the Latin word "cereus," which means "waxen." Dragon fruit is cultivated in Peru, The plant prefers warm conditions with mild winters and no frost. They can tolerate the occasional short burst of cold weather but ideally they should be kept above 10 degrees. They can handle heat, drought, humidity and poor soils but will grow tastier fruit with regular watering and rich soil. Choose a sunny spot and then boost the soil with compost, manure and certified organic pelletized fertilizer. A dose of lime is also beneficial. Soil needs to be free draining as these plants rot easily if the roots sit in water. Clay soils are not ideal. Alternatively they will happily grow in a large pot. Plants are grown with a thick support and tie one or two main stems to the support to encourage vertical straight growth. Trim away any other side shoots. Once the stems have reached the desired height cut off their ends to encourage new branching shoots. These can then be allowed to spread out and hang downwards. Commercial growers often use a rose wheel and encourage side way branches to grow over the wheel making it easier for picking fruit. Every two three weeks fertilizer is applied. Periodically remove some of the longer shoots to keep it under control and allow space for new growth to develop. This is important because flowers form on the ends of new season growth so each year you need new growth to get fruit. Less congestion tends to lead to bigger fruit as well. Plants are grown in Kuchh area in Gujarat. Fruit is ripe approximately a month after flowering but this can vary with local conditions. Fruit will not continue to ripen once picked so you need to look for other signs before picking. Check that the colour of the fruit is bright and even all over and the small "wings" on the fruit are starting to wither. Lightly press the fruit in your hand and if ripe it will give just a little. Pick by twisting the fruit off the plant or cut with secateurs.

More watering will cause splitting of outer skin hence avoided. The skin is not edible but the seeds are just like a kiwi fruit. Dragon fruit grow easily from seed or cuttings. To grow from seed, squash some flesh onto paper towel and keep moist in a warm position but away from direct sunlight. Seeds will sprout 2-3 weeks later and can be potted up into punnets. Water weekly with fertilizer to develop strong seedlings and pot into individual pots once large enough. Seedlings will take several years to reach fruiting size.To take a cutting simply break off a segment 30-50cm long and leave in a dry shady spot for a week. This allows the cut end to seal and prevents rotting. Plant into a pot and keep in a bright shady spot while roots are forming before then moving into the sun. Don't over water during this time. Cuttings can be taken at any time but will grow faster if done during the warmer months.

Figures 2.1. (A) Plants of Dragon Fruit; (B and C) Red Coloured Fruits of Dragon Fruits (D) A Cut Dragon Fruit with White Flesh and Small Black Seeds.

On the outside, the fruit has the appearance of a hot pink or yellow bulb with spike-like green leaves shooting up like flames around it. Fleshy white fruit has dotted black seeds. This fruit comes in red- and yellow-skinned varieties. The seeds have a nutty flavour. Peel and cut it into sections when you're ready to eat it. Much like a guava, you eat the flesh and discard the skin. You could also cut it in half and scoop out the flesh with a spoon. Dragon fruit is best eaten raw, but you can throw it on the grill like some other fruits. Enjoy it on its own or add it to: Cocktails, desserts, fish especially cod, tuna and mahimahi, salads, smoothies. Toss it into a fruit salad along with other tropical fruits like pineapple and mango.

2.3 Phytochemicals Present in Different Varieties

Dragon fruits may have following varieties:

- **Pink skin with white flush** (*Hylocereus undatus*): This is well known variety, it is the least sweet. It's sold under the names including Alice, Cosmic Charlie and Guyute.
- **Pink skin with red or pink flush** (*Hylocereus costaricensis, H. polyrhizus)*: It is bigger and sweeter than its white-fleshed variety. It is sold in stores under names such as Red Jaina and Bloody Mary.
- **Pink skin with purple flesh:** It is known as American beauty.
- **Yellow skin with white flesh** (*Hylocereus megalanthus*)**:** Yellow dragon fruit is uncommon to find, but it's also the sweetest.

The flesh of the dragon fruit is low in calories and fat-free. It also contains plenty of fibre. Dragon fruit is an excellent source of fibre. The daily recommendation for adults is at least 25 g and dragon fruit packs 7 g in a single 1-cup serving. It may benefit gastrointestinal and cardiovascular health. Fibre helps in weight loss. Dragon fruit has a ton of beneficial vitamins and minerals, including:

- Carotenoids are present, which may reduces cancer risk.
- Lycopene - improves heart health and reduces cancer risk.
- Magnesium is important for cell function, and dragon fruit provides 18 per cent of the recommended daily amount.
- Iron is important component of blood and energy, and dragon fruit contains 8 per cent of the recommended daily intake.
- Vitamin C helps to boost immune system.
- Dragon fruit improves gut flora and improves colon cancer risk.

Table 2.1. Presence of Nutrients per 100g Fruit Pulp

Nutrient	*Amount per 100 g Fruit*	*Nutrient*	*Amount per 100 g Fruit*
Water	87 g	Protein	1.1g
Fat	0.4g	Carbohydrates	11 g
Fibre	3g	Vit. B1 (Thiamine)	0.04mg
Vit. B2 (Riboflavin)	0.05mg	Vit. Niacin	0.16 mg
Vit. C (Ascorbic acid)	20.5 mg	Calcium	8.5 mg
Iron	1.9mg	Phosphorus	22.5 mg
Magnesium	68mg	Zinc	ND
Calories	102		

ND: Not detected.

2.4 Medicinal and Nutritive Value

The heavenly dragon fruit has a lot of medicinal values. According to the paper, previous animal studies had shown a potential link between dragon fruit consumption and better control of diabetes. The consumption of dragon fruit encourages the growth of pancreatic cells that produce insulin. The analysis examined multiple clinical trials comparing the effects of dragon fruit with a placebo in people with type 2 diabetes or prediabetes. The dragon fruit's affect on fasting plasma glucose in people with pre-diabetes. They also reported that the more dragon fruit a person eats, the greater is the control of blood sugar, However, the study did not find a correlation between eating dragon fruit and improved control of type 2 diabetes. The researchers recommended need of more studies regarding prediabetes and dragon fruit consumption to determine whether eating dragon fruit can help stabilize blood sugar levels (Soryono 2006; Sonawane 2017).

Dragon fruit lowers the cholesterol concentration, to balance blood sugar concentration, to prevent colon cancer, to strengthen kidney function and bone, to strengthen the brain working, increasing the sharpness of the eyes as well as cosmetic ingredients (Suryono 2006). The fruit is rich in antioxidants and vit. C, B vitamins, carotene, polyunsaturated fatty acids, protein and minerals like Ca, Fe, K, Na *etc.*

Super red dragon fruit from Pasuruan district contains the highest Vitamin C compared to the three other locations. Kristanto (2003) suggests that the content of Vitamin C in dragon fruit ranges from 8-9 mg/100 g. The result of this study shows that Vitamin C content is lower. This is possible because the fruit is too ripe, resulting in lower number of Vitamin C. According to Winarno (1995) and de Man (1999), the content of Vitamin C in raw fruit is higher than in a mature one, and the more mature the fruit is the lesser the Vitamin C content (Rahmawati and Mahajoeno 2009). It can boost your iron levels. Iron is very important for moving oxygen through your body and giving energy, and dragon fruit has high iron content. And the vitamin C in dragon fruit helps your body take in and use the iron. It's rich in antioxidants like flavonoids, phenolic acid, and betacyanin. These natural substances protect your cells from damage by free radicals - molecules that can lead to diseases like cancer and premature aging. The dragon fruit helps in getting rid of toxic heavy metals. It helps in improving eye sight. Lycopene, responsible for the red color in dragon fruit has shown to be linked with a lower prostate cancer risk. Vitamin C accumulates in phagocytic cells, such as neutrophils, and can enhance chemotaxis, phagocytosis, generation of reactive oxygen species, and ultimately microbial killing. It is also needed for apoptosis and clearance of the spent neutrophils from sites of infection by macrophages, thereby decreasing necrosis/NETosis and potential tissue damage. The role of vitamin C in lymphocytes is less clear, but it has been shown to enhance differentiation and proliferation of B- and T-cells, likely due to its gene regulating effects. Vitamin C deficiency results in impaired immunity and higher susceptibility to infections (Carr and Maginni 2017). In turn, infections significantly impact on vitamin C levels due to enhanced inflammation and metabolic requirements. Furthermore, supplementation with vitamin C appears to be able to both prevent and treat respiratory and systemic infections. Prophylactic prevention of infection requires dietary vitamin C intakes that provide at least adequate, if not saturating plasma levels (*i.e.*, 100-200 mg/day), which optimize cell and tissue levels. In contrast, treatment of established infections requires significantly higher doses of the vitamin to compensate for the increased inflammatory response and metabolic demand.

The seeds of dragon fruits are high in polyunsaturated fats (omega-3 and omega-6 fatty acids) that reduces triglyceride and lower the risk of cardiovascular disorders (Sonawane 2017).

2.5 Associated Health Risk

Studies conducted on dragon fruits have revealed that it can cause allergic reactions. The symptoms caused as swelling of the tongue, and vomiting. Excessive use of fruits may turn color of urine red or pink. The color returns normal once the fruit is out of our system.

2.6 Molecular Markers

Selenicereus megalanthus H. is a tropical fruit belonging to the family Cactaceae, is rich in essential nutrients, antioxidants and bioactive components. It presents wide variability in different characteristics and a great demand in the market; however, genetic studies in Colombia are scarce. The main aim of this study was to characterize the genetic diversity of 76 yellow pitahaya genotypes with eight ISSR markers. Genetic parameters expected average heterozygosity (He), percentage of polymorphic loci, genetic distances and Fst were estimated with TFPGA. As a result, 225 alleles were generated and the number of polymorphic loci ranged 85 (CT, AG) to 90 (GT). High genetic diversity was found, with an average value of heterozygosity of 0.34 with a genetic differentiation coefficient (Fst) of 0.26, indicating that there was a great genetic diversity, similar values than those reported in other studies of pitahaya genetic diversity in Colombia. The 76 genotypes were grouped into K=3 according to geographic location, however, in some groups a mixture of individuals from different origins was observed. The analysis of molecular variance (AMOVA) showed higher variation (75 per cent) within groups than among groups (25 per cent). These results provide information that can be used to develop conservation strategies for dragon fruit and breeding programs to obtain more productive pitahaya genotypes with superior quality, high yield and with resistance to biotic and abiotic factors.

2.7 Commercial Applications

2.7.1 Use of Pectin

Dragon fruit is rich in pectin. It contains 14.96 to 20.14 per cent pectin of the dried peels (Izalin *et al.,* 2016). About 7.5 per cent of the pectin is extracted by microwave assisted extraction method (Thirugnanasambandham *et al.,* 2014).

It was revealed that dragon fruit could be substantial source of pectin in fruit production. Izalin *et al.* (2016) recommended using dragon fruit peel pectin as a thickener in food products such as low viscous food and beverages.

2.7.2 Use of Betalins as Natural Food Colour

Betalains are plant derived natural pigments that are presently gaining popularity for use as natural colorants in the food industry (Gengatharan *et al.*, 2015). The growing interest of consumers in the aesthetic, nutritional and safety aspects of food has increased the demand for natural pigments. These are used as alternative colorants in food products. Although betalins from red beetroot are one of the most widely used food colorant, betalins are not as well studied as compared to other natural pigments such as anthocyanins, carotenoids or chlorophylls. Pigments betalins comprising of betacyanins and betaxanthins are present in sufficient amount in peel of dragon fruit. These pigments were extracted by Rebecca *et al.* (2008). These pigment are tolerant to colour loss during processing (Harivaindaran *et al.*, 2008; Ruzainah *et al.*, 2009; Woo *et al.*, 2011). Dragon fruit peel dye can be preserved at 4° C without light for 3 weeks. Rodriguez *et al.* (2016) showed that the antioxidant, anti-inflammatory, antiangiogenic and GST inducing activities of betalins from peels were enhanced through carbohydrate encapsulation. Pharmacological properties, such as antioxidant, anti-cancer, anti-lipidemic and antimicrobial activity of betalins are reported from sources such as red beetroot, amaranth, prickly pear and red pitahaya, for potential application as functional foods.

2.7.3 Use of Dragon fruit as Natural Probiotics

Dragon fruit contains small seeds which have been made to determine the composition of linoleic acid. Xu *et al.* (2016) found flesh rich in polysaccharides which help in growth of bacteria Lactobacilli. The gastrointestinal microflora helps to suppress the pathogens and act as natural probiotic (*Sonawane 2017)*. Having more prebiotics in your system can improve the balance of good to bad bacteria in our intestine. Specifically, dragon fruit encourages the growth of the probiotics lactobacilli and bifido-bacteria. In your gut, these and other helpful bacteria can kill disease-causing viruses and bacteria, and help in digestion of food.

REFERENCES

Carr A C, Maggini S (2017) Vitamin C and Immune Function. Nutrients 9(11): 1211. doi: 10.3390/nu9111211.

de Man JM. (1999) Principles of food chemistry. 3rd ed. Aspen Publishers. Maryland

Gengatharan A, Dykes G A, Choo W S (2015) Betalains: Natural plant pigments with potential application in functional foods.LWT - Food Science and Technology. 64(2): 645-649, ISSN 0023-6438, https: //doi.org/10.1016/j.lwt.2015.06.052.

Harivaindaran K V, Rebecca O P S, Chandran S (2008) Study of optimal temperature, pH and stability of dragon fruit (*Hylocereus polyrhizus)* peel for use as potential natural colourant. Pakistan J. Biological Sci. 11(18): 2259-2263.

Izalin N M Z, Kharidah M, Jamilah B, Noranizan M A (2016) Functional properties of pectin from dragon fruit (*Hylocereus polyrhizus*) peel and its sensory attributes J. Trop. Agaric. and Food sci. 44(1): 95-101.

Kristanto D (2003) Dragon fruit cultivation in pots and in the garden. Penebar Swadaya. Jakarta. [Indonesia].

Rahmawati B, Mahajoeno E (2009) Variation of morphology, isozymic and vitamin C content of dragon fruit varieties. Nusantara Bioscience 1(3): 131-137.

Randhawa G.S. and Chadha K.L. (1994) Fruit drop and its control in mango and citrus ICAR, New Delhi, India 64 pp.

Rebecca O P S, Zuliana R, Boyce A N, Chandran S (2008) Determining pigment extraction efficiency and pigment stability of dragon fruit (*Hylocerus polyrhizus*). J. Biological Sci. 8(7): 1174-1180.

Rodriguez E B, Vidallon M L P, Mendoza D J R (2016) Health-promoting bioactivities of betalains from red dragon fruit (*Hylocereus polyrhizus* (Weber) Britton and Rose) peels as affected by carbohydrate encapsulation. J.Sci. Food and Agric. 96(14): 4679-4689 DOI: 10.1002/jsfa.7681

Ruzainah A J, Ridhwan A R, Nor Zaini C M, Vasudevan R (2009) Proximate analysis of dragon fruit (*Hylocereus polyrhizus*) Amrican J. Appl. Sci. 6(7): 1341-1346.

Soryono J (2006) Consuming dragon fruit to treat various diseases. Sinar Tani. 15-21 February 2006.

Sonawane M S (2017) Nutritive and medicinal value of dragon fruit. The Asian Journal of Horticulture 12(2): 267-271. DOI: 10.15740/HAS/TAJH/12.2/267-271.

Thirugnasambandham K, Sivakumar V, Prakash M J (2014) Process optimization and analysis of microwave assisted extraction of pectin from dragon fruit peel. Carbohydrate Polymers. 112(4): 622-626.

Winarno FG. (1995) Food enzymes. Gramedia. Jakarta. [Indonesian]

Woo KK, Ngou F H, Ngo L S, Soong W K and Tang P Y (2011) Stability of betalin pigment from red dragon fruit (*Hylocerus polyrhizus*). Am. J. Food Technol. 6(2) 140-148

Xu L., Zhang Y. and Wang L. (2016) Structural characteristics of water soluble polysaccharide purified from dragon fruit (*Hylocereous undatus*) pulp. Carbohydrate Polymers. 146 (1): 224-230

Chapter 3

Nutritive and Healthy Fruit: Pineapple Guava [*Feijoa sellowiana* (O. Berg) Burret]

Pineapple guava has a wide range of health effects, including adding weight-loss goals, improving digestion, lowering cholesterol levels, boosting the immune system, increasing bone strength, decreasing blood pressure, optimizing nutrient uptake, balancing the metabolism, increasing circulation, stimulating cognitive function and regulating blood sugar levels, among others. In native countries, Feijoa fruits are used to treat thyroid gland diseases due to high content of iodine. Because of the high Vitamin C content, it is often used as a kitchen remedy to cure common cold, and fever. The fruit is suggested to be consumed by people suffering from blood pressure irregularity. It is said that by regular consumption it prevents people from cancer. Fiejoa extract is used in many personal care products because of its nutrient contents. Pulp of Feijoa with the seeds is used as a natural exfoliant on skin. The extract of Feijoa exhibits anti-cancer effect on haematological and solid cancer cells.

***Keywords**: Pineapple guava, Feijoa sellowiana,Vitamin C, Inflammation, Oxidation, Atherosclerosis, Heart disease, Digestion, Cognitive function, Gastrointestinal.*

3.1 Introduction

Pineapple guava in many parts of the world is called– *Feijoa.* Scientifically it is known as *Acca sellowiana*. The plant that bears this fruit is a shrub or small tree native to regions in South America, including Argentina, Brazil and Colombia. This fruit was named after Joam da Silva feijo who was a Brazilian botanist. This fruit was collected by the German explorer named Freidrich Sellow in Southern Brazil in

1815. It was introduced in France in 1890 by the French botanist and horticulturist named Dr Edouard Andre. It was introduced in New Zealand in 1920's. It is now widely cultivated for its sweet fruit, as well as for ornamental purposes. The fruit is green and ellipsoid-shaped, and roughly the size of a plum or a small avocado. The unique flavor and impressive supply of nutrients make pineapple guava highly sought after, as it can have many different culinary applications, from an ingredient in smoothies to cocktails, desserts, chutneys and stewed fruit dishes. The fruit has tart and strongly sweet aromatic flavor. The taste of the fruit is like the combination of strawberry, guava and pineapple, lemon or quince. The texture of the peel is rough and thin. The seeds contain volatile oil (Shaw *et al.,* 1983).

The fruit weighs about 50 g. *Feijoa* is available during the autumn season. When the fruit got ripen, it drop from the tree or is picked from the tree directly. According to USAD major nutrients contained in this fruit are Vitamin C, Total dietary fiber, Carbohydrate, Vitamin B9 Vitamin B6 (Pyridoxine), Vitamin B5 and Copper. The *Feijoa* (*Acca sellowiana*) is a tree native to Brazil, belonging to family Myrtaceae and is also cultivated in New Zealand and California.

This plant is usually grown for its sweet and aromatic fruits. The plant has pale gray bark with greenish gray leaves. There are 21 varieties of *Feijoa* such as Apollo, Choiceana, Coolidge, Edenvale Improved Coolidge, Edenvale Late and so on. *Feijoa* or pineapple guava also called guavasteen, pineapple guava or New Zealand Banana. It has got various names such as *Feijoa* in Japanese, faiijo in Hindi, feiyueguo in Chinese, feykhoa in Russian, Ananasguave or *Feijoa* in Dutch. The fruit is oval in shape and about the size of chicken's egg which is about 2.5 to 8cm in length and 1 – 5 cm in diameter. They have lime green waxy skin, jelly textured seed pulp which is encircled by soft and cream colored flesh. The fruit has tart and strongly sweet aromatic flavor.

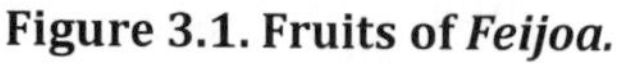

Figure 3.1. Fruits of *Feijoa.*

Figure 3.2. A Flowering Twig of *Feijoa*. Note the Numerous Red Coloured Stamens

Pineapple guava has a very unique flavor, with sweet, tangy and bitter elements, which many people compare to guavas and pineapples, as the common name implies, but there is also a slight strawberry flavor. In some cultivars, there are very subtle notes of mint, which may increase as the fruit ripens. To ensure the best flavor and taste of pineapple guava, fruits should be collected the day they drop from the tree, as this indicates ideal ripeness. Before that, the taste is more bitter, whereas after the fruit falls, it can quickly become overripe and unpleasant to consume.

3.1.1 Types of Pineapple Guava

- ☆ ***Acca sellowiana* 'Apollo'**: This variety is self-fertile and will pollinate with different varieties. It produces medium-sized fruit that ripens at the end of the season. This fruit is described as having a slight gritty texture.
- ☆ ***Acca sellowiana* 'Coolidge'**: This variety produces large, early fruits and is popular in cooler growing zones. It is considered self-fertile.
- ☆ ***Acca sellowiana* 'Nikita'**: This variety is among the earliest to produce fruits. It has a small, compact growing habit perfect for containers or small spaces.

3.2 Nutrition present in Pineapple Guava

The calorie content per serving (100 g) is only 55, which is unusually low. This tropical fruit also contains significant levels of vitamin C (more than 50 per cent of your daily recommendation per serving), as well as a diverse selection of B vitamins and trace amounts of vitamins E, K and A. In terms of mineral content, pineapple guava contains moderate levels of copper, manganese, magnesium, potassium, iron and calcium. A single serving of this fruit also delivers more than 15 per cent of your daily recommended dietary fiber, in addition to various phytochemicals, phenols and antioxidants (Ebrahimzadeh *et al.,* 2008).

3.3 Therapeutic Potential

Pineapple guava has a wide range of health effects, including weight-loss, improving digestion, lowering cholesterol levels, boosting the immune system, increasing bone strength, decreasing blood pressure, optimizing nutrient uptake and balancing the metabolism. It increases circulation, stimulating cognitive function and regulating blood sugar levels, among others. There are very few reported side effects, although allergies to this fruit do exist, and some reports of gastrointestinal distress and low blood sugar levels have been documented. It contains oil (Fernandez *et al.,* 2004).

Usually Crohn's disease and ulcerative colitis are referred as inflammatory bowel disease. Crohn's disease affects all digestive systems whereas ulcerative colitis affects large intestines. The investigation shows the capability of hydrophilic *Fejioa* fraction for modulation of autophagy by TLR2 with the use of IBD cell lines.

Results show that extracts of *Feijoa* is effective to manage inflammatory bowel disease in the patients (Monforte *et al.*, 2014).

The antibacterial activities were evaluated of an aqueous extract from *Feijoa* fruit. The extract was tested against Gram positive and Gram negative bacteria with the use of broth dilution test on whole blood leukocytes of humans and isolated neutrophils which is based on chemiluminescence assay. Experiment shows that an extract helps to inhibit the growth of bacteria. Thus result shows that *Feijoa* possesses antibacterial properties.

The juice extracted from *Feijoa* is used to ascertain anti-inflammatory activities with the use of carrageenan induced paw edema test. In vitro test is involved to find out the effects on release of superoxide anion from neutrophils in human blood. The juice was analysed with the use of high performance liquid chromatography method. It identifies the chemical compounds such as quercetin, eriocitrin, syringic acid, ellagic acid, pyrocatechol, gallic acid, catechin, rutin and eriodictyol. Results showed meaningful anti-inflammatory activity in fruit juice that helps to inhibit neutrophils basal activity in all experiments. It lowered edema significantly and inhibits O_2 production during experiment which make sure the anti-inflammatory effect of *Feijoa* juice.

The extract of *Feijoa* exhibited anti-cancer activities on haematological and solid cancer cells due to presence of flavons. It is observed that the extract did not have toxic effects on normal myeloid progenitors that suggested that extract of *Feijoa* possess tumor selective activities (Paola *et al.*, 2007). The purification, fractionation and analyses of extract showed flavone as an active component. Dietary fiber is well-known for its ability to provide relief from constipation. An average person requires half of the dietary fiber regularly. Women under 50 years of age should intake 25 g of fiber per day and about 38 grams for men. For children, the recommended intake of dietary fiber is 14 g/1000 Kcal. One *Feijoa* provides ¼ of the daily requirement of dietary fiber. Fiber is essential for our good gut bacteria. When those bacteria are healthy and strong then nutrients in food could be absorbed properly. In addition, healthy gut bacteria help to protect lining of gut wall.

Feijoa has high content of Vit. C which is associated with prevention of flu and colds. Antioxidants found in *Feijoa* assist to eliminate free radicals from blood. It lowers oxidative stress in lining of arteries. An inflammation oxidation is the main cause for atherosclerosis, heart disease and buildup of plaques (Weston 2010). *Feijoa* has healthy amounts of potassium which is known as electrolytes that assist in regulation of muscle functions in the body. Potassium influence the smooth function of muscles which allows arteries dilation as well as contraction as needed. Potassium plays a vital role in flowing of blood.

Study shows that an extract of *Feijoa* showed antidepressant effect in mice. *Feijoa* has high content of folate which is helpful in treating depression. The deficiency of folate is associated with increasing chances of depression. Folate helps to neutralize metabolite known as homocysteine. It activates homocysteine which could interfere with hormones such as serotonin that assist in regulation of mood and have become a target of most pharmaceutical antidepressants. *Feijoa* possess antimicrobial and antibacterial properties. It contains adequate amounts of minerals and vitamins so its regular consumption provides boost to immune system. Vitamin C assists in stimulation of white blood cells production which acts as the first line of defense of the body. It also contains antioxidant that helps to eliminate free radicals. A single serving (200g) of *Feijoa* provides more than 50 per cent daily recommended vitamin C.

The foods with high content of potassium are vital for the people suffering from high blood pressure. They have high chances of cardiovascular disease, strokes and atherosclerosis. Potassium acts as a vasodilator which helps to lower tension in blood vessels as well as arteries and also eases the strain on cardiovascular system. Usually Crohn's disease and ulcerative colitis are referred as inflammatory bowel disease. The investigation shows the capability of hydrophilic *Fejioa* fraction for modulation of autophagy by TLR2 with the use of IBD cell lines. Results show that extracts of *Feijoa* is effective to manage inflammatory bowel disease in the patients (Nasef *et al.,* 2015).The antibacterial activities were evaluated of an aqueous extract from the fruit. The extract was tested against Gram positive and Gram negative bacteria with the use of broth dilution test on whole blood leukocytes of humans and isolated neutrophils which is based on chemiluminescence assay. Experiment shows that an extract helps to inhibit the growth of bacteria. Results shows that *Feijoa* possess antibacterial properties which is used to counteract bacteria.

The juice extracted from *Feijoa* is used to ascertain anti-inflammatory activities with the use of carrageenan induced paw edema test. In vitro test is involved to find out the effects on release of superoxide anion from neutrophils in human blood. The juice was analyzed with the use of high performance liquid chromatography method. It identifies the chemical compounds such as quercetin, eriocitrin, syringic acid, ellagic acid, pyrocatechol, gallic acid, catechin, rutin and eriodictyol. Results showed meaningful anti-inflammatory activity in fruit juice that helps to inhibit neutrophils basal activity in all experiments. It lowered edema significantly and inhibits O_2 production during experiment which make sure the anti-inflammatory effect of *Feijoa* juice.

It is observed that the extract did not have toxic effects on normal myeloid progenitors that suggested that extract of *Feijoa* possess tumor selective activities. The purification, fractionation and analyses of extract showed flavone as an active component. Results showed that flavone is an active principle and the extract of *Feijoa* possesses anti-cancer activities. Dietary fiber is well-known for its ability to provide relief from constipation. An average person requires half of the dietary fiber regularly. Women under 50 years of age should intake 25 g of fiber per day and about 38 grams for men. For children, the recommended intake of dietary fiber is 14 g/1000 Kcal. One *Feijoa* provides ¼ of the daily requirement of dietary fiber.

Fiber is essential for our good gut bacteria. When those bacteria are healthy and strong then nutrients in food could be absorbed properly. In addition, healthy gut bacteria help to protect lining of gut wall.

Feijoa has high content of folate which is helpful in treating depression. The deficiency of folate is associated with increasing chances of depression. *Folate* helps to neutralize metabolite known as homocysteine

It activates homocysteine which could interfere with hormones such as serotonin that assist in regulation of mood and have become a target of most pharmaceutical antidepressants. Vitamin B are essential for the overall body functions and metabolic activities such as synthesizing red blood cells and proteins, stimulating nervous system function, directing hormone production and generating energy within cells. Favorably, *Feijoa* has balanced levels of vitamin B.

3.4 Culinary and Traditional Uses

Feijoa fruit is a delicious fruit and it finds its use in culinary delight. The yummy, egg-shaped fruit is grown in many parts of the world and is very popular for its taste and nutrition content. Pulp of the fruit is eaten by scooping it out from within the skin. For the sweet aroma of the pulp of *Feijoa* it is used in smoothie and even as a flavour in pie, cakes, ice-creams, yogurt, and cookies. Dried *Feijoa* is cooked in stew and even curry. It makes a great side dish. It is a common ingredient for chutneys and pickles. *Feijoa* wine, or *Feijoa* juice tinctured in vodka, gin, or other cocktails and mocktails make a great appetizing and yummy drink. *Feijos* is also used in fruit salad and other desserts. The tea prepared from fruit and flowers is helpful for treating diarrhea and dysentery.

In Paraguay, crushed flowers are applied to mild burns, rashes, stings, insect bites and itchy as well as inflamed areas. Lotions prepared from flowers are useful to ease sunburn. The fruit slices are used in the form of poultice. It is also useful for treating pyelonephritis, atherosclerosis and cardiovascular diseases. Because of the higher Vit. C content in *Feijoa,* it is usually utilized like a kitchen cure to heal common cold, and fever. The fruit is usually recommended to be eaten by individuals struggling with blood pressure. level irregularity. It will help in regularizing the blood pressure level. It is considered that frequent usage of *Feijoa* helps prevent from cancer. There are very few reported side effects, although allergies to this fruit do exist, and some reports of gastrointestinal distress and low blood sugar levels have been documented. Research has shown that eating pineapple guava fruit is able to help regulate blood sugar levels, due to its low level of calories and carbohydrates.

3.5 Genetic Diversity and Molecular Markers in *Feijoa*

There are two types or biotypes in the *A. sellowiana*, previously classified as the Uruguay and Brazil types. Also, the present study determined that there is significant genetic diversity in the *A. sellowiana* accessions composing the AGB-Feijoa, maintained by Epagri, in São Joaquim, Brazil (Saifert *et al.,* 2020).

Currently, the majority the AGB-Feijoa accessions were collected in the state of Santa Catarina or from altitudes of 900–1100 m. Taking into account the effects of climate change, a priority should be to collect accessions from altitudes and latitudes not yet represented in the collection. The genotypic data analyses revealed accessions originating from TIs, QLs, UCs, farmland linked to PGB, and other natural populations which are not represented in the AGB Feijoa.

In general, this study showed that some alleles of all eight loci are missing from the AGB-Feijoa. These results suggest the need for new collection expeditions of *Feijoa* germplasm to reduce the detected gap. The largest number of missing alleles (99) is concentrated in the areas of TIs, QLs, and UCs scattered throughout southern Brazil. These populations, which were characterized by Borsuk (2018), are extremely important for the conservation and improvement of *A. sellowiana*. In addition, we verified the population that had the largest number of alleles per locus, which would facilitate the organization of a collection expedition to incorporate these accessions in the AGB-Feijoa. In the TIs, Terra Indígena Cacique Doble and Terra Indígena Monte Caseiros, there are 19 alleles missing from the AGB-Feijoa. The QLs, Quilombola Faxina and Rincão do Couro and Quilombola Moçambique, have a total of 23 and 22 alleles, respectively, which are not represented in the AGB-Feijoa. In the UCs, the most relevant for conservation purposes would be the Refugio da Vida Silvestre dos Campos de Palmas with 31 alleles and the Floresta Nacional de Passo Fundo with 24 alleles missing. In situ on farm, the populations of the participatory genetic breeding (Donazzolo *et al.,* 2020) are of greater interest for conservation since they are mainly cultivated in agro-forestry systems in Ipê, Antônio Prado, and Monte Alegre dos Campos-RS. Overall, those populations store 26 alleles, most of which are missing from the AGB-Feijoa. In addition, we detected that 25 alleles present in natural populations and farmland of Rio Grande do Sul (Donazzolo 2012) are missing from the AGBFeijoa, which are also relevant for conservation (Saifert *et al.,* 2020).

Amplified fragment length polymorphisms (AFLP), intersimple sequence repeat (ISSR), and a small number of simple sequence repeats (SSR) markers were used in the construction on the first genetic map of the species (Quezada *et al.,* 2014). This map along with those developed for the species *Psidium guajava* (Padmakar *et al.,* 2015) represent the only ones developed for fruit species within the Myrtaceae family. The 641 common markers in the composite map presented all five possible joint marker configuration types defined considering the segregation pattern in both populations simultaneously (Quezada *et al.,* 2021). However, markers heterozygous in the maternal genotype and homozygous in both paternal genotypes ("ab" × "aa"/"ab" × "aa," "D1.10"-"D1.10") were largely the most frequent, with 449 (70.05 per cent) of the 641 common mapped markers. The configuration type "D2.15"-"D2.15," homozygous in the female parent and heterozygous for the male parent in both populations, was the second most frequent marker type, with 70 (10.92 per cent) mapped markers. A high agreement was observed between LG assignment of the Discovery-identified SNPs, mapped in the composite map and the physical positions of these markers among the *Eucalyptus grandis* chromosomes. This same result was observed in the single population maps. Out of 1,897 markers mapped in the composite map, 474 (24.98 per cent) SNPs corresponded to markers identified by the Discovery pipeline, and of these, 437 (92.19 per cent) showed correspondence between the assignment to a LGs in *A. sellowiana* and *E. grandis* chromosomes. Other 32 markers were mapped in discordant reference chromosomes and the remaining five were assigned to *E. grandis* genome scaffolds. Despite the high level of synteny observed, the marker order (colinearity) in the *A. sellowiana* composite map was not consistent with the physical positions in the E. grandis genome. This map will be also a useful tool to guide the assembly of the *A. sellowiana* genome sequence, as a reference to anchor and orient the sequence scaffolds. With the development of composite genetic maps, the occurrence of genome structural variation or conserved synteny can be evaluated across divergent species (Quezada *et al.*, 2014).

3.6 Conclusion

People who regularly consume pineapple guava will receive health benefits related to blood pressure, cholesterol, obesity, immune health, oxidative stress, metabolism, osteoporosis, indigestion, diabetes, circulation, cognitive function, and nutrient deficiencies. With a strong supply of vitamins and minerals in the pineapple guava fruit, regular consumption can give your immune system a much-needed boost. Vitamin C is able to stimulate the production of white blood cells, the body's first line of defense, while also acting as an antioxidant to seek out free radicals. A single serving of pineapple guava has more than 50 per cent of your daily recommended vitamin C. Potassium-rich foods are important for people who are suffering from high blood pressure, and are therefore at high risk of cardiovascular disease, atherosclerosis and strokes. Potassium is a vasodilator, meaning that it can reduce the tension in blood vessels and arteries, and generally ease strain on the cardiovascular system. High levels of dietary fiber (roughly 17 per cent of your daily recommended fiber per serving) means that this fruit is able to optimize digestion by stimulating peristaltic motion and improving nutrient uptake. This can help soothe symptoms of indigestion, constipation, bloating, cramping and general stomach upset. Besides improving digestion, dietary fiber is also directly linked with lower levels of cholesterol, particularly "bad" cholesterol that can increase your risk of heart disease. By scraping this cholesterol out of the arteries and blood vessels, you decrease your risk of blood clots, heart attacks and strokes. The antioxidants present in pineapple guava have been associated with increased memory and retention, better focus, and a reduced risk of neuro-degenerative diseases, such as Alzheimer's disease and dementia.

The antioxidants can seek out and neutralize free radicals in neural pathways before they can cause the accumulation of plaque. B vitamins present in it are incredibly important for the overall functioning of the body, particularly when it comes to metabolic activities like synthesizing proteins and red blood cells, directing hormone production, stimulating nervous system function, generating energy within the cells. This can increase your energy levels and keep you more active and fit in daily routine. Although there is a relatively small amount of iron in pineapple guava, it can still aid your red blood cell production and circulation, while the metabolic boost of Vit. B can also stimulate blood flow. This means increased oxygenation to critical areas of the body, and higher energy levels.

REFERENCES

Borsuk, L. J. (2018) Avaliação da diversidade genética e morfológica da goiabeira-serrana [*Acca sellowiana* (O. Berg) Burret] em Terra Indígenas, Áreas Quilombolas e em Unidades de Conservação no Sul do Brasil e acesso ao conhecimento tradicional associado ao uso e manejo da espécie. Ph.D. diss., Federal Univ. of Santa Catarina, Florianópolis.

Donazzolo, J. (2012). Conservação pelo uso e domesticação da Feijoa na serra gaúcha-RS. Ph.D. diss., Florianópolis: Federal Univ. of Santa Catarina.

Donazzolo, J., Stefenon, V. M., Guerra, M. P., and Nodari, R. O. (2020). On farm management of *Acca sellowiana* (Myrtaceae) as a strategy for conservation of species genetic diversity. Scientia Horticulturae, Amsterdam 259, 108826. https: //doi.org/10. 1016/j.scienta.2019.108826

Ebrahimzadeh, M. A., Hosseinimehr S. J., Hamidinia A. and Jafari M. (2008), Antioxidant and free radical scavenging activity of *Feijoa sellowiana* fruits peel and leaves, Pharmacologyonline 1: 7-14.

Fernandez, X., Loiseau, A. M., Poulain, S. and Monnier, Y. (2004), Chemical Composition of the Essential Oil from Feijoa (*Feijoa sellowiana* Berg.) Peel, Journal of Essential Oil Research 16(3): 274-275.

Monforte, M. T., Fimiani, V., Lanuzza, F., Naccari, C., Restuccia, S. and Galati, E. M. (2014), Feijoasellowiana Berg Fruit Juice: Anti-Inflammatory Effect and Activity on Superoxide Anion Generation, Journal of Medicinal Food, 17 (4): 455-461.

Nasef, N. A., Mehta, S., Powell, P., Marlow, G., Wileman, T. and Ferguson, L. R. (2015), Extracts of *Feijoa* Inhibit Toll-Like Receptor 2 Signaling and Activate Autophagy Implicating a Role in Dietary Control of IBD, PLoS ONE 10(6): e0130910. doi: 10.1371/journal.pone.0130910

Padmakar, B., Kanupriya, C., Latha, P. M., Prashant, K., Dinesh, M., Sailaja, D. (2015). Development of SRAP and SSR marker-based genetic linkage maps of guava (Psidium guajava L.). Sci. Hortic. 192, 158–165. doi: 10.1016/j. scienta.2015.04.031

Paola, B., Luigi, M., Marco, M., Antonella, D. and Angela N. (2007), *Feijoa sellowiana* derived natural Flavone exerts anti-cancer action displaying HDAC inhibitory activities, The international journal of biochemistry and cell biology, 39 (10) doi: 10.1016/j.biocel.2007.05.010

Quezada M, Amadeu R R, Vignale B, Cabrera D, Pritsch C, Garcia A A F (2021) Construction of a High-Density Genetic Map of *Acca sellowiana* (Berg.) Burret, an Out crossing Species, Based on Two Connected Mapping Populations. Front. Plant Sci. 12: 626811. doi: 10.3389/fpls.2021.626811

Quezada, M., Pastina, M. M., Ravest, G., Silva, P., Vignale, B., Cabrera, D. (2014). A first genetic map of *Acca sellowiana* based on ISSR, AFLP and SSR markers. Sci. Hortic. 169, 138–146. doi: 10.1016/j.scienta.2014. 02.009

Saifert L, Sánchez-Mora F D, Borsuk L J, Donazzolo J, da Costa N C F, Ribeiro H N, Nodari R O (2020) Evaluation of the genetic diversity in the *Feijoa* accessions maintained at Santa Catarina, Brazil. Plant Genetic Resource. Crop Science. 60: 345–356.

Shaw GJ, Ellingham PJ and Birch EJ. 1983. Volatile constituents of feijoa-headspace analysis of intact fruit. *J.Sci.Fd.Agric.* 34: 743-747.

Weston RJ. (2010). Bioactive products from fruit of the feijoa (*Feijoa sellowiana*, Myrtaceae): A review FoodChemistry 121(4): 923-926.

Chapter 4

Multipurpose Butternut Tree: Mahua [*Madhuca longifolia* (J.Konig) J.F.Macbr.]

Mahua is one of the most common trees available in the forests of Gujarat, U.P., M.P. and Chhattisgarh region, and many tribal people earn their employment by collection, processing and shelling the flowers, fruits, and seeds of this plant. Mahua [Madhuca longifolia (J. Konig) J.F.Macbr.] belongs to the dicot family of Sapotaceae. This is a wild plant, generally grown without investing anything in forests, this tree is approximately 20 m in height and is evergreen or semi-evergreen in nature. It has short, whole and large rounded crown throughout the green forest part of India. Mahua is of an Indian origin tree which has several undiscovered marvelous properties. Besides food value limited data and survey has been documented regarding non-medicinal utilization of Mahua. Due to presence of high sugar content flowers are edible. They also contain good amount of minerals like Calcium and Phosphorus. Small amount of protein and fat are also present in flowers. Trifed, a website of the Ministry of Tribal Affairs, Government of India reports: "Mahua oil is edible and has medicinal properties. It is emollient and is used in skin disease, rheumatism and headache. It is also a laxative and considered useful in habitual constipation, piles and haemorrhoids and as an emetic. Native tribes also used it as an illuminant and hair fixer." It has also potential to be used as biodiesel.

Keywords: *Multipurpose tree, Butternut, Mahua, Flowers, Seed, Therapeutic uses, Madhuca longifolia.*

4.1 Introduction

Nature has blessed us extensively with the wide range of diversified plants used for various purposes like decoration, flowering, fruiting and medicinal, *etc.* India is considered as a treasure of various medicinal and aromatic plant species which are being utilized since ancient time. According to WHO (2003) approximately 65 per cent of the world's population integrate the medicinal plant for treatment. *Madhuca* is one of the naturally occurring tree with numerous health benefits. Tribal people use its flowers for curing of skin diseases, headache, pitta and bronchitis (Sinha *et al.,* 2017). Flower juice is supplemented to lactating women for augmentation of breast milk (Sunita and Sarojini 2013). Flowers of plant are edible and have high nutritive value majorly high amount of sugars and subsequently having good amount of vitamins, proteins, minerals and fats (Patel and Naik 2010; Patel *et al.,* 2011). Because of the higher amount of sugar, the flowers are utilized as a sweetener in preparation of numerous traditional dishes like *halwa, kheer, meethi puri* and *barfi etc.*

Due to the availability of the flower for a shorter time and highly perishable nature of the flower, it is not yet fully explored, except a few for its value addition. The flowers as food and medicine and its future prospective regarding its value addition (Behera *et al.,* 2016). It is a multipurpose tree which fulfil three fundamental needs of tribal individuals *i.e.* Food, Fodder and Fuel. It is cultivated in warm and humid regions for its oleaginous seeds, producing between 20 and 300 kg of seeds annually per tree, depending on maturity. The fat is used for the care of the skin, to manufacture soap or detergents, and as a vegetable butter. The wood is also important. India is known for wide diversity of such plants, which are utilized traditionally and have significance of being commercialized *Mahua* is one of those plants which is occupying novel space in the ethnic as well as economical life of the traditional people.

4.1.1 Area and Production

Mahua is a frost resistant tree that can grow in marginal areas of dry tropical and subtropical forests up to an altitude of 1200-1800 m, in India. It requires mean annual temperature of 2-46°C, mean annual rainfall ranging from 550-1500 mm and mean annual humidity from 40-90 per cent (Kureel *et al.,* 2009). In India, large quantities of deciduous *mahua* trees are found in the 12 states and the annual production of flowers is 45000 Million tonnes during 2016 (Behera *et al.,* 2016). Madhya Pradesh is the most astounding *mahua* developing state with average trade volume of 5,730 metric tonnes and worth about Indian rupees 8.4 million (Thakor and Babu 2008).

4.1.2 *Madhuca* a Sacred Tree

Madhûca tree is the sacred tree of various temples in South India, including Irumbai Mahaleswarar Temple, Iluppaipattu Neelakandeswarar Temple, Tirukkodimaada Senkundrur at Tiruchengode, and Thiruvanathapuram (www.shaivam.org). The Tamil saint philosopher Valluvar is believed to have born under an Iluppai tree within the Ekambareshwarar Temple at Mylapore, and hence mahua remains the sanctum tree of the Valluvar shrine built within the Ekambareshwarar temple complex (Ramakrishnan 2019).

4.2 Constituents Present in Flowers and Fruits

Mahua flowers are rich source of sugars which is responsible for its sweet taste and can be utilized to make indigenous or modern alcoholic beverages. *Mahua* flowers contains good amount of Vit. C which is responsible for its antioxidant activity (Indu and Annika 2014). *Mahua* flower contains carotene which is precursor of Vit. A. Flowers also contain good amount of minerals like Calcium and Phosphorus. Small amount of protein and fat are also present in *mahua* flowers. Brief composition of *mahua* flower is given in Table 4.1. Various researches have been done to find out medicinal properties of *mahua* flowers like antihelmenthic, antibacterial, analgesic, hepatoprotective, antioxidant, and anticancer (Table 4.2).

Table 4.1. Composition of *Madhuca* Flower

Sl.No.	*Constituents*	*Fresh Flowers*	*Dry Flowers*
1	Moisture	73.6-79.82 (per cent, d.b.)	11.61-19.8 (per cent, w.b)
2	pH	4.6	—
3	Starch (g/100 g)	0.94	—
4	Ash (per cent)	1.5	1.4-4.36
5	Total sugars (g/100 g)	47.35-54.06	41.62
6	Total Inverts (per cent)	54.24	—
7	Cane sugars (per cent)	3.43	—
8	Reducing sugars (g/100 g)	36.3-50.62	28.12
9	Proteins (per cent)	6.05-6.37	5.62
10	Fats (per cent)	1.6	0.09-0.06
11	Fibers (per cent)	10.8	—
12	Calcium (mg/100 g)	45	0.14-8
13	Phosphorus (mg/100 g)	22	0.14-2
14	Carotene (μg/100 g)	307	—
15	Vitamin-C (mg/100 g)	40	7

Source: Hiwale (2015); Patel *et al.* (2011).

Seed oils and fats are increasingly being utilized for biofuels, making these oils potentially more costly and a less plentiful and reliable supply for food needs. Oil and fats market is interested in solutions that can either address these agricultural supply chain challenges. There are few plants that produce oil and fat in sufficient commercial quantity. Vegetable oils contribute about 85 per cent of the visible oil or fat available for consumption (Lawson 1995; Yadav *et al.,* 2012a). Several plants are grown not only for food and fodder but also for amazing variety of novel

Table 4.2. Medicinal Properties of *Madhuca* Flowers

Medicinal Properties	*Type of Extract*	*Remarks*	*References*
Hepato-protective activity	Methanolic	Methanolic extract of flower showed potential protective effect by lowering the levels of SGOT, SGPT, ALP and total bilirubin by increasing serum level of total proteins and albumins.	Umadevi *et al.* (2011); Patel *et al.* (2012); Yadav *et al.* (2012) (a); Mishra and Pradhan, (2013); Sinha *et al.* (2017)
Antihel-menthic activity	Both methanolic and ethanolic	Among both extracts methanolic extract of flower demostrated best anti helmenthic activity against Indian earth worm.	Katiyar *et al.* (2011); Yadav *et al.* (2012); Sinha *et al.* (2017)
Antibacterial activity	Both aqueous and methanolic	Aqueous extract showed more antibacterial activity than methanolic one for *Bacillus subtilis* and *Klebsiella pneumonia*	Verma *et al.* (2010); Patel *et al.* (2012); Yadav *et al.* (2012); Sinha *et al.* (2017)
Analgesic activity	Both aqueous and alcoholic	Analgesic effect was studied through tail flick, hot plate and chemical graded doses on mouse which shows analgesic effect as per dose value.	Chandra, (2001); Saluja *et al.* (2011); Patel *et al.* (2012); Yadav *et al.* (2012) (a); Amia and Ekka, (2014); Verma *et al.* (2010); Sinha *et al.* (2017)
Antioxidant activity	—	As concentration of flower extract and ascorbic acid increases, the ferric reducing antioxidant power increases.	Indu and Annika, (2014)
Anticancer activity	—	Cell viability was found to decreases as the concentration of floral extract increases and cytotoxic effect was found to increase.	Indu and Annika, (2014)

SGOT: Serum glutamic oxaloacetic transaminase; SGPT: Serum glutamic pyruvic transaminase; ALP: Alkaline phosphatase.

industrial products. Search is necessary of new sources of oil plants which provide high recovery of oil or fat for nutritional, pharmaceutical applications and other uses. The genus *Madhuca*, belonging to the family of Sapotaceae, is a multipurpose tree with its species, *Madhuca longifolia, M. latifolia* and *M. butyracea* being the most prevalent. Buttercup or mahua, *Madhuca longifolia* (Koenig) (Synonyms, *Madhuca indica* Gmelin; Family, Sapotaceae), is a large, shady, deciduous tree dotting much of the central Indian landscape, both wild and cultivated. The tree is economically important because of widespread uses of its flowers, fruits, seeds and timber (Ramadan *et al.,* 2006). The tree wins in fame due to the liquor distilled from the flowers, which are used to make vinegar. The expectorant flowers are used to treat chest problems such as bronchitis. They are also taken to increase production of breast milk. The distilled juice of the flowers is considered a tonic, both nutritional and cooling. The leaves are applied as a poultice to relieve eczema. In Indian folk medicine, the leaf ash is mixed with ghee to make a dressing for wounds and burns. Mahua preparations are used for removing intestinal worms, in respiratory infections and in cases of debility and emaciation. The astringent bark extract is used for dental-related problems, rheumatism, and diabetes. *Madhuca longifolia* flowers seasonally and produces green-fleshy fruits containing three to four seeds. The buttercup fruit-seeds, generally ellipsoidally shaped (Figure 4.1a), measured from 1.5 to 2.0 cm and from 1.3 to 1.6 cm across the length and breadth, respectively. The medicinal properties attributed to this plant are stimulant, demulcent, emollient, heating, and astringent (Ramadan and Moersel 2006; Yadav *et al.,* 2011a, b). Previous phytochemical studies included characterization of sapogenins, carbohydrates, triterpenoids, steroids, saponins, flavonoids, and glycosides (Yosioka *et al.,* 1974; Yoshikawa *et al.,* 2000). In view of the attributed medicinal properties, new components including: madhucic acid (a pentacyclic triterpenoid), madhushazone (an untypical isoflavone), madhusalmone [(a bis(isoflavone)] (Yoshikawa *et al.,* 2000) and four new oleanane-type triterpene glycosides (madlongisides A-D) (Siddiqui *et al.,* 2004). Madhucosides A and B were isolated from *M. longifolia* and showed significant inhibitory effects on both superoxide release from polymorphonuclear cells and hypochlorous acid generation from neutrophils (Pawar and Bhutani 2004). The fruit is valued for its seed which contains high quantity of lipids (ca. 50–61 per cent), commercially known as mahua or mowrah butter, and it has many edible, medicinal and non-food applications (Ramadan *et al.,* 2006). The whole mahua seeds contain 50–61 per cent oil, 16.9 per cent protein, 3.2 per cent fiber, 22 per cent carbohydrates, 3.4 per cent ash, 2.5 per cent saponins, and 0.5 per cent tannins. Oil represents the major component which is thrice the amount of protein. The deoiled seed cake contains 30 per cent protein, 1 per cent oil, 8.6 per cent fiber, 42.8 per cent carbohydrates, 6 per cent ash, 9.8 per cent saponins, and 1 per cent tannins (Singh and Singh 1991). Defatting of mahua seed increased the protein, saponin and tannin levels. The levels of saponins could be reduced by treatment with isopropanol. The deoiled mahua seed cake showed good oil absorption and emulsification properties. The in vitro digestibility of mahua seed cake after treatment with isopropanol was found to be 81 per cent. Detoxified mahua seed flour appears to be a good source of protein for food and feed products (Singh and Singh 1991; Ramadan and Moersel 2006). The cake is a rich source of sugars, nitrogen and proteins. The cake has been used as a low grade fertilizer, bio-pesticide, included in animal feed (up to 20 per cent) and in dye removal from waste waters. Various detoxification methods have also been tried for the use of cake as an improved animal feed only. Gupta *et al.* (2012) evaluated the use of raw as well as detoxified cake (produced from water treatment) for biogas production and mushroom cultivation. Significant enhancement in the biogas (93 per cent) and the mushroom yield (128 per cent) was obtained. In a recent study, Gupta *et al.* (2013) evaluated the conditions affecting biogas production from raw and detoxified mahua seed cake. Detoxified mahua seed cake produced better results compared to raw cake. Significant reduction in celluloses (34.4 per cent) and hemicelluloses (29.7 per cent) and an increase in the nutrients (NPK) of the digested slurry were obtained.

4.3 Utilization of Mahua Flowers

Due to its valuable composition, the flower is used traditionally and is still utilized in value-addition of different products.

4.3.1 Traditional Uses

Flower bearing period of *mahua* is March-April, as it is an annual bearing tree. Flower sheds when it gets mature at dawn. Fresh *mahua* flowers are sweet in taste and contain different phytochemicals. Traditionally, the fresh flowers are collected and dried under direct sunlight for 2-3 days and stored in gunny bags in normal environment.

4.3.2 Utilization in Food Production

Due to high content of sugar (sucrose, glucose, fructose, arabinose, few amount of maltose and rhamnose) tribal people are utilizing *mahua* flowers as a sweetening agent in numerous local and traditional dishes like *halwa, meethi puri, kheer* and *burfi* (Patel and Naik 2010). Behera *et al.* (2016) referred that tribal people are also utilizing the *mahua* flowers along with some grains (rice, ragi, jowar) or root crops (sweet potato) for preparing cake. Sun-dried flowers are boiled with seeds of tamarind and sal, taken as substitute of grain staples by poor tribal people. Apart from this, flowers, fruits are likewise utilized as cattle feed which helps in improvement of health of cattle and increase in milk production (Sinha *et al.*, 2017).

In term of fermented products flowers are also used as crude material for production of alcohol and alcoholic beverages. Local people of North-West India used to collect and dry the *mahua* flowers for preparation of "*mahua daaru*" which contains 20-40 (per cent)alcohol. *Mahua* lowers are mixed with water and kept aside for fermentation. During the fermentation *Navshar* (Ammonium chloride) and jaggery are added. Sometime black pepper is added to develop a strong hot flavour. After fermentation, the mixture is kept in a container having traditional distillation setup. It is reported that one kilogram of dried flowers yields 300-400 ml of *daaru* by this method (Kumari *et al.*, 2018).

Mahua flowers are also utilized by the tribal people of Gujarat in making liquire and Orissa to make a Country liquor called *mahuli* (Behera *et al.*, 2018). The method of *mahuli* preparation is similar to earlier described method of *mahua daaru* but only the difference found in distillation process and ingredients *i.e.* at the time of fermentation *bakhar tablets* are added. These tablets consists of *Asparagus racemosus* (roots), *Cissampleos pareira* L (roots), *Clerodendrum serrautum* (roots), *Dipteracanthus suffruticosus* (roots), *Elephantopus scaber* L. (roots), *Lygodium flexuosum* (roots), *Ochna obtisata* (roots), *Phoenix acaulis* (roots), *Holarrhena pubescens* (bark), *Homalium nepalense* (bark), *Woodfordia fruticose* L (flower), *Xantolis tomentosa* (fruit), *M. indica* Gmel (seeds)] is added. Behera *et al.* (2016) fermmental fermented flowers are kept for distillation and alcohol content which found in distillate is about 30-40 per cent. A brief description of various non-fermented and fermented traditional products is described in Table 4.3. In Madhya Pradesh, the Gonds make a drink sulfi from sulfi tree. From ancient times *Mahua* flowers are used by tribals and also Britishers in past produced alcoholic drink (Sen 2015).

Table 4.3. Traditional Uses of *Mahua* Flowers

Sl.No.	*Uses*	*Key Points*	*Remarks*	*References*
	Non-fermented flowers			
1	As a sweetener	*Mahua* flower used as a sweetener in many dishes like halwa, meethi puri, barfi.	Due to presence of high amount of sugars (sucrose, fructose, arabinose, maltose, rhamnose.	Patel (2008)
2	Preparation of cake	It is made from *mahua* flowers rice or other cereals or root crops.	Pre-soaked rice and *mahua* flowers are mixed and grinded, paste is covered with Sal leaves and burned on fire to make cake.	Behera *et al.*, 2016
3	As a substitute for staple grains	It is generally used by poor tribal people.	Sundried flowers are boiled with seeds of tamarind and Sal and stored.	Amia and Ekkka 2014
4	As a cattle feed	Spent flowers (Flowers left after fermentation and distillation) are used.	Spent flowers are fed to cattle, reported improvement in cattle health and increase in milk production.	Sinha *et al.*, 2017
	Fermented flowers			
5	Preparation of "*mahua daaru*"	Produced from dried *mahua* flowers by tribal people,	Alcohol content of "*mahua daaru*" ranges from 20-40 (per cent).	Kumari *et al.*, 2016
6	"*Mahuli*" preparation	Traditionally made by local people of Orissa.	Alcohol content of "*mahuli*" is reported between 30-40 (per cent).	Kumari *et al.*, 2016; Behera *et al.*, 2016

Food and health-promoting applications *Mahua* butter has many edible and medicinal applications. *Mahua* butter has been used as substitute for cocoa butter. The semisolid *mahua* fat is used in cooking, adulteration of ghee, and

Figure 4.1A-D. (A) Fresh Flowers; (B) Fruit and (C) Seeds of *Madhuca indica;* (D) Sun Drying of *Madhuca* using Traditional *Supa* Prepared from Bamboo in Chhattisgarh Village, India.

manufacturing chocolates. The seed fat has emuluscent property; it is used for skin disease, rheumatism, headache, laxative, piles and sometimes used as galactogogue (Ramadan and Moersel 2006; Yadav *et al.,* 2011a). Khatoon and Reddy (2005) prepared plastic fats with no trans fatty acids suitable for use in bakery and as vanaspati by interesterification of blends of palm hard fraction (PSt) with mahua fats. Blends of PSt/mahua (1:1 and 1:2) showed three distinct endotherms, indicating heterogeneity of TAG, the proportions of which altered after inter esterification. After inter esterification for 1 h at 80°C blends showed an SFC similar to those of commercial hydrogenated shortenings and vanaspati. Therefore, they could be used in these applications in place of hydrogenated fats as they are free from trans fatty acids. Jeyarani and Yella Reddy (2010) prepared specialty fats using mahua and kokum by enzymatic interesterification (IE) using 1,3-specific lipase (Lipozyme TL IM). There was a significant change in the solids fat content with time of IE and this is attributed to the decrease in monounsaturated and disaturated types of TAG and increase in trisaturated TAG. The melting points of the blend subjected to IE for 60 min resembled that of commercial milk fat and the one inter esterified for 6 h showed a wider melting range, similar to that of hydrogenated fats used bakery purposes. Besides its edible and medicinal uses, mahua fats can also be utilized in the manufacture of laundry soaps and lubricants (Parrota 2001; Ramadan and Moersel 2006).

4.4 Utilization for Medicinal Purposes

The universal role of plants in the treatment of disease is exemplified by their employment in all the major system of medicine, irrespective of the underlying philosophical premise (Evans 2009). Plants are having a great importance to pharmaceutical industry, because these are rich source of drugs and a vast reservoir of chemical diversity for screening programs aimed at new drug discovery. Most of the drugs which are mention in the Indian medicinal system are from plant source (Hoffman and Leaders 1996). The duration of the mediaeval period is known as between 8th century to 18th century AD (Md. Alam 2008). In Ayurveda, *mahua* flowers are considered as to be cooling agent, carminative, galactagogue, and astringent. It is also reported to be beneficial for heart, skin, and eye diseases. *Mahua* flowers are used traditionally as a remedy of many diseases by tribal people. Fresh juice of flower is utilized as tonic and also cure skin diseases, eye diseases, raktapitta and headache due to "pitta". Sunita and Sarojini, (2013) referred that tribal people offer raw flowers to lactating mothers for increasing their lactation. Roasted flowers are consumed to cure cough and bronchitis by local people. Acharya and Srivastava, (2008) reported that *mahua* flowers can cure impotency and general debility when consumed with milk. Flowers fried in ghee act as a cooling agent and help to cure piles. Table 4.4 provides brief description of utilization of *mahua* flower as remedy for various diseases by traditional people. Charak Samhita (Saxena 2023 p. 742) mentioned use of *Madhuca* leaves and bark when used after heating helps in gouty arthritis.

a) Anti-Inflammatory Effect

Inflammation is a result of the defensive mechanism of our immune system. When any foreign particle enters into our body, the defensive mechanism starts working, and this initiates with the migration of WBC towards the injury site, and phagocytic action takes place. The inflammation is a stepwise process which starts with phosphide which forms arachidonic acid followed by the production of prostaglandin which starts pain. Almost all anti-inflammatory drugs in nature act by inhibiting the COX, either selectively or non-selectively. The anti-inflammatory action of Mahua has been proved by a few researchers. They induced inflammation in the hind paw of the rat using carrageenan and formaldehyde, and the extract of *Madhuca indica* seed in petroleum ether as a solvent was compared with the standard drug, diclofenac sodium in response to inflammation. The overall test was performed and evaluated in the plethysmograph apparatus.

Table 4.4. Traditional Medicinal Uses of *Mahua* Flowers

Medicinal Uses	*Way of Consumption*	*Remarks*	*References*
Used as tonic	Flower juice	Flower juice having high amount of protein so it is used as tonic.	Mishra and Pradhan 2013; Amia and Ekka 2014; Sinha *et al.*, 2017
Cure skin diseases		Flower juice rubbed on skin for oleation to relieve from itching	Mishra and Pradhan, 2013; Amia and Ekka, 2014; Sinha *et al.*, 2017
Cure eye diseases		Flower juice is used for treatment of eye diseases.	Amia and Ekka 2014
Cure Raktapitta		Flower juice is used to arrest bleeding	Sinha *et al.*, 2017
Cure headache due to "pitta"		Flower juice is used as nasal drops	Sinha *et al.*, 2017
Cure diarrhoea and colitis	Flower powder	Flower act as an astringent to cure diarrhoea and colitis	Amia and Ekka *et al.*, 2014; Sinha *et al.*, 2017
Increases lactation	Raw flowers	Flowers act as a galactogouge which can help in augmentation of breast milk.	Mishra and Pradhan, 2013; Sinha *et al.*, 2017
Cure cough and bronchitis	Roasted flowers	—	Palani *et al.*, 2010; Chandra 2011; Mishra and Pradhan, 2013; Sinha *et al.*, 2017
Cure impotency and general debility	Flower mixed with milk	—	Acharya and Shrivastava, 2008; Amia and Ekka 2014
Cure piles	Flowers fried in ghee	*Mahua* flower act as a cooling agent to cure piles	Sinha *et al.*, 2017

b) Antidiabetic Activity

Madhuca indica has also shown its effectiveness as an antidiabetic. Methanolic extract of the leaves and bark of this plant has been studied in streptozotocin-induced diabetes in male albino Wistar rat, and the result was compared with the standard insulin-treated rat. After 30 days of evaluation periods, the result showed a markedly decrease in blood sugar level of the rat who had received the extract, orally. This result confirms the antidiabetic activity of *M. indica* (Swain *et al.*, 2007).

c) Anthelmintic Activity

The anthelmintic activity of the ethanolic extract of the flower and bark of this plant has been proved by a few researchers. They used the ethanolic and methanolic extracts of *M. indica* and compared their anthelmintic activity by in vivo method and used metronidazole as the standard drug. It was considered that the anthelmintic activity may be because of the presence of tannins. The ethanolic, methanolic and hydroalcoholic extracts of *M. indica* have tannins as one of the phytoconstituents with anthelmintic activity (Patel *et al.,* 2012).

d) Antimicrobial Activity

In many regions of Chhattisgarh tribal people used to consume the leaves and bark of Mahua to treat their wounds (Yadav *et al.,* 2012b). They firstly crush the fresh and young leaves and once their active constituents come out they apply this juice cum extract on their wounds, and after regular interval of application of the extract, they find a good result and the wound started to heal and thus the activity of Mahua as an antimicrobial came into light. Several investigators performed this activity in the laboratory using aqueous, ethanolic and methanolic solution to extract the phytoconstituents of leaves and bark parts of Mahua (*Madhuca indica*) by using cup plate methods (Katiyar 2011). They have taken Ampicillin as a standard drug and after comparison; they found satisfactory response against *Staphylococcus* and *Escherichia coli* as a sample bacteria.

4.5 Value addition of *Mahua* Flowers

Nowadays the researchers and food processors are getting attracted toward the underutilized crops for the development of new products. Because of this few products are developed from *mahua* flowers also. But these are very limited at limited research institutes, universities, and industries, which might be due to limited availability in limited space and perishability. Some of the possible products which have been prepared from *mahua* flowers by researchers and processors to utilize this flower commercially for its value addition in term of fermented and non-fermented foods (Pinakin 2018).

Sinha *et al.* (2017) reported that fresh flower juice is concentrated and used as liquid sweetener in bakery and confectionery products. Fresh flowers are also utilized for making puree and sauces by crushing the flowers. Pulp of ripe flowers can be utilized to make intermediate moisture foods (IMF) like jam, jelly and marmalade (Chandra 2011; Patel *et al.,* 2008).

Recently Orissa University of Agriculture and Technology, Bhubaneshwar has developed many value-added products from dry *mahua* flowers like candy, cake, RTS, toffee, squash and ladoo (Dash 2017). Mishra *et al.* (2013) reported the storage study of *mahua* RTS in combination of ginger and fennel extracts. Other value-added products like candied flower, glazed flower and *mahua* bar are also developed from dried *mahua* flowers (Bakhara *et al.,* 2016). Because of having high amount of fermentable sugars, *mahua* flowers are utilized for making wine by various researchers scientifically using *Saccharomyces cerevisiae* (Kumari *et al.,* 2016). Freshly prepared *mahua* wine has been fortified with traditional Indian herbs (Black pepper, cinnamon, clove, cumin, fenugreek, nutmeg, fennel and Indian cassia) for development of new value-added product, called *mahua* vermouth (Yadav *et al.,* 2012). Malavade and Jadav, (2000) reported that dry flowers are also utilized to make fermented products like brandy, acetone, and lactic acid. Beside that all *mahua* flowers can be successfully used as a substrate for surface fermentation using *Aspergillus niger* for production of citric acid. But in these aspects the availability of the documentation is still very low which need to be explored by the researchers and food processors. The complete detail of the utilization of *mahua* flower recently by various researchers for preparation of value added products is given in Table 4.5 along with some specific remarks.

Table 4.5: Value-Added Products of made from *Madhuca* Flowers

Sl.No.	*Products*	*Way of Utilization*	*Remarks*	*References*
	Non-fermented flowers			
1	Puree and sauces	Fresh flowers	Fresh flowers are crushed into puree after removal of stamens manually and processed to make puree.	Patel 2008; Sinha *et al.,* 2017
2	Juice Concentrates	Fresh flower juice	Used as a sweetener in bakery and confectionary.	Sinha *et al.,* 2017
3	*Mahua* Jam	Pulp of ripe flowers	Jam is made with addition of citric acid.	Patel,2008; Jha *et al.,* 2013; Bakhara *et al.,* 2016; Sinha *et al.,* 2017
4	Jelly		Combined with guava to reduce astringency of *mahua* flower.	Sinha *et al.,* 2017
5	Marmalade		By addition of citrus peels.	Sinha *et al.,* 2017
6	Candied flowers	Dry flowers	—	Bakhara *et al.,* 2016; Dash, 2017.
7	Glazed flowers			
8	*Mahua* bar			
9	*Mahua* candy			
10	*Mahua* toffee			
11	*Mahua* cake			
12	*Mahua* squash			
13	*Mahua* ladoo			
14	*Mahua* RTS		RTS blended with ginger extract @ 10 (per cent) have TSS of 18° Brix and with fennel extract @ 5 (per cent) have TSS of 14.8° Brix.	Mishra *et al.,* 2013

Sl.No.	*Products*	*Way of Utilization*	*Remarks*	*References*
15	Antioxidant rich beverage	*Mahua* flower and Amla juice	The blend showed TPC of 15.94 (mg GAE/ml) and 91.22 (per cent) DPPH radical scavenging activity.	Patel *et al.* (2016)
	Fermented flowers			
16	*Mahua* wine	Fermentation of flower juice	Fermentation at 16° C favours wine quality and increase alcohol content (up to 9.9 per cent). Sensory evolution reported that addition of yeast during fermentation is acceptable but tannin addition is not required.	Yadav *et al.* (2009 a; b); Kumari *et al.* (2016)
17	*Mahua* Vermouth	Fortification of young *mahua* wine with spices	*Mahua* vermouth is having 18.4 (per cent) alcohol and 1.26 (mg/100 g) tannin. There is no deterioration found during one year of storage.	Yadav *et al.,* 2012 b; Kumari *et al.,* 2016)
18	Brandy	Fermentation of dry flowers	—	Malavade and Jadav, (2000); Sinha *et al.* (2017)
19	Acetone			
20	Lactic acid			
21	Citric acid	Surface fermentation of *mahua* flowers using *Aspergillus niger*	At 14 (per cent) sugar, 0.07 (per cent) Nitrogen level and 4 (per cent) methanol level, citric acid production was found highest. At this condition, high production of CA found by *A. niger* NCIM-595 than NCIM-545.	Thorat and Patil, (2016)

The major factors affecting the quality of *mahua* flowers and their food products, are poor post-harvest storage and lack of modern technologies for the value addition of *mahua* flowers (Awasthi *et al.,* 1975). As a result, the poor tribal people and small local entrepreneurs are facing lots of economic problems due to low profit market of *mahua* flowers and their products (Patel and Naik 2010).

4.6 Use as Biodiesel

Biodiesel is a renewable, biodegradable and non-toxic fuel. Biofuel may be derived from edible oil obtained from rapeseed, soybean, palm, sunflower, tree borne oilseed like jatropha (*Jatropha curcas*), neem (*Azadirachta indica*), castor

(*Ricinus communis*), karanj (*Pongamia pinnata*) and kokkam (*Garcinia indica*) (Carlos *et al.,* 2010; Yadav *et al.,* 2012a). For use as bio diesel, fatty acid methyl esters (FAME) of vegetable oils have been found suitable for use as fuel in diesel engine. Saponification value (SV), iodine value (IV) and Cetane number (CN) were used to predict the quality of FAME of oil for use as biodiesel. The SV depends upon the molecular weight and the percentage concentration of fatty acid components, IV depends upon three variables percentage concentrations of unsaturated fatty acids, their molecular weight and the number of double bonds present in them while CN gives the indication of ignition quality of the fuel, a higher value indicates better quality of fuel (Azam *et al.,* 2005). These parameters varied from 198.3–202.8, 52.0–68.6 and 58.0–61.6, respectively, in FAME of mahua fat (Yadav *et al.,* 2012a, b; Singh *et al.,* 2021).

4.7 Carbon Sequestration

A total of 17.98,30.25,133.1,519.2 and 811.7 kg for the trees belonging to 10-30, 31-60,61-90,91-200cm and more than 200cm girth classes respectively C was sequestered by *M. indica* in Gujarat (Pandey *et al.,* 2013).

4.8 Conclusion

Due to its multipurpose use it is satisfying the basic need of tribal people in the form of 3F *i.e.* food, fodder and fuel because of its vital components and abundant availability in that particular areas. But due to its limited availability for the short span only it is still untouched by the processors, researchers and consumers beside those places and particular time. Based on the current review knowledge the quality attributes of the flower are getting deteriorated because of the malpractices followed by tribal people for its preservation. To overcome these issues, there is a strong need for the commercial utilization of this flower along with advance technologies for the development of various valuable food products along with their availability throughout the year, which will definitely help in upliftment of tribal people economy and their sustainable development. A development of efficient technologies and awareness among the tribal people will definitely help in improving the quality attributes of the flower which may help in enhancement of employment and income of those societies. The various value added products included in this review, will play an important role for the sustainable development of tribal economy and wellness of local folks.

REFERENCES

Acharya D and Shrivastava A. (2008) Indigenous Herbal Medicines: Tribal Formulation and Traditional Herbal Practices. Avishkar Publishers Distributors, Jaipur, 2008.

Alam MD. Zulfeequar (2008) Herbal Medicine, APH Publishing Corporation, New Delhi Edition 3, 2008: 11-15

Amia E and Ekka N, (2014) *Madhuca longifolia* latifolia: an important medicinal plant used by tribes of north-east part of Chhattisgarh. OIIRJ. 4: 227-231.

Awasthi Y C Bhatnagar S C and Mitra C R (1975) Chemistry of sapotaceous plants: Madhuca species of India. Econ. Bot. 29: 380-389.

Behera S, Ray R C, Swain M R, Mohanty R C, Biswal A K. (2016)Traditional and Current Knowledge on the Utilization of Mahua (L.) Flowers Madhuca latifolia by the Santhal Tribe in Similipal Biosphere Reserve, Odisha, India. Ann Trop Res. 38: 94-104.

Bakhara C K, Bal L M, Pal U S, Sahoo N R, Panda M K. (2016) Post-Harvest practices and value addition of Mahua (*Madhuca longifolia*) flower in Odisha. Agric Engine Today. 40: 22-8

Chandra D. (2011) Analgesic effect of aqueous and alcoholic extracts of *Madhuca longifolia* (Koeing). Indian J Pharmacol. 33: 108-111.

Dash, S. K. (2017) Post-Harvest Management and Value Addition for Tribal Areas. Citied from: http: //zpd7icar.nic.in/Dr per cent 20S per cent 20K per cent 20Das.pdf Accessed on: 09/10/2017.

Dave Jaydeep Pinakin D. J, Kumar V, Kumar A, Gat Y, Suri S, Sharma K. (2018) Mahua: A boon for Pharmacy and Food Industry. Curr Res Nutr Food Sci 2018; 6(2). doi : http: //dx.doi.org/10.12944/CRNFSJ.6.2.12

Dutt D, and Pandey A. (2009) Mahua: A potential tree borne oilseed. National oilseeds and vegetable oils development board, Gurgaon.

Evans WC, Trease and Evans (2009) Pharmacognosy. Saunders Publication, Edition 16, (2009): 03-04.

Gopalan C, Ramasastri B V, Subramanian S C. (2007) Nutritive Value of Indian Food. National Inst. Nutrition (ICMR) Press, Hyderabad.

Gupta A, Chaudhary R, Sharma S. (2012) Potential applications of mahua (*Madhuca indica*) biomass. Waste Biomass Valoriz. 2012; 3: 175–189. doi: 10.1007/s12649-012-9107-9.

Gupta A, Kumar A, Sharma S, Vijay VK. (2013) Comparative evaluation of raw and detoxified mahua seed cake for biogas production. Appl Energy. 2013; 102: 1514–1521. doi: 10.1016/j.apenergy.2012.09.017.

Hoffman FA, Leaders FE(1996) Botanical (herbal) Medicine in Heath Care. Regulatory Perspective Pharm New 1: 23-25.

Hiwale S. (2015) Sustainable Horticulture in Semiarid Dry Lands, Mahua (*Bassia latifolia* Roxb.). Springer, New Delhi. 2015; 255-261.

Indu S. and Annika D. (2014) Cytotoxic and antioxidant potential of *Madhuca indica* flowers. World J Pharm Pharm Sci. 3: 2108-2114.

Jeyarani T, Yella Reddy S. (2010) Effect of enzymatic inter esterification on physicochemical properties of mahua oil and kokum fat blend. Food Chem. 2010; 123: 249–253. doi: 10.1016/j.foodchem.2010.04.019.

Jha S, Vaibhav V, Suneetha V. A Culinary(2013) Mahua (*Madhuca indica*) flower from Bihar, India A potential in Production of Jam, Alcohol for Pharmacological benefits with Fertilizer value. Int J Drug Dev Res. 5: 362-367.

Khatoon S, Reddy SRY. (2005) Plastic fats with zero trans fatty acids by interesterification of mango, mahua and palm oils. Eur J Lipid Sci Technol. 2005; 107: 786–791. doi: 10.1002/ejlt.200501210.

Kumar Patel P K, Janghel V, Chandel S S, Jyoti Sahu (2019) *Madhuca indica* (Mahua) - Pharmaceutical, nutraceutical and economical importance for tribal people of Chhattisgarh state. International Journal of Pharmaceutical and Phytopharmacological Research (eIJPPR). 9 (3): 16

Kureel R.S., Kishor R. Dutt S. Pandey A. (2009) Mahua: a potential tree borne oilseed. National oilseeds and vegetable oils development board, Gurgaon.

Katiyar S, Tandon M, Chandekar A, Upamanyu N. (2011) Pharmacognostic standardization, phytochemical investigation and the anthelmintic evaluation of the extract of *Madhuca indica* JF GMEL flowers. Pharmacologyonline. 3: 892-903.

Kumari A, Pandey A, Ann A, Raj A, Gupta A, Chauhan A. (2016) Indigenous Alcoholic Beverages of South Asia, in: Joshi, V.K. (Ed.), Indigenous Alcoholic Beverages of

South Asia. CRC Press, New York 2016, 501-566.

Malavade D M, Jadhav B L. (2000) Alcohol production from *Madhuca indica* flower. Trends. Life. Sci. 15: 59-65.

Mishra S and Pradhan S. (2013) *Madhuca lonigfolia* (Sapotaceae): A review of its traditional uses and nutritional properties. Int J Human Soc Sci Invent. 2013; 2: 30-36.

Mishra B K, Jain N K, Kumar S, Sharma K C. (2013) Storage stability of ready-to-serve beverage from mahua (*Madhuca indica*) flowers. Indian J Hortic. 2013; 20: 459-62

Palani S, Raja S, Karthi S, Archana S, Kumar B S.(2010) In vivo analysis of nephro and hepato protective effects and antioxidant activity of Madhuca longifolia against acetaminophen- induced toxicity and oxidative stress. *J Pharm Res.* 2010; 3: 9-16.

Pandey C.N. Pandey R. Bobda R. (2013) Carbon stock assessment of selected Tree species in urban and suburban areas of Gujarat. Forest dept. Govt. of Guarat 82-63pp.

Patel M. and Naik S.N. (2010) Flowers of *Madhuca indica* J.F. Gmel; Present status and future perspectivesIndian J. Nat. Prod. and Resources 1(4): 438-443.

Patel M, Pradhan R C, Naik S. (2011) Physical properties of fresh *mahua*. Int Agrophys 2011; 25: 303=306.

Patel M. (2008) Biochemical investigations of fresh mahua (*Madhuca indica*) flowers for nutraceuticals. Doctoral dissertation. Indian Institute of Technology, Delhi, Hauz Khas, New Delhi.

Patel P K, Prajapati N K, Dubey B K. (2012) *Madhuca indica*: a review of its medicinal property. Int. J Pharma Sci Res. 2012; 3: 1285-1293.

Patel M, Jaiswal S, Naik B, Naik M, Saxena D, Naik S N. (2016) Development of antioxidant rich beverage from Mahua (*Madhuca indica*) and Amla (*Emblica officinalis*). J. Sci. Ind. Res. 2016; 75: 35-39.

Pinakin D.J. Kumar V.Kumar A.Gat Y.Suri Sheenam, Sharma K. (2018) Mahua: a boon for Pharmacy and food industry. 6(2).https: //dx.doi.org/10.12944/ CRNFSJ.6.2.12.

Ramadan MF, Moersel J-T. (2006) Mowrah butter: nature's novel fat. *Inform.* 2006; 17: 124-126.

Ramadan M F, Mohdaly A A A, Assiri A M A, Tadros M, Niemeyer B (2006) Functional characteristics, nutritional value and industrial applications of *Madhuca longifolia* seeds: an overview. J Food Sci Technol. 53(5): 2149–2157. DOI 10.1007/s13197-015-2095-6

Ramakrishnan, D. H. (2019). As a war of words rages outside, peace reigns inside this temple . The Hindu. Chennai: Kasturi and Sons. 15th Nov. 2019; p. 3.

Saluja M S, Sangameswaran, B, Hura, I S, Sharma A, Gupta S K, Chaturvedi M. (2011) In vitro cytotoxic activity of leaves of *Madhuca longifolia* against Ehrlich Ascites Carcinoma (EAC) cell lines. Int J Drug Dev Res. 2011; 1: 55-57.

Saxena O.P. (Nidar) (2023) Charak Samhita in Hindi by DPB Pub. Delhi pp.742

Sen C.T. (2015)A history of food in India. Speaking Tiger Pub. New Delhi 349 pp.

Shekhawat N. and Vijayvergia R. (2010) Investigation of Anti inflammatory, Analgesic and Anti Pyretic properties of *Madhuca indica* Int J Mol Med Adv Sci. 2010; 6: 26-30.

Singh C. Cahuhan N. Rani A. Kishir K. Singh K. Sigh R. (2021) Ethnobotanical and pharmacological studies of pteridophytes (Ferns) from Mussoorie region, Dehradun district, Uttarakhand. International Journal of Botany Studies. ISSN: 2455-541X 6(6): 694-702. www.botanyjournals.com

Singh A, Singh IS. (1991) Chemical evaluation of mahua (*Madhuca indica*) seed. Food Chem. 1991; 40: 221–228. doi: 10.1016/0308-8146(91)90106-X.

Sinha J, Singh V, Singh J and Rai A K. (2017) Radiochemistry, Ethnomedical Uses and Future Prospects of Mahua (*Madhuca longifolia*) as a Food: A Review J Nutr Food Sci. 2017; 1.

Sunita M. and Sarojini P. (2013) *Madhuca lonigfolia* (Sapotaceae): A review of its traditional uses and nutritional properties. Int J Human Soc Sci Invent. 2: 30-36.

Swain M R, Kar S, Sahoo A K, Ray R C. (2007) Ethanol fermentation of mahula (*Madhuca latifolia* L.) flowers using free and immobilized yeast *Saccharomyces cerevisiae*. Micro Res. 162: 93-98.

Thakor R and Babu M S (2008). Engendering the market with mahua a community based initiative in Mandla. Citied from: Website-www.thelivelihood school.org.

Thorat S S, Patil G. (2016) Standardization of Process Parameters for Production of Citric Acid from Mahua Flowers (*Madhuca indica*) by Surface Fermentation using *Aspergillus niger* NCIM-545 and NCIM-595. J. Food. Ferment. Tech. 2016; 6: 111-120.

Umadevi M, Maheswari C, Jothi R, Paleti S K, Reddy Y S, Narayanan R V.(2011) Hepatoprotective Activity of Flowers of *Madhuca longifolia* (Koen.) Macbr. Against Paracetamol-Induced Hepatotoxicity. *Res J Pharm Tech.* 2011; 4: 259-262.

Verma KS, Saxena N, Sinha R, Agarwal A.(2010) Phytochemical Screening and Therapeutic Profiling of *Madhuca indica*. J.F. Gmel. IJOR. 2010; 23: 109-115.

Verma N, Jha K K, Kumar U, Deepak K, Singh N K, Singh A K (2014) Biological Properties,Radiochemistry and Traditional Uses of Mahua (*Madhuca longifolia*): A Review. 2: 630- 638.

World Health Organization (WHO), (2003). Traditional Medicines Citiedfrom:

http: //apps.who.int/gb/archive/pdf_files/WHA56/ea5618.pdf Accessed on: 09/10/2017.

Yadav P, Singh D, Mallik A, Nayak S, (2012a) *Madhuca longifolia* (Sapotaceae), a review of its traditional uses, phytochemistry and pharmacology. *Int J* Bio *Med Res.* 2012a; 3: 291-305.

Yadav P, Garg N, Diwedi D H. (2009) Standardization of pre-treatment conditions for mahua wine preparation. J. Ecofriend. Agric. 2009; 4: 88-92.

Yadav P, Garg N, Diwedi D H. (2009) Effect of location of cultivar, Fermentation temperature and additives in the physico-chemical and sensory qualities on mahua (*Madhuca latifolia* JF Gmel) wine Preparation. *Natural Prod. Rad.* 2009; 8: 419-425.

Yadav P, Garg N, Diwedi D. (2012b) Preparation and evaluation of Mahua (*Bassia latifolia*) Vermouth. J. Food. Ferment. Tech.; 2: 57-61.

Yoshikawa K, Tanaka M, Arihara S, Chandra BP, Roy SK, Matsumura E, Katayama S. (2000) New oleanene trirerpenoid saponins from *Madhuca longifolia.* J Nat Prod. 2000; 63: 1679– 1681. doi: 10.1021/np000351r. [

Yosioka I, Inada A, Kitagawa I. (1974) Soil bacterial hydrolysis leading to genuine aglycone-VIII structures of a genuine sapogenol protobassic acid and a prosapogenol of seed kernels of *Madhuca longifolia* L. Tetrahedron. 1974; 30: 707–714. doi: 10.1016/S0040- 4020(01)97155-8.

Chapter 5

Gorgeous and Tangy Plums: Aloo Bukhara (*Prunus domestica* L.)

A member of Rosaceae, Plum or Aloo Bukhara is very attractive fruit tree of hilly areas in India. It contains beta carotene in sufficient amount which is very essential phytonutrients required to fight against various types of cancer. The anti-aging fruit boots the power of both body and mind thus enhances the brain capacity to function actively for long time. Plum enhances the level of vitality among people is consumed on regular basis. Prunus species is the healthiest fruit having capacity to keep heart healthy by reducing the bad blood cholesterol, boosting blood health, and keeps arteries healthy, if consumed regularly. Plum helps in removing unnecessary bile from liver so that extra cholesterol can be used for addressing the bile deficiency. Its high level of potassium (K) and low level sodium (Na) plays vital role in normalizing the blood pressure and lowers down the water retention in body. It is supposed to be having good source of vitamin K which helps in protecting from the unnecessary blood clotting thus prevents from high blood pressure and stroke. The plum is very low in saturated fat, and cholesterol and is a good source of dietary fiber, and vitamins A, C and K. These delicious, succulent fruits provide high calories.

Keywords: *Plum, Aloo Bukhara, Delicious, Nourishing, β Carotene anticancer, antiaging.*

5.1 Introduction

The plum is a wonderful, juicy fruit that belongs botanically to the Rosaceae family. This fruit belongs to the genus *Prunus*, which also includes peaches, nectarine, and almonds. Other local names of plum are 'Aloo Bukhara' in Hindi, 'Alpagoda Pandu' in Telugu, 'Alpakoda' in Tamil and 'Onagida Draakshe' in Kannada. It is one of the few fruits that come in a panorama of colors. The plant which bears this fruit is a short tree cosmopolitan in distribution, including China, United States, Japan, and Europe. The plum fruits differ in color and size according to the region of origin. The plant bears the plum fruits mostly during the months between May and September. Dried plums are known as prunes. There are around 2000 varieties of plums available in the world, and 100 are available in the United States alone. The country grows approximately 12 varieties of plum. Their primary producers are Punjab, Himachal Pradesh and Jammu and Kashmir. The world crop is estimated to be 6.6 Mt for plum, these are treated fresh or in dried form.

Dried plums are considered as healthy food in the West and used as medicine in India, contain phenolic compounds with protective actions against age-related diseases. The plum is very low in saturated fat, cholesterol, and Na. It is also a good source of dietary fiber, and vitamins A, C and K. These delicious, succulent fruits are quite low in calories. It even contains several health-promoting compounds, minerals, and vitamins. These compounds include sorbitol and isatin (Stacewicz-Sapuntzakis *et al.,* 2001).

Aloo Bukhara is very attractive, healthy, tasty and antioxidants rich fruit packed with the vitamins (especially vitamin A), minerals, iron and *etc.* it boosts the mental and emotional health of the people as it contains all the essential properties of boosting the health if consumed on daily basis. It can be used as a wonderful vitamins capsule of boosting the health. It has been used for years (Nakatani *et al.,* 2000) for both health and beauty benefits by the people. Aloo Bukhara is the rich source of vitamin A and beta-carotene which are the most essential elements for the skin health thus it helps in nourishing the proper skin health if eaten on regular basis. It prevents skin from various skin problems like blemishes, eczema, acne and *etc.* It reduces the risk of osteoporosis, bone demineralization and other problems among women after the menopause. Plum boosts the proper growth and development of the bones and joints (by enhancing the secretion of enzymes and growth factors for bone growth) among menopausal women if consumed in adequate amount (Cobianchi and Watkins 1984).

It is the rich source of iron and folic acid thus reduces the chance of early menopause by maintaining the proper flow of blood and haemoglobin count among women. It provides relief from the semen leakage problem among men if the mixed water of soaked plum and one tea spoon full (tsf) aniseed (saunf) is taken regularly. It is the richest source of iron and folates, which are most required elements to enhance the blood haemoglobin content thus it is most beneficial to get prevented from the anemia. Aloo Bukhara contains high level of potassium which boosts the functioning of nerves and muscles thus prevents from many health problems related to the nerves and muscles during old age. Its high level of insoluble dietary fibers provides stomach the sense of fullness for longer period of time thus prevents from overeating and involves in the weight management.

Figures 5.1A-D. Plum or Aloo Bukhara Attached to a Tree and Picked up for Marketing.

It contains vitamin A and β carotene which helps in improving eye sight and preventing from the various eye disorders. It also helps in improving the appetite and normalizing the digestive system functioning as it is the best natural home remedy for all. It provides instant relief from the weakness and fatigue if taken on daily basis on empty stomach in the morning after routine walk. Its regular consumption is very effective home remedy to get relief from the problem of enlarged spleen and get normal sized spleen. It provides relief from the foot inflammations if plum is rubbed on the foot palm for around 15 mins. It helps in treating the sleeping disorders among people if it (6-8) is taken regularly before sleep time.

Its antioxidants richness makes it able to boost weak memory power during old age. Plum also provides relief from the problems of throat swelling, soreness, osteoarthritis, rheumatoid arthritis, scanty urination and prostate gland problem. It enhances metabolism by promoting the enzymatic actions in body. Its high level of antioxidants, carotenoids and phytonutrients helps in getting prevented from the diseases caused by free radicals.

It contains good level of magnesium mineral which has ability to promote muscles, nerves and blood vessels relaxation. It also protects from the asthma, muscle soreness, migraine, headaches and *etc.* It naturally assists in treating the problem of constipation, irregular bowel movements and *etc.* disorders of digestive tract due to the availability of insoluble dietary fibers. Its regular consumption prevents from the age-related macular degeneration, infections as well as enhances the iron synthesis and absorption in the body. Prune contains lutein, beta carotene, cryptoxanthin, zeaxanthin and *etc.* which actively help in fighting against the free radicals and early ageing problem.

5.1.1 Historical Background

Genus *Prunus* is the part of subfamily Prunoideae of the family Rosacea with basic chromosome number x=8. Rehder (1940) classified *Prunus* into 77 spp. Plums hold the center of *Prunus* stage. The centers of origin for plums include *P. domestica* (European), *P. insititia* (Western Asia), *P salicina.* (Japanese), *P americana* (North America). *P. domestica*, the European plum is the most recent crop species to have arisen within Prunophora (Watkins 2014). The five main groups of plum evolved in three distinctly different areas of the world. All five groups can be easily hybridized. The dominant commercial group is the hexaploidy European plum *P. domestica.* European breeding program focuses on fruit character while North American breeding has more concerned to the adaptation of environment (Cobianchi and Watkins 1984). Genetic engineering can be used to create new fruit crops by protoplast fusion.

5.2 Benefits of Leaves

It provides relief from the intestinal worm problem if paste of the plum leaves is applied to the lower abdominal area means below the belly point. Drinking water or tea prepared by boiling some plum or Aloo Bukhara leaves provides instant relief from the excessive running nose problem. It provides skin a radiant glow, soft and supple touch. Prune prevents skin from the dryness and roughness including other facial skin problems by providing it a natural freshness. It also provides relief from the dark and black lips problem if fresh peeled plum is rubbed on the lips and gives lip a soft and pink look. Plum prevents skin cells damage by protecting it from the harmful UV sun rays. Vitamin C found in it protects skin from various seasonal infections.

5.3 Health Benefits of Plums

Plums neutralize the harmful effects of free radicals and protect against the development of health conditions like asthma, arthritis, heart stroke and cancer. The antioxidants found in this fruit protect the body from free radicals that speed up the aging process. Zeaxanthin, an important dietary carotenoid, selectively absorbs light into the retina where it provides antioxidant and protective UV light filtering functions. These antioxidants are particularly effective in neutralizing the destructive oxygen radical called superoxide anion radical and are known to prevent oxygen-based damage to fats, such as fats that comprise a substantial portion of the brain cells or neurons. The antioxidants in this fruit also improve the memory by neutralizing the cell-damaging radicals.

The compounds sorbitol and isatin, found in plums, help in regulating the functioning of the digestive system and in relieving constipation (Lee *et al.,* 2009). The vitamin C in the plum helps the body develop resistance against infectious agents, counter inflammation and free radicals (Stacewicz-Sapuntzakis *et al.,* 2001). Plums also help in maintaining the heart health. They contain antioxidants that help to get rid of the free radicals and prevent cholesterol oxidation. Cholesterol oxidation is a threat to healthy blood vessels, which may eventually lead to a number of heart diseases such as heart stroke and high cholesterol. The vitamin C in plums is extremely helpful for controlling heart diseases in diabetic patients (Kuo *et al.,* 2015).

Figures 5.2. (A) Dried Plums; (B) A Glass of Delicious Juice Obtained from Plums.

Some studies have connected the plum to the prevention of cancer, especially breast, gastrointestinal and respiratory tract cancers (Tsuji *et al.,* 2011). Anthocyanins, which are reddish blue pigments in this fruit, protect against cancer by mopping the free radicals. Vitamin A found in plums is also known to protect from oral and cavity cancer. Plums are particularly good in preventing breast cancer. The antioxidant power of plums and their phytonutrients cease the development of the breast cancer cells without killing the healthy cells. Plums contain potassium that is an important component of the cell and body fluids that help in controlling the heart rate and the blood pressure (Chai *et al.,* 2012). It also prevents platelet clotting which may result in atherosclerosis, high blood pressure, stroke and coronary heart disease. It also maintains a healthy level of blood pressure. The fiber in plums helps to lower the cholesterol levels (Groathouse *et al.,* 2010).

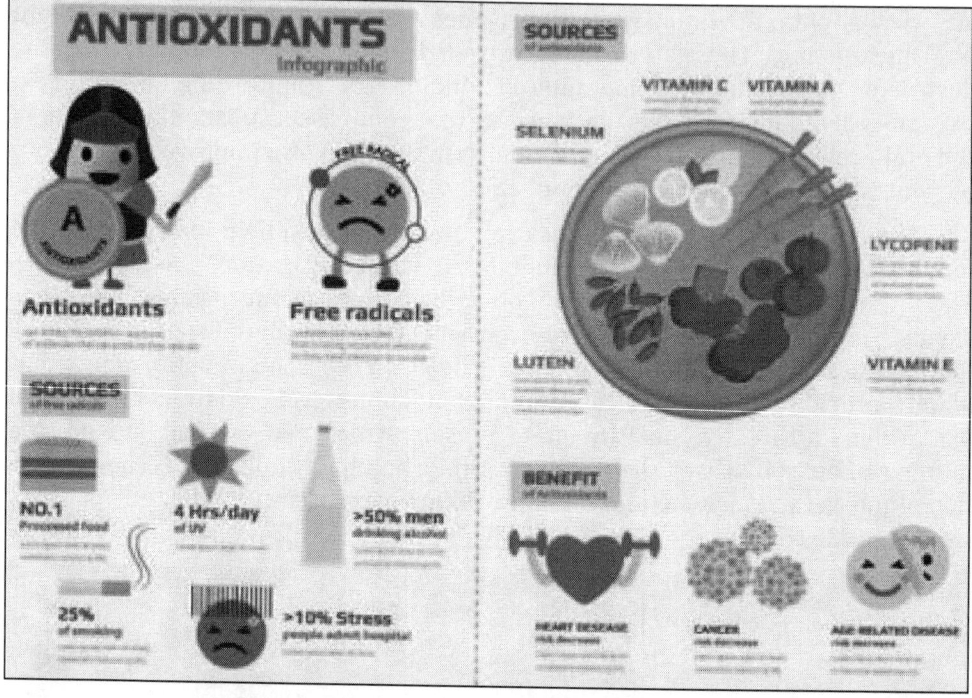

Figures 5.3. Sources of Antioxidants and their Uses (www.facebook.com/Doctor adviceforhealthandfitness).

The plums have great richness of health promoting components like antioxidants and phenolics (Cao *et al.,* 1997; Nakatani *et al.,* 2000; Auger *et al.,* 2004). Predominant phenolic compounds in plums are caffeic acid and its derivatives (Nakatani *et al.,* 2000; Tomas *et al.,* 2001). According to literature data, depending on the variety, environmental conditions and used analytical methods, phenolic components in plums fall within a wide range of values, *i.e.*

Acids	*Quantity (mg/kg)*
Neochlorogenic acid	85-1300
Chlorogenic acid	13-430
Cryptochlorogenic acid	956

Source: Kayano *et al.,* 2002.

From data it can be concluded that phenolic acid content in plums are higher than grapes, kiwi, apples, pears, melons, bananas, while in dried form it exhibits antioxidant capacity higher than that of black currants, blueberries, strawberries, *etc.* (Wang *et al.,* 1996; Kayano *et al.,* 2002).

Plums also contain vitamin B6, which prevents an increase in the homocysteine levels and reduces the possibility of a heart attack. Both plums and prunes are known to be effective natural laxatives and can help to maintain regular bowel habits (Stacewicz-Sapuntzakis *et al.,* 2001). Plums can also protect against macular degeneration. Eating three servings of this fruit daily can lower the risk of developing age-related macular degeneration (Anonymous 2006). Try adding diced plums to your morning cereals or salads.

Plum and prunes have the increased ability to absorb iron in the body. This ability is due to the high vitamin C content in this fruit. The iron in the plum is highly required for the formation of red blood cells. It improves the blood circulation in the body and prevents anemia. It also removes toxins and prevents constipation, leading to weight loss naturally. Plum is rich in fiber and antioxidants, which aid proper digestion and good metabolism. It contains citric acid, which prevents tiredness and cramps. It also improves liver and gastrointestinal function. Thus, eating plums or drinking plum juice regularly helps in reducing weight.

Plums are quite low in glycemic index, so eating plums controls the blood sugar level and reduces the risk of developing type 2 diabetes. The phytonutrients found in plums control the release of glucose into the blood after you have just eaten your meal and regulate the insulin spike between the meals. So, it is recommended that diabetic patients should incorporate plums in their diet as it will not worsen the situation and will keep them healthy (Kuo *et al.,* 2015).

The soluble fiber in plums also helps to lower cholesterol by soaking up the extra bile in the intestine and then excreting it. Bile is made in the liver to digest fat. When the body excretes bile along with fiber, the liver makes use of the cholesterol in the body to make more bile, thereby lowering the amount in circulation in the body. Soluble fiber also inhibits the amount of cholesterol manufactured by the liver in the first place. Plum also assists the body in making healthy tissues and a strong immune system. It prevents cold and flu and is helpful to people who suffer from recurrent infections.

Several researches have concluded that eating plums boosts the bone health, especially of post-menopausal women. It reverses bone loss due to the rich amount of phenolic and flavonoids compounds. It also contains boron which plays an important role in ossification and actively participates in the preservation of the bone density and the general upkeep of your bones.

Magnesium in plums is necessary for muscle contraction and nervous impulses. These have a good amount of folic acid and calcium which contribute to the supply of folates. Folate is very essential for pregnant women and the growth of the fetus. Plums also have diuretic and detoxifying effects on the body.

Additionally, high potassium content of prunes (745 mg/100 g) might be beneficial for cardiovascular health. Dried prunes are an important source of boron, which is postulated to play a role in prevention of osteoporosis. A serving of prunes (100 g) fulfills the daily requirement for boron (2 to 3 mg).

5.3.1 Uses of Plum in Antiaging and for Cosmetics

The vitamin E and beta-carotene, an antioxidant in plums, protect the fatty acids and the cell membranes from the attack of the free radicals, which allows the body to fight the signs of aging skin such as loss of elasticity and wrinkles. The plum contains a high amount of antioxidants and high anti-inflammatory properties, which increase the circulation and cell metabolism, reducing inflammation (Nakatani *et al.*, 2000), which promotes the elasticity of the skin and keeps acne and wrinkles away.

Drinking plum juice can also help to even out the skin tone and firm the skin by improving the skin health and rejuvenating its appearance. Collagen deficiency often makes the skin dull and lifeless. The plum contains a high amount of vitamin C that is required for the production of hydroxyproline and hydroxylysine, which are needed to bind the molecules that produce collagen. This rejuvenates the skin from the roots and reduces wrinkles and other signs of aging. Plums can also reduce the appearance of scar tissue by increasing the blood flow to the scar tissue, minimizing the look of the scar and blending in the development of new skin.

Plums also help to replace damaged tissue and help to heal the wounds at a faster pace. They improve the skin elasticity and create scar tissue and ligaments for a quick recovery. Applying plum pulp on the face lightens dark spots and treats skin discoloration, freckles and age spots to give you a younger looking and healthy skin. Plum can help to maintain the hair follicles, connective tissue or collagen of the scalp and protects the cells both in the follicles and in the blood vessels of the scalp, thus promoting hair growth. Free radicals often make our hair weak, brittle and thin. These conditions often hinder proper hair growth. Vitamin E in the plum strengthens the cell membranes and prevents free radicals from attacking the follicles of the scalp. Thus, people who eat plums and prunes regularly have healthy, strong and thick hair. The vitamin C in the plum can also help you get rid of dandruff. Plums can help fight the bacteria on the scalp and soothe the skin to ward off dandruff. This will get rid of the follicles debris and encourage new hair growth. Plum can also help reverse adrenal gland fatigue. This is very crucial for stopping hair loss and encourages more hair growth since the adrenal glands are very important for maintaining proper hormonal imbalance.

The high iron content in this fruit improves the blood circulation and repairs the capillaries to promote stronger and thicker hair. Plums can also help to treat damaged hair and aid in the treatment of a variety of hair disorders that can damage the follicles and prevent proper hair growth. It also helps combat alopecia, male baldness in men. The plum is a highly pigmented fruit that not just combats hair loss problems but also retains the natural color of the hair. It also prevents premature graying.

5.4 Nutritional Benefits of Dried Aloo Bukhara

Dried plums, are considered as healthy food in the West and are usually used as medicine in India. It contains phenolic compounds with protective actions against age-related diseases. The plum is very low in saturated fat, cholesterol, and sodium. It is also a good source of dietary fiber, and vitamins A, C and K. These delicious, succulent fruits are quite low in calories. It even contains several health-promoting compounds, minerals, and vitamins. These compounds include sorbitol and isatin. It contains minerals like potassium, fluoride, and iron. The total antioxidant strength of plums measured in terms of ORAC (Oxygen Radical Absorbance Capacity) is 6259.

5.4.1 Contents Per 100 g Dried Fruit

Energy: 240 kcal; Carbohydrate: 63.88 g; Sugar: 38.13 g; Dietary fibers: 7.1 g; Fat: 0.38 g; Protein: 2.18 g; **Vitamins** - Vitamin A: 148 µg; Thiamine: 0.051 mg; Riboflavin: 0.186 mg; Niacin: 1.882 mg; Pantothenic; cid: 0.422 mg; Vitamin B6: 0.205 mg; Folate: 4 µg; Choline: 10.1 mg; Vitamin C: 0.6 mg; Vitamin E: 0.43 mg; Vitamin K: 59.5 µg; **Electrolytes-** K: 732 mg; Na: 2 mg; **Mineral-** Ca: 43 mg; Fe: 93 mg; Mg: 41 mg; Mn: 0.299 mg; P: 69 mg; Zn: 0.44 mg; Se: 0.66 mcg; Cu: 150 mcg; Fl: 4 µg; **Phytonutrients-** β-carotene: 39 µg; β-crypto-xanthin: 35 µg; Lutein zeaxanthin: 394 µg. (Source: USDA Nutrient Database)

5.5 Molecular Data

Prunus domestica is a hexaploid and has low genetic diversity.The number of Genes are SNPs. The 7 ssR markers generated a total of 140 alleles in European plum.

5.6 Conclusion

The fruit of the plum tree (*Prunus domestica*) has been used as a traditional medicinal food. It enhances immunity against infectious agents and treats cancers. However, limited information exists on the mechanisms responsible for its immune enhancing properties. Plums have a high content of antioxidants which function to protect against harmful free radicals that could lead to heart disease or cancer. Plums are also a good source of vitamin C, and A (in the form of β-carotene), and fiber while being very low in calories. Studies have also shown that including fruit in your diet, such as plums or prunes, can help your vision. A recent study published in the Archives of Ophthalmology showed that including 3 or more servings of fruit everyday could decrease your risk of developing age-related macular degeneration, the leading cause of blindness in older adults. This extensive literature review summarizes the current knowledge of chemical composition of prunes and their biological effects on human health. Because of their sweet flavor and well-known mild laxative effect, prunes are considered to be an epitome of functional foods, but the understanding of their mode of action is still unclear. Dried prunes contain approximately 6.1 g of dietary fiber per 100 g. The laxative action of both prune and prune juice could be explained by their high sorbitol content (14.7 and 6.1 g/100 g, respectively). Prunes are good source of energy in the form of simple sugars, but do not mediate a rapid rise in blood sugar concentration, possibly because of high fiber, fructose, and sorbitol content. Prunes contain large amounts of phenolic compounds (184 mg/100 g), mainly as neochlorogenic and chlorogenic acids, which may aid in the laxative action and delay glucose absorption.

REFERENCES

Anonymous (2006) European Food Safety Authority, Panel on Dietetic Products, Nutrition and Allergies. "Scientific Opinion on the substantiation of health claims related to dried plums of 'prune' cultivars (*Prunus domestica* L.) and maintenance of normal bowel function (ID 1164, further assessment) pursuant to Article 13(1) of Regulation (EC) No 1924/2006". EFSA Journal. 10 (6): 2712. doi: 10.2903/j.efsa.2012.2712.

Auger C, Al-Awwadi N, Bornet A, Rouanet JM, Gasc F, Cros G, Teissedre PL (2004) Catechins and procyanidins in Mediterranean diets. Food Res. Int. 37: 233-245

Cao G H, Sofic E, Prior L R (1997) Antioxidant and pro-oxidant behavior of flavonoids: structure–activity relationships. Free Radical Bio. Med. 22: 749–760.

Chai S C, Hooshmand S, Saadat R L, Payton M E, Brummel-Smith K, Arjmandi B H (2012) Daily apple versus dried plum: impact on cardiovascular disease risk factors in postmenopausal women. J Acad Nutr Diet. 112(8): 1158-68. doi: 10.1016/j.jand.2012.05.005.

Cobianchi D, Watkins R (1984) Plum descriptors. IBPGR/CEC (AGPG/IBPGR/84/92. Pp. 1-31

Groathouse J, Barcena H, Gustafson C, Wilson M V, Symonsbergen D J, Lucas K A, Holmes A E (2010) Wild Plum: novel particles of improved optical brightness and fluorescence. J Cosmet Sci. 61(4): 279-288.

Kayano S, Kikuzaki H, Fukutsaka N, Mitani T, Nakatani N (2002) Antioxidant activity of prune *Prunus domestica* L. constituents and a new synergist. J. Agr. Food Chem., 50: 3708- 3712.

Kuo P H, Lin C I, Chen Y H, Chiu W C, Lin S H (2015) A high-cholesterol diet enriched with polyphenols from Oriental plums (*Prunus salicina*) improves cognitive function and lowers brain cholesterol levels and neurodegenerative-related protein expression in mice. Br J Nutr. 113(10): 1550-7. doi: 10.1017/S0007114515000732.

Lee S H, Lillehoj H S, Cho S M, Chun H K, Park H J, Lim C I, Lillehoj E P (2009) Immunostimulatory effects of oriental plum (*Prunus salicina* Lindl.). Comp Immunol Microbiol Infect Dis. 32(5): 407-17. doi: 10.1016/j.cimid.2007.12.001.

Nakatani N, Kayano S, Kikuzaki H, Sumino K, Mitani T (2000) Identification, quantitative, determination, and antioxidative activities of chlorogenic acid isomers in prune *Prunus domestica* L. J. Agr. Food Chem. 48: 5512-5516

Rehder A (1940) Manual of cultivated trees and shrubs Dioscorioides Press

Stacewicz-Sapuntzakis M, Bowen P E, Hussain E A, Damayanti-Wood B I, Farnsworth N R (2001) Chemical composition and potential health effects of prunes: a functional food? Crit Rev Food Sci Nutr. 41(4): 251-86.

Tomas B F A, Gil M I, Cremin P, Waterhouse A L, Hess-Pierce B, Kader A A (2001) HPLC- DAD-ESIMS analysis of phenolic compounds in nectarines, peaches, and plums. J. Agr. Food Chem. 49: 4748-4760.

Tsuji R, Koizumi H, Fujiwara D (2011) Effects of a plum (*Prunus mume* Siebold and Zucc.) ethanol extract on the immune system *in vivo* and *in vitro*. Biosci Biotechnol Biochem. 75(10): 2011-2013.

Utsunomiya H, Yamakawa T, Kamei J, Kadonosono K, Tanaka S (2005) Anti-hyperglycemic effects of plum in a rat model of obesity and type 2 diabetes, Wistar fatty rat. Biomed Res. 26(5): 193-200.

Wang H, Cao G, Prior R L (1996) Total antioxidant capacity of fruits. J. Agr. Food Chem., 44: 701-705.

Watkins R (2014) Cherry, plum, peach, apricot and almond. In: Evolution of crop plants. Eds J. Smartt and N.W. Simmonds. Pub by Wiley India Pvt. Ltd. 423-428.

www.facebook.com/Doctoradviceforhealthandfitness

www.hashmi.com

Chapter 6

Nutritious and Healthy Dry Fruit: Pistachio Nut (*Pistacia vera* L.)

Pistachios are prized for their unique, slightly sweet flavor as well as for their intriguing viridescent hues. Pistachios are one of the world's oldest nuts, being grown in the Middle East for thousands of years. It's a nut with a hard outer egg-shaped shell and a tasty kernel inside that's commonly loved as a snack food. The health benefits of pistachios include a healthy heart, weight management, protection against diabetes and hypertension, and improved digestion. The vitamins, minerals, fats, and protein found in pistachio are all good for health.

Keywords: *Pistachio nut, Pistacea, Diabetes, Minerals, Anemia, LDL, HDL, ALA, DHA, EPA, Glycation.*

6.1 Introduction

Pistachios are one of the oldest edible nuts They are native to Asia, particularly in Iran and Iraq. Archaeological evidence dates the association of pistachios and man as early as 6,000 BC. They were cultivated in Iran, Iraq, and Syria and were introduced to the Romans only in 100 AD. Today, apart from Iran, Iraq, and Syria, pistachios are being produced in many countries such as the United States, Australia, Turkey, and China, among others. Pistachio nuts belong to the *Anacardiaceae* family from the genus *Pistacia*. A pistachio tree takes about 10 to 12 years to produce the first crop. They are drupe, where the fruit has a large seed in the center. This seed is edible. The nuts of pistachio are not seasonal and are available all year long. They can be bought from almost any grocery store. Pistachios are available today in various forms such as shelled, unshelled, sweetened or salted. The shell of the nut protects it from physical damage and various infections (AL-Saghir and Porter. 2012).

On the other hand, salted and sweetened pistachios may not be a good choice because of their high sodium and sugar content. The ideal way is to buy shelled nuts, as they are in their natural form. They must be stored in airtight containers in cool and dry places so that they can last for many months without going bad. Pistachios are prized for their unique, slightly sweet flavor as well as for their intriguing viridescent hues. Pistachios are one of the world's oldest nuts, being grown in the Middle East for thousands of years. They are even mentioned in the Old Testament of the Bible, which indicates their rich history as a prized food source. Like all nuts, pistachios are rich in beneficial nutrients. However, pistachios have some unique properties that set them apart. Their versatility means that you can eat pistachios alone as a snack, on top of a salad, mixed with dried fruits in trail mix, in baked goods, or as a crunchy coating for fish or meat. In addition to these distinctive traits, consider the following nutritional benefits of pistachios as an added perk to their particularly pleasant palate (Parfitt 2003). Pistachio, the second non-oil and the third exported good of Iran, is one of the most delicious and most expensive snacks of all other nuts and protein materials. In a study Ardakani (2006) calculated that the added value of 300,000 tons of pistachio production annually (with value of 933 million dollars), including only four products of pistachio processing industry (split pistachio, pistachio kernel, pistachio shred and pistachio powder) is about 58 million dollars. This shows that processing industry is very important in the improvement of current situation in Iran (Dreher, 2012).

6.2 Nutritional Importance

Most nuts contain large amounts of protein relative to their size, and pistachios are no exception. Approximately 49 pistachio kernels contains 6 g of protein. Your body breaks this protein down into its constituent amino acids, which can be used to repair tissues or to create new molecules. Plus, eating protein-rich foods helps you feel satiated, reducing your cravings for unhealthy foods later on. This makes pistachios the perfect mid-afternoon snack (Okay 2002).

Pistachios are a high-fat food, but that is not a bad thing. Per serving, pistachios contain 13 total grams of fat. However, only 2 grams of fat are saturated fats, the unhealthy fats that are associated with higher risk of cardiovascular disease (American Heart Association, 2016). The rest of the fats are polyunsaturated and monounsaturated fats, which actually protect the heart (Ardakani *et al.,* 2009).

This includes omega-3 fatty acids, a type of fat that has been associated with lower levels of "bad" cholesterol and higher levels of "good" cholesterol. Pistachios contain alpha-linoleic acid (ALA), a beneficial type of omega-3 fatty acid that can also be converted to DHA and EPA, the two other forms of omega-3s that are only found in animal sources. Thus, vegetarians and vegans may eat pistachios as a source of the omega-3 fatty acids their bodies need.

As our cells grow older, they accumulate oxidative damage. This can be related to the presence of free radicals, which can cause considerable damage to cells. A class of molecules called antioxidants can sweep up these free radicals and reverse some of the cellular damage.

Pistachios are an excellent source of antioxidants, including lutein, beta-carotene, and gamma-tocopherol used as a precursor to vitamin E. Both vitamin A and vitamin E themselves have very high antioxidant activity, making pistachios a great way to reap some of the oxidative damage-fighting effects of these vitamins. In a randomized study of the effects of pistachios, researchers found that incorporating these nuts into the diet was associated with lower levels of harmful LDL cholesterol, possibly because of the antioxidants present in pistachios.

Vitamin B6 has myriad health benefits. Failure to get enough vitamin B6 has been associated with elevated risk of cardiovascular disease, certain types of cancer, and cognitive dysfunction. Thus, getting vitamin B6 through your diet may improve your cardiovascular health and keep your brain sharp. Women need 1.5 mg of vitamin B6 per day, while men need 2 mg per day. A 1-ounce serving of pistachios contains 0.4 mg of vitamin B6, helping you reach your daily recommended intake of this beneficial vitamin. It contains 20 per cent zinc.

Pistachios are prized for their unique, slightly sweet flavor as well as for their intriguing viridescent hues. Pistachios are one of the world's oldest nuts, being grown in the Middle East for thousands of years (Chauhan and Kevalia 2014).

Phosphorus is an element that is essential for our proper physiological functioning. Not only does phosphorus make a structural component of all cells, but it also regulates a variety of physiological reactions. Getting enough phosphorus ensures that your cells can continue to produce energy and also strengthens the bones (Table 6.1).

Table 6.1. Nutritional Value of Pistachio Nut

Principle	*Nutrient Value*	*Percentage of RDA*
Energy	557 Kcal	29
Carbohydrates	27.97 g	21.5
Protein	20.60 g	37
Total Fat	44.44 g	148
Cholesterol	0 mg	0
Dietary Fiber	10.3g	27
	Vitamins	
Folates	51 µg	13
Niacin	1.3 mg	8
Pantothenic acid	0.520 mg	10
Pyridoxine	1.7 mg	131

Principle	*Nutrient Value*	*Percentage of RDA*
Riboflavin	0.160 mg	12
Thiamin	0.870 mg	72.5
Vitamin A	553 IU	18
Vitamin C	5 mg	8
Vitamin E-γ	22.60 mg	150
	Electrolytes	
Sodium	1 mg	0
Potassium	1025 mg	22
	Minerals	
Calcium	107 mg	11
Copper	1.3 mg	144
Iron	4.15 mg	52
Magnesium	121 mg	30
Manganese	1.2 mg	52
Phosphorus	376 mg	54
Selenium	7 μg	13
Zinc	2.20 mg	20
	Phyto-nutrients	
Carotene-β	332 μg	–
Crypto-xanthin-β	0 μg	–
Lutein-zeaxanthin	0 μg	–

Data modified from Drecher (2012).

6.3 Health Benefits of Pistachio

Pistachios help in reducing bad cholesterol, LDL and increases good cholesterol, HDL in the body thus preventing heart diseases. It also increases strength of the nerves making heart stronger. Pistachio benefits include the presence of vitamin A, vitamin E and anti-inflammatory properties reducing the inflammation in the body caused due to any problems. 60 per cent of the daily required phosphorous is present in a cup of pistachios which can prevent even type-2 diabetes. Also phosphorous in pistachios helps in breaking down proteins into amino acids resulting in glucose tolerance. Vitamin B6 is the protein that helps in carrying oxygen in the blood. The high quantity of B6 in pistachios if taken daily can increase the oxygen quantity in the blood and increase the haemoglobin count in it (Bulló *et al.,* 2015).

Vitamin B6 is present in high concentration in pistachios which is very beneficial to nervous system. Amines are the messaging molecules in the nervous system. In order to develop, they need amino acids which are dependent on vitamin B6 presence in the body. This vitamin builds a sheath called myelin around the nerve fibers. And messages pass via Nerve fibers from one nerve to another. Vitamin B6 helps in building many amino acids which help in proper transmission of nerve impulses. Macular degeneration is an age related eye disease which slowly reduces the eye vision making adults unable to read and work properly. It makes them unable to recognize people as well. Free radicals attack the cells and damage them resulting in macular degeneration. Lutein and Zeaxanthin are two anti-oxidants that are found in Pistachios which fight against these free radicals and destroy them preventing the cells from damage and thus preventing macular degeneration.

Vitamin B6 is very important for a healthy immune system. This helps in building blood and proper blood supply throughout the body. Vitamin B6, that is present in high quantity in pistachios helps in increasing the haemoglobin count in the blood. This oxygen rich blood is then supplied to the brain making it even more active. Glands like spleen, thymus, *etc.* become healthier and work properly with good supply of blood containing more white blood cells (WBC) that is required to fight against infections.

Vitamin E is very essential for a healthy skin and this anti-oxidant is present in pistachios. It integrates the cell membranes of the mucous membranes of the skin membranes. It protects skin from harmful UV rays, prevents from skin diseases and makes skin healthy and more beautiful. Vitamin E present in pistachios is helps fight ageing process of skin and making you look younger. oil present in this nut has emollient properties helping skin to moisturize and keep it from drying. It is used as aroma oil, medicinal massage oil *etc.* Vitamin B6 helps in increasing the blood count. The WBC or white blood cells help in preventing infections and various cancers.

One of the biggest health benefits of pistachios is that they are heart-friendly nuts. Research suggests that regular pistachio consumption can decrease the levels of bad LDL cholesterol in the body. The antioxidants, phytosterols, and unsaturated fatty acids (both polyunsaturated and monounsaturated fatty acids) are great for promoting a healthy heart.

Pistachios are also the first choice of people who are dieting. Evidence suggests that pistachios, with their lower calories, high protein, low saturated fat, and high unsaturated fats, are a perfect contender for ideal weight management. Pistachios contain carotenoid antioxidants such as lutein and zeaxanthin. These antioxidants help in reducing the risk of contracting an age-related macular disease in the elderly. They can be included, along with vegetables, for an increased supply of lutein and zeaxanthin.

Pistachios also help in eliminating dryness from the skin. The saturated fat plays a major role in this aspect. The fat also acts as a good carrier or base oil as it is used as an ingredient in traditional therapies like aromatherapy and traditional massage therapy. One of the other health benefits of pistachios is that it is a good source of dietary fiber, which helps in the smooth digestion of food. A serving of about 30 grams of pistachios contains about 3 grams of dietary fiber, more than enough to get your digestive tracts working at optimum capacity.

According to research, pistachios greatly influence sexual vitality in men. Studies have shown that those men who had a daily intake of about 100 grams of these nuts for three weeks improved their erectile function by around 50 per cent. This was confirmed by ultrasound scans of blood flow in the penis. Even the shells of pistachios are found to contain a considerable amount of antioxidants. Antioxidants neutralize the free radicals that form in the body after cellular metabolism and can cause conditions like cancer and heart disease.

One of the final health benefits of pistachios is that they function in controlling diabetes. In people suffering from diabetes, the sugars form inappropriate bonds with proteins and make them unusable. This process is known as glycation. The antioxidants present in them help in reducing the process of glycation and thus can help in controlling diabetes. Pistachios have a high copper content, which helps in the higher absorption of iron into the body from food sources. This can help combat conditions like anemia.

6.4 Uses in Cosmetics

Pistachio oil is an excellent natural moisturizer. It happens to be naturally infused with demulcent properties which tend to have highly hydrating and smoothening effects on the skin. One can consider substituting the regular moisturizer with Pistachio oil in order to get that smooth and silky skin.

They contain a variety of potent antioxidants that tend to neutralize free radicals, a chief culprit behind premature aging and in ensuring soft and silky skin for a longer time. Pistachios contain a considerable amount of essential fatty acids which are vital for maintaining a healthy, glowing complexion. Including green apples in your routine diet help you ensure a robust, radiant skin all the time. Being a rich source of vitamin E, a fat soluble antioxidant, Pistachios play a vital role in maintaining hale and hearty skin. They protect your skin from sun damage, thereby reducing the odds of skin cancer and sunburns. Pistachios help promoting healthy vision and are highly recommended to those struggling with any vision related issues. A hair mask using pistachios deeply nourishes and moisturizes your hair while improving the flexibility of hair fibers.

6.5 Conclusion

Besides minerals they also provide the essential vitamins like vitamins A, K, C, E, and vit B6. According to Ros (2010), pistachios contain a lower calorie content of only 160 calories per one ounce in comparison to other nuts and dried fruits such as cashew nuts, walnuts, almonds, pecans, hazelnuts, Brazil nuts, and chestnuts. Pistachios contain a higher amount of protein in comparison with other nuts such as almonds, cashews, hazelnuts, macadamias, pecans, and walnuts. The amount of protein found in pistachio is highest in comparison to other nuts. The fat content in pistachios is also the lowest as compared to the other nuts. They are also a good source of mono-unsaturated fatty acids (MUFAs) such as oleic acid and antioxidants. Taking them on a regular basis helps in decreasing low-density lipoprotein (LDL cholesterol).

REFERENCES

AL-Saghir, M.G., and D.M. Porter. (2012) Taxonomic revision of the genus *Pistacia* L. (Anacardiaceae). American Journal of Plant Sciences, 3: 12-32.

Ardakani A.S.H. (2006). The vital role of pistachio processing industries in development of Iran non-oil exports. Acta Horticulturae. 726: 579–581.

Ardakani, A. S. H.; Shahedi, M.; Kabir, G. (2009). "Optimizing Formulation of Pistachio Butter Production". Journal of Science and Technology of Agriculture and Natural Resources. 13 (47): 49–59.

Bulló, M; Juanola-Falgarona, M; Hernández-Alonso, P; Salas-Salvadó, J (2015). "Nutrition attributes and health effects of pistachio nuts". The British Journal of Nutrition (Review). 113 (2): S79-93. doi: 10.1017/S0007114514003250. PMID 26148925.

Chauhan M.G. and Kevalia J.K. (2014) Microscopic profile of drugs used in Indian systems of medicine. Vol. 3 Seed drugs Part II, Gujarat Ayurvedic University, Jamnagar. 299pp.

Dreher, M.L. (2012). "Pistachio nuts: composition and potential health benefits". Nutrition Reviews (Review). 70 (4): 234–40. doi: 10.1111/j.1753-4887.2011.00467.x.

Okay Y. (2002). "The comparison of some pistachio cultivars regarding their fat, fatty acids and protein content". Die Gartenbauwissenschaft. 67 (3): 107–113. JSTOR 24137567.

Parfitt, D.E.; Arjmand, N.; Michailides, T.J. (2003) Resistance to *Botryosphaeria dothidea* in pistachio. Hort Science. 38 (4): 529.

Ros E. (2010) Health benefits of nut consumption. Nutrients. 2010 Jul; 2(7): 652-682. doi: 10.3390/nu2070652. Epub 2010 Jun 24.

Chapter 7

A Powerhouse of Widespread Nourishment: Wheatgrass (*Triticum aestivum* Linn.)

Wheatgrass, shoot of Triticum aestivum Linn. (Hindi Name- gehun, kanak, Sanskrit name- godhuma) belongs to family:- Poaceae and consists of fat, carbohydrate, sugar, protein, vitamins viz. thiamine, riboflavin, niacin, pantothenic acid, folate, choline, Vitamin E, Vitamin K, amino acid, enzymes, minerals such that Ca, K, Zn, Na, Mg, B, Mo, chlorophyll which is most important of the therapeutic activity of wheatgrass, and some trace elements. The fresh juice prepared from cotyledons of wheat is rich in chlorophyll (pH 7.4) and is extensively used for therapeutic purposes. The various enzymes responsible for its pharmacological actions are protease, amylase, lipase, cytochrome oxidase, trans hydrogenase and super oxide dismutase. The other notable feature of wheatgrass is its high proportion of amino acids such as aspartic acid, glutamic acid, arginine, alanine and serine. Wheatgrass juice gets easily assimilated and quickly absorbed in the blood and is often referred to as Green Blood and is used therapeutically for the treatment of various problems and is recommended to the patients suffering from cancer, ulcer, disease related to joints such as gout, osteoarthritis, thalasemia, eczema, diseases related to digestive system, circulatory system, reproductive system, respiratory system, tooth, gum and migraine. An additional enzyme in wheatgrass is P4D1. This enzyme stimulates the renewal of nucleic acid and acts as an anti-inflammatory. In the present review attempts have been made to provide detailed information on the importance of wheatgrass, its chemical composition, pharmacological activities emphasizing the multitude potential of wheatgrass.

Keywords: *Chlorophyll, SOD, GBT, P4D1, Gluten, Amino acids, Anti-inflammatory, Wheat grass (WG), Wheatgrass Juice (WGJ).*

7.1 Introduction

Wheatgrass (WG) is obtained from the mature sprouts of wheat grains (*Triticum aestivum* Linn.) belonging to Poaceae family. It has many therapeutic values *viz.* anti-inflammatory, antioxidant, anti-carcinogenic, immune-modulatory, laxative, astringent, diuretic, antibacterial, anti-aging and remarkable healing properties. Its use in acidity, colitis, kidney malfunctions, atherosclerosis and swelling has been shown to be beneficial. The name "green blood" of WG is attributable to its high chlorophyll content which accounts for 70 per cent of its total chemical constituents. WG Juice is a house of nutrients, minerals and vitamins. It contains plenty of minerals like calcium, phosphorus, magnesium, alkaline earth metals, potassium, zinc, boron, and molybdenum *etc.* It's rich source of Vitamins A, C, E and B complex. As WGJ bears a closer resemblance to the "haemoglobin" in our blood, the juice is often called as "green blood" and the therapy using it is called as GBT "green blood therapy" WGJ is complete food (Wakeham 2013; Roshan *et al.*, 2016). WG consist of 16 amino acids in which 8 are essential. Although the therapeutic activity of WG has approved for last four decades but it was not used before because of the lack of powerful clinical data to support the therapeutic activity. But few experiments has evaluated the efficacy of WG in the form of powder or juice in the treatment of chronic disease like cancer, rheumatoid arthritis, and ulcer, *etc.* Coeliac is a condition that is caused by an adverse immune system reaction to gliadin, a gluten protein found in wheat. Upon exposure to gliadin, the enzyme tissue trans-glutaminase modifies the protein, and the immune system cross-reacts with the bowel tissue, causing an inflammatory reaction. That leads to flattening of the lining of the small intestine, which interferes with the absorption of nutrients. The only effective treatment is a lifelong gluten-free diet (Ashok 2011).

WGJ is very safe and the chances of the side effects are very low or minimum. It may cause nausea and headache if excessive quantities are taken. Throat swelling may occur in hyper sensitive to individual. 100 g of WG powder is equal to 23 kg of fresh vegetables. Ideally, WG should be taken about an hour prior to meal. This allows the body to fully metabolize it without competing with other foods, and it may also curb hunger. Wheatgrass consumption throughout the world, especially in western countries, has sky rocketed in the recent years. Wheatgrass is high in chlorophyll and minerals. A benefit to wheat grass is that it can easily be found in its "live" form. The enzyme content of fresh WG is impressive. The benefits of these enzymes directly affect the functions of the intestine and bowel. They aid in the assimilation of food and loosen debris in the lower bowel. Wheatgrass contains an antioxidant enzyme, SOD (superoxide dismutase) (Bodla and Mujoriya 2011). Wheatgrass in general is an excellent blood purifier as well as an energy booster. Wheatgrass contains glycogen (Thammana *et al.*, 2016), which is quickly translated into energy. Muco-polysaccharides found in the carbohydrate structure of, are found to strengthen bodily tissue (heart and arteries) and lower cholesterol. This is especially A benefit to wheat grass is that it can easily be found in its "live" form (Yang *et al.*, 2001). The enzyme content of fresh wheat grass is impressive. The benefits of these enzymes directly affect the functions of the intestine and bowel. They aid in the assimilation of food and loosen debris in the lower bowel.

Figures 7.1A-D: Young Seedlings of Wheatgrass, Powder, Juice and Tablets.

7.2 Pharmacokinetics and Phytochemistry of WGJ

The carbohydrates, protein and lipid in WG are digested, absorbed, and metabolized by normal physiological process. Wheat grass juice is extracted from the cotyledons of the common wheat plant *Triticum aestivum*. It has been used therapeutically from years all around the world because of its healing properties. The juice from WG is high in chlorophyll, active enzymes, vitamins and other important nutrients like Iron (Lai *et al.,* 1978).

The chlorophyll (Rana *et al.,* 2011) present in wheatgrass and haemoglobin has similar structures except for the central moieties. The compounds in chlorophyll and magnesium depict a striking similarity in having a tetra pyrrole ring structure, the only difference between the two being the nature of the central metal atom Mg in chlorophyll and Fe in haemoglobin (Polshetiiwar and Khorate 2016). Also the pH of WGJ is 7.4 which equals same as that of blood. Therefore, it is absorbed quickly in the blood and is often referred to as Green Blood. This juice is extensively used therapeutically for the management of diseases like Anemia, Thalassemia, Inflammatory Bowel diseases, *etc.* Nutritional deficiency anaemia refers to a decrease in the concentration of haemoglobin in the blood due to poor dietary habits that causes deficiency of iron, proteins, vitamins like vitamin C, vitamin B_{12} along

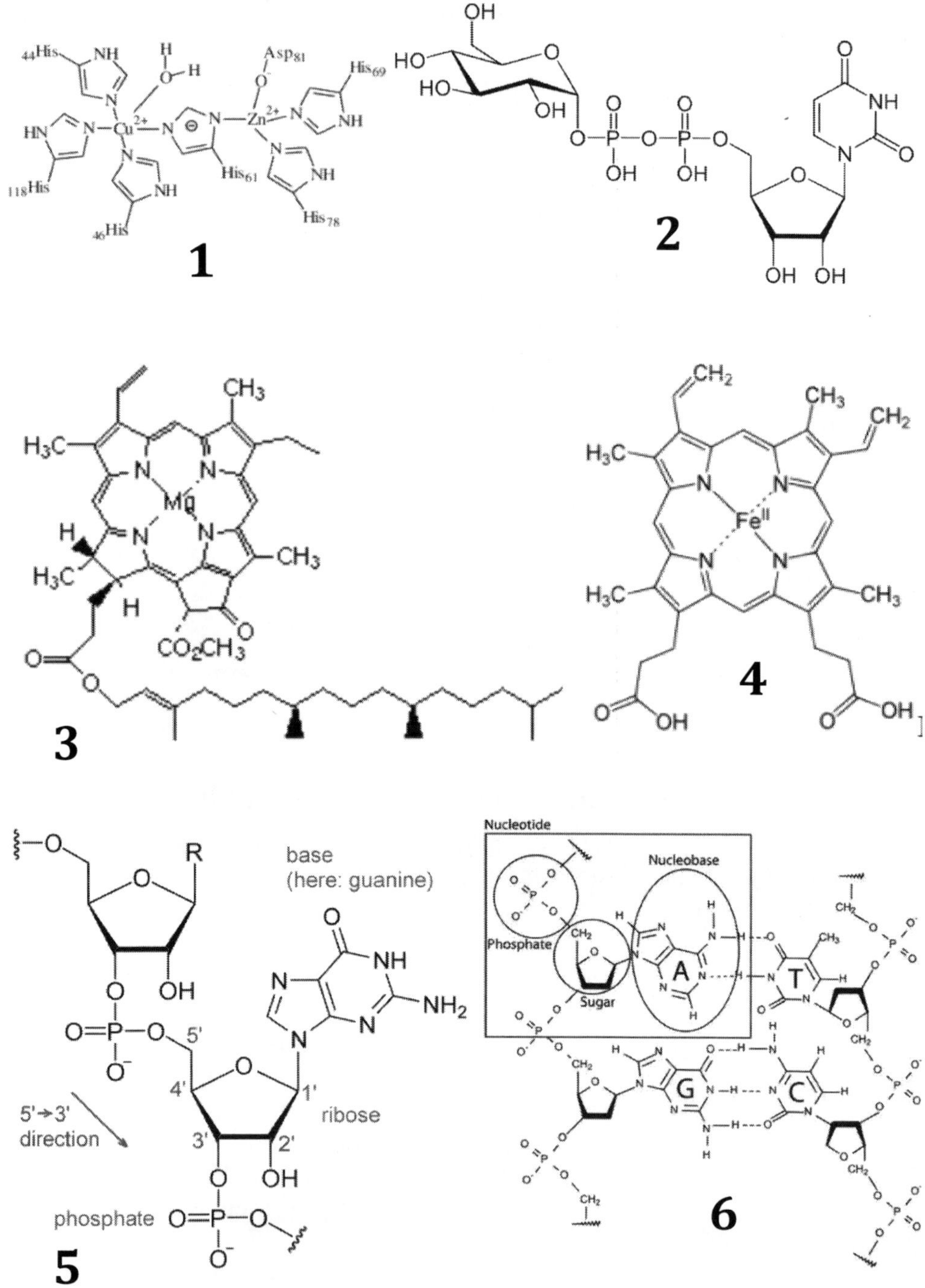

Figure 7.2. (1-6) Structural Formula of Chemicals Present in Wheat Juice.

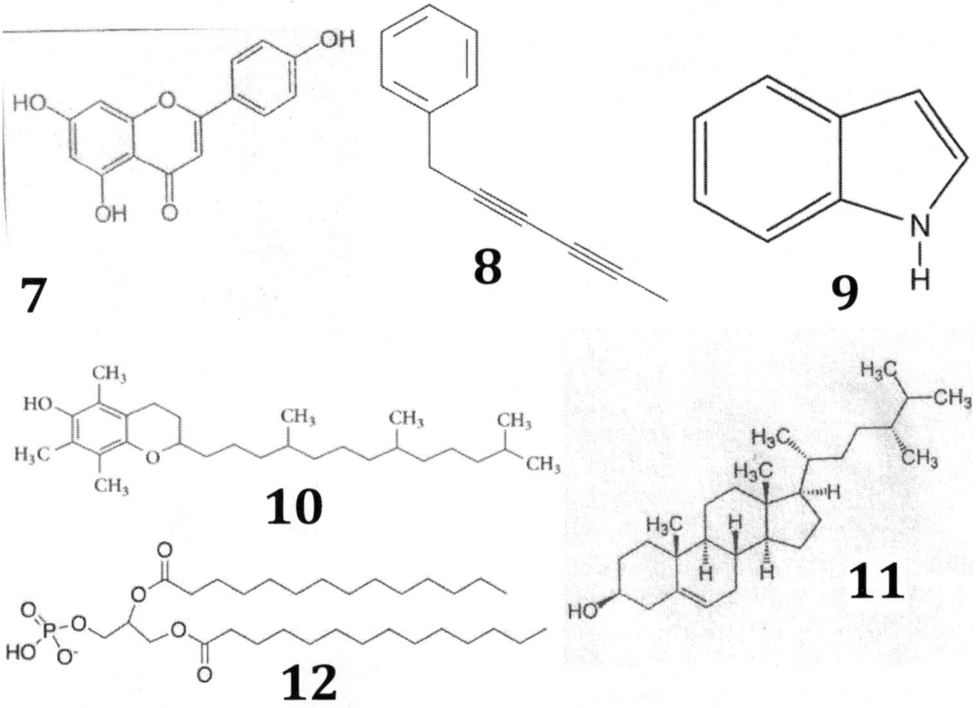

Figure 7.2. (7-12) Structural Formula of Chemicals Present in Wheat Juice.

with folic acid. Deficiency of the above nutrients can affect the production of red blood cells. The major clinical utility of WGJ is due to its antioxidant action which is derived from its high content of bioflavonoid. Other compounds present, which make this grass therapeutically effective is choline. Being highly anti-bacterial, consuming WG will help to alkalise and detoxify the lymph and blood cells, helping the body to rebuild and transport toxins out of the body quickly and effectively. The analogy between chlorophyll and haemoglobin can be demonstrated with respect to the structure of their porphyrin heads. The structure of both WG consumption throughout the world, especially in western countries, has sky rocketed in the recent years. WG is high in chlorophyll and minerals. A benefit to wheatgrass is that it can easily be found in its "live" form.

Table 7.1. Nutritional Value of Wheat Grass Juice per 100g WG

Nutritional Value Concerning to Wheatgrass	
Energy	35 calorie
Carbohydrate	1 per cent
Dietary fiber	8 per cent
Fat	0
Protein	4 per cent

Table 7.2. Amount of Amino Acid per 100 g Wheat Juice

Quantity of Amino Acids in WGJ μg/ml			
Aspartic acid	510.3	Threonine	105.8
Serine	201.8	Asparagine	3039.6
Glutamine	200.6	Proline	33.6
Glycine	20.6	lanine	166.4
Valine	272.1	Methionine	14.0
Isoleucine	145.1	Leucine	101.0
Tyrosine	121.8	Phenylalanine	200.9
Lycine	174.5	Histidine	232.2
Tryptophan	160.1	Arginine	252.9

Wheat grass contains glycogen, which is quickly translated into energy. An additional enzyme in wheat grass is P4D1. This enzyme stimulates the renewal of nucleic acid (RNA/DNA) (Shirude 2011; Padelia *et al.,* 2010) and acts as an anti-inflammatory. This is especially beneficial for those suffering from arthritis and other inflammatory diseases. The other notable feature of wheat grass is its high proportion of amino acids like arginine, serine, lysine, aspartic acid, glycine, alanine, methionine, leucine, tryptophane, phenylalanine, and valine.

Table 7.3. Minerals Present in per 100 g Wheat Grass Juice

Quantity of Minerals in Wheatgrass per 100 ml WGJ			
Minerals	57	K (Potassium)	57
P (Phosphorus)	8.2	Ca (Calcium)	2.4
S (Sulphur)	2,37	Mg (Magnesium)	1.7
Na (Sodium)	1.42	Al (Aluminium)	0.31
Zn (Zinc)	0.02	Cu (Copper)	0.007

WG has been shown to have potential anti-inflammatory and anti-aging properties. WG is promoted to treat a number of conditions including the common cold, cough, bronchitis, fever, infections, inflamed mouth and throat, and skin disorders like hemorrhoids, psoriasis, ivy, eczema, burns and thalassemia. The extract of WG has shown significant improvement in rectal bleeding and abdominal pain associated with ulcerative colitis, a type of inflammatory bowel disease. Flavonoids, anti-inflammatory agents- apigenin (Rathore *et al.,* 2016) and agropyrene (Lam and brush 1950) contained in WG, play a role in alleviating symptoms. Another flavonoid present in WG is Indole (Gahan *et al.,* 1943), which helps in synthesis of enzymes and deactivating carcinogens in the liver. Other benefits of WG are: removal of toxins from the body, improvement of blood sugar balance, prevention of tooth decay, maintenance of healthy hair, aiding digestion, and reducing high blood pressure levels. Chlorophyll present in WG can protect us from carcinogens; it strengthens the cells, detoxifies the liver and blood stream, and chemically neutralizes the polluting elements.

Wheat germ oil is rich in and essential fatty acids. Wheat bran oil is also high in Tocopherols, 68 per cent of which is Epsilon-Tocopherols. α-Tocopherol (Chernomorsky and Segelman 1988), Sitosterol, ergosterol (Mancinelli *et al.,* 1998), and campesterol, phospatidic acid (Calzuola *et al.,* 1998) and glyceroinositophosphatidic acids, phytoglycolipid, serine, *etc.*, are also reported. Leaf protein isolate contains methionine, tryptophane, histidine, arginine and total lysine. WG has the ability to increase the amount of oxygen in the blood, making it a great way to stimulate circulation. Instead of reaching for antacids to relieve heartburn or indigestion, introduce WG into your daily regimen. WG contains several elements that can boost digestion, including a great deal of fiber, and B complex vitamins, which boost the function of the muscles of the digestive system. Thiamine helps convert carbs into energy, and riboflavin keeps the mucosal lining of your digestive tract healthy; WG contains both. WGJ is effective in treating arthritis. Chlorophyll is thought to fight inflammation, which is associated with joint pain and may be the reason WG is so helpful in these cases. Not only does chlorophyll boost the immune system, it also helps to increase oxygen supply in your body's cells and tissues, contributing to cell regeneration, which heals the body and reduces fatigue symptoms. WG improves lipid levels. As WGJ supports immunity and reduces inflammation, it can also be an excellent tool for reducing congestion. Some tout the benefits of snorting WGJ through the nose to clear the sinuses by pulling toxins and breaking up mucus–which we don't recommend. Fortunately, you can just as easily reap these benefits by drinking WG. Wheatgrass' antibacterial (Kakkar *et al.,* 2012; Dubey *et al.,* 2012) benefits and its ability to reduce chronic inflammation combine to make WG an excellent tool to reduce acne and occasional breakouts. Wheatgrass's anti-cancer benefits stem from its blood oxygenating ability; cancer thrives in a low-oxygen environment (Kulkarni *et al.,* 2006; Kapil 2012), so WG may contribute to cancer prevention in this way. WG extract proved successful in topically treating acute radiation-induced skin reactions. WG was shown to be a powerful anti-hyperglycemic agent. WG could be beneficial for those suffering from diabetes or other hyperglycemic issues.

7.3 Common Ailments Cure

Wheat protein, which comprises up to eight per cent of the grain, has a special benefit as it has eight of the essential amino acids in delicately balanced proportions. A complete internal rejuvenation takes place when Wheat protein is metabolized into health-building amino acids. These amino acids build a resilient muscle that comes back to its original form after stretching and bending, healthy skin and hair and clearer eyesight and nourish the heart and lungs, tendons and ligaments, brain, nervous system and glandular network. The B-complex vitamins, especially thiamine, riboflavin and niacin offered by natural brown Wheat promote youthful energy and nourishment to the skin and blood vessels. An abundance of minerals in natural brown Wheat helps to nourish the hormonal system, heal wounds and regulate blood pressure. Wheat also offers iron to enrich the bloodstream and phosphorus and potassium to maintain internal water balance along with other nutrients. Wheat, thus helps restore internal harmony.

Wheat is valuable in the prevention and cure of pyorrhea. It takes time to eat wheat and as it is generally taken with other foods, it compels the chewing of other foods also. This not only provides the needed exercise for the teeth and gums but also a great aid to digestion. WGJ acts as an excellent mouth wash for sore throats and pyorrhea. It also prevents tooth decay and tooth aches. Therefore, it is beneficial to chew wheat grass which draws out toxins from the gums and thus checks bacterial growth.

The bran of wheat, which is generally discarded in milling of the flour, is more wholesome and nourishing than the flour itself. It is an excellent laxative. The laxative effects of bran are much superior to those fruits or green vegetables as cellulose of the latter is more easily broken down by bacteria while passing through the intestine. The bran is highly beneficial in the prevention and treatment of constipation due to its concentration of cellulose, which forms a bulk-mass in the intestines and helps easy evacuation due to increased peristalsis. It has been scientifically proved that chlorophyll arrests growth and development of harmful bacteria. Wheat grass therapy can be effectively used for skin diseases and ulcerated wounds as by retarding bacterial action (Singh and Thalwal 2013), it promotes cell activity and normal re-growth by drinking WGJ regularly, an unfavorable environment is created for bacterial growth. Poultice of WGJ can be applied on the infected area, as it is an able sterilizer. Externally, wheat flour is useful as a dusting powder over inflamed surface as in burns, scalds and various itching and burning eruptions, Whole wheat flour, mixed with vinegar, boiled and applied outwardly removes freckles. Wheat grass juice used as an enema helps detoxify the walls of the colon and is very helpful in mucous and ulcerative colitis, chronic constipation and bleeding piles. The chlorophyll content present in wheat enhances heart and lung functions. Capillary activity also increases while toxemia or blood poisoning is reduced. Due to increased Iron content in the blood and hemoglobin, lungs function better. Oxygenation improves and the effect of carbon dioxide is minimized. It is for this reason that WGJ is prescribed for circulatory disorders.

7.4 Discussion and Conclusion

A wide range of health benefits have been attributed to WG. This review concluded that the wheat grass juice (WGJ) has higher degree of curative index and which is useful for various pharmacological activities such as anticancer, antibacterial, antimicrobial, antidibetic, antilukemic, antioxidant, antiulcer *etc.* (Peryt *et al.,* 1992; Reynolds 2005; Swati *et al.,* 2021).

The isolation of properties in WG is becoming more prevalent. Superoxide dismutase is gaining interest as an antioxidant as is apigenin, both of which are present in WG. There is evidence to suggest that it is the combination of these properties which make WG activity potentially useful as a medical treatment.

The name "green blood" is because of the structural homology of chlorophyll with hemoglobin indicates the role of chlorophyll as a blood builder in various clinical conditions involving hemoglobin deficiency. The chlorophyll present in WG helps in combating inflammation and eases joint pain, which is linked with curing arthritis. It helps in detoxifying the liver by ridding it of excess toxins such as heavy metals. WG contains a compound known as P4D1, which increases fertility levels in the human body. The chlorophyll present in WG helps increase the red blood cell count and normalizes blood pressure. WG is an incredible source of vitamin A, C, phenolic compounds, antioxidants, minerals and most importantly, chlorophyll. It's great for your heart, your brain, skin and your entire body. It is a nutritionally complete source of food which has shown to sustain and promote growth and development. WG consists of all essential amino acids, making it a complete protein source.

Wheatgrass contains selenium, which facilitates the healthy functioning of the thyroid gland. The chlorophyll present in WG contains enzymes, amino acid chains and antioxidants which helps in repairing damaged DNA and decomposing the superoxide radicals in the body and slows down the ageing process (Shirude 2009; Rajesh and Ramesh 2012; Chauhan 2014). This results in a better health overall. The antibacterial and antimicrobial property of WG boosts oral hygiene and reduces the risk of cavities and gum inflammation. WG help in keeping common diseases such as flu, cough and cold, and other infections at bay. WG is used to treat skin diseases such as eczema, psoriasis, acne, skin wounds by regenerating the skin cells. It also helps in converting carbohydrates into much-needed energy. Wheatgrass is widely believed to be a remedy for dandruff and reverses graying hair. WG is proven to be an effective anti-hyperglycaemic agent, which makes it beneficial for those suffering from diabetes. Being chock full of vitamins, minerals and other nutrients, it reduces the craving for other foods to compensate for the intake of such nutrients. By reducing such cravings, it helps in preventing overeating. By clearing out the harmful, toxic substances, the body can focus its energy better on other forms of physical activity and in performing day-to-day activities. By balancing the iron levels in the blood, WG helps in fighting depression and improve mental health. WG has been proven to reduce bowel inflammation and ease movement. By compensating for any lack of vitamins and minerals, it helps in tiding over the menstrual cycle better. In view of the above benefits, it is recommended to consume WGJ every morning, for a healthy day ahead. It can contribute towards the overall well-being of the people. Health experts vouch for the immense beneficial effects of WG, but at the same time caution that it should be treated as an addition to our everyday diet, and not as a supplement or replacement. Widespread data from a number of studies has made known the multitude effects of WG is known to help diminish fatigue, improve sleep, increase strength, naturally regulate blood pressure and blood sugar, support weight loss, improve digestion and elimination, support healthy skin, teeth, eyes, muscles and joints, improve the function of our heart-lungs and reproductive organs, heal ulcers and skin sores, slow cellular aging, improve mental function, and is beneficial in arthritis and muscle cramping, thalassemia, hemo-lytic anemia, cancer, asthma, allergy, inflammatory bowel disease and detoxification. Thus, it should be made part of daily dietary intake in order to explore its utmost benefits. The structural homology of chlorophyll with hemoglobin indicates the role of chlorophyll as a blood builder in various clinical conditions involving hemoglobin deficiency - thus the name "green blood". To conclude WG seems to be very promising herbal drug and extensive research work is needed in order to explore its therapeutic application in various diseases.

Consumption of freshly squeezed WGJ resembles living food which has bio-electricity. This high vibration vitality is actually the life power inside the living juice. This asset of life-power vitality can conceivably unleash effective recharging vibrations and more prominent availability to one's inward being. These intense supplements can likewise prevent DNA destruction and shield the humans from the continuous impacts of preadult maturing and cell breakdown. Late research demonstrates that exclusive living foods and juices can reestablish the electrical charge between the vessels and the cell dividers which supports the resistant framework. WGJ is especially high in chlorophyll along with some other constituents. WG washes down and manufactures the blood because of its high substance of chlorophyll. Chlorophyll, the life blood of plants is the main result of light and in this manner contains more mending properties than some other component. Drinking WGJ resemble drinking fluid daylight. Chlorophyll conveys elevated amounts of oxygen (in addition to other things) which is particularly intense in helping the body to reestablish irregularities. The chlorophyll which has more oxygen conveys more oxygen to the blood. Red Blood cells numbers will increase and blood oxygen levels raises rapidly with the drinking of WG squeezed juice and utilizing WGJ. WG found to break down scars formed in the lungs also serves to remove various drug deposits from the body, filters the blood and organs and counteracts acids and poisons in the body. It builds the compound level in our cells, supporting in the revival of the body and the digestion system of supplements. Researches found that insertion of WG chlorophyll packs into the sinuses clears the congestion. It also helps to cure chronic sinusitis. WG is one of green substances that are esteemed by well-being of people as an incredible source of nutritious supplements. WG is known as complete nourishment. Researches should focus on this exceptionally encouraging natural medication, endeavors are expected to direct broad studies on the wheat grass both in test models and human subjects to create wheat grass treatment with no side effects in prevention, cure and treatment of many harmful health conditions for which the current medical sciences did not discover any fine solution.

REFERENCES

Ashok SA. (2011). Phytochemical and Pharmalogical Screening of Wheatgrass Juice (*Triticum aestivum* L.) Int. J. Pharmaceut. Sci. Rev. Res., 9: 159-164.

Bodla RB., Mujoriya R. (2011). A study on wheat grass and its Nutritional value Food Sci. Qua. Manag., 2(1): 1-8.

Calzuola I., Valeria M., Gianfranceschi GL. (2004). Synthesis of antioxidants in wheat sprouts. J. Agri. Food Chem., 52: 5201–5206.

Chauhan M. (2014). A pilot study on wheat grass juice for its phytochemical, nutritional and therapeutic potential on chronic diseases. International Journal of Chemical Studies 2(4): 27-34.

Chernomorsky S., Segelman A. (1988). Review Article: Biological activities of chlorophyll derivatives. N. J. Med., 85(8): 669-673.

Dubey PK., Bais N., Kakkar A., Netam R., Dubey S., Verma DR. (2012). Studies on antimicrobial and anthelmintic potential of wheatgrass. J. Glob. Pharma. Technol. 11(4): 08-11.

Gahan E., Kline P., Finkle T. (1943). Chlorophyll in the treatment of ulcers. *Derm. Syph.,* 47: 849-851.

Kakkar A., Dubey PK., Dubey S., Khare P., Bias N., Netam R. (2012). Studies on synergistic antimicrobial potential of aloe-wheatgrass extract combination. *Asian J. Pharma. Lif. Sci.,* 2(2): 101-110.

Kapil S. (2012) The Elixir Of Life, Wheatgrass Juice. Popular Book Depot.

Kulkarni SD., Tilak JC., Acharya R., Rajurkar NS, Devasagayam T, Reddy AV. (2006). Evaluation of the antioxidant activity of wheatgrass (*Triticum aestivum* L.) as a function of growth under different conditions. Phytother. Res., 20: 218-227.

Lai CN., Dabney BJ., Shaw CR. (1978). Inhibition of in vitro metabolic activation of carcinogens by wheat sprout extracts. Nut. Canc., 1: 27– 30.

Lam C., Brush B. (1950). Chlorophyll and wound healing. Experimental and clinical study. Am, J, Surg., 8: 204-210.

Mancinelli L., Marsili V, Gianfranceschi G. (1998). Presence of biologically active substances in the wheat bud. Gastroenterol. Int. 11: 50–51. 51.

Padalia S., Sushma D., Indira R., Gupta A., Mamta D. (2010). Review Article Multitude potential of wheatgrass juice (Green Blood): An overview *Chroni. You. Scient.,* 1(2): 23- 28.

Peryt B., Szymczy KT., Lesca P. (1992). Mechanism of antimutagenicity of wheat sprout extracts. Mut. Res., 269: 201–205.

Polshettiwar S., Khorate SS. (2016). *Triticum aestivum*-A Green Gold. World J. Pharma. Pharmaceut. Sci., 5(4): 636-651.

Rajesh M., Ramesh B. (2012). Therapeutic Potential of Organic *Triticum aestivum* Linn.(Wheat Grass) in Prevention and Treatment of Chronic Diseases: An Overview. Int. J. Pharmaceut. Sci. Dru. Res., 4(1): 10-14.

Rana S., Kamboj JK., Gandhi V. (2011). Living life the natural way – Wheatgrass and Health. *Func. Foods Heal. Dis.,* 1(11): 444-456.

Reynolds CA. (2005). DNA technology-based cellular assay used to measure specific biological activity in a wheatgrass extract. J. Austral. Integ. *Med.* Asso., 1-3.

Roshan K., Rathore KS., Bharkatiya M., Goel PK., Naruk PS., Saurabh SS. (2016). Therapeutic potential of *Triticum asetivum* Linn. (Wheatgrass or Green blood therapy) in the treatment and prevention of chronic and acute diseases: An Overview, PharmaTut., 2016; 4(2): 19-27.

Shirude AK. (2011). Phytochemical and pharmacological screening of wheatgrass juice (*Triticum aestivum* L.). Int. J. Pharmaceut. Sci., Rev. Res., 9(1): 159- 164.

Singh J., Thalwal S. (2013). Comparative analysis of Medicinal Plants for their Antimicrobial Potential and Phytoconstituents Screening. Int. J. Dru. Dev. Res., 5(3): 447-453

Swati P., Sushma D., Indira R., Alka G., Mamta D. (2021). Multitude potential of wheatgrass juice (Green Blood): An overview. Chron. you. scient., 1(2): 23-28.

Thammana M., Sreerangam S., Nambaaru S. (2016). A Mini Review on Wheatgrass. Res. Rev.: J. Pharmacog. Phytochem., 4(3): 13-19.

Wakeham P. (2013). The medicinal and pharmacological screening of wheatgrass juice (*Triticum aestivum* L.): an investigation into chlorophyll content and antimicrobial activity. The Plymou. Stud. Scient., 6(2): 20-3

Yang F., Basu TK., Ooraikul B. (2001). Studies on germination conditions and antioxidant content of wheatgrass. Int. J. Food Sci. Nut., 52: 319–30.

Chapter 8

Irresistible Health Benefits of Super Food: Rajgira (*Amaranthus* spp.)

Grain amaranth is a pseudo-cereal with unique agricultural, nutritional, and functional properties. Amaranth is a healthy grain often used as an alternative to gluten-rich grains. As amaranth is gluten free, it is a great replacement grain for all people who suffer from any form of gluten intolerance, especially those who suffer from celiac diseases. Amaranthus grain or Ramdana is the perfect example of the nutraceutical. Amaranth oil with to cotrienol and squalene has potential to be used as medicinal foods. Amaranth oil has antibacterial, anti-tumor, and burn and wound-healing properties. It also lowers blood serum cholesterol. Amaranth is an excellent source of protein which contains those amino acids that are usually found only in animal foods. It is loaded with fiber, iron, calcium, vitamins and minerals, and is significantly more nutritious than whole wheat. It is also the best plant source of squalene, a powerful antioxidant used as a dietary supplement for diabetics and those suffering from hypertension and metabolic disorders. Amaranth grain that is becoming rapidly popular has many nutritious benefits and having applicability treating many disorders with remarkable "Nutraceutical Properties". One of the nutritional powerhouse that we seem to forget about the plain and simple amaranth or otherwise known as rajgira or ramdana.

Keywords*: Amaranth grain, Squalene, Tocopherols, Phytosterols, Lysine, Gluten, Micronutrients.*

8.1 Introduction

Amaranth (*Amaranthus* spp., Greek "eternal") originated in South America, specifically Ecuador, Peru and Bolivia. Botanically, Amaranth belongs to the family of dicotyledonous plants Amaranthaceae, which includes more than 60 species, most of which are weeds. They can grow in width and in height, and there are both low and varieties that can grow up to 4 meters high. Three varieties of amaranth are the most important: *A. hypochondriacus* (Mexico), *A. cruentus* (Guatemala) and *A. caudatus* (Peru and other countries in the Andes). The variety *A. hypochondriacus* is examined the most widely and thoroughly. Amaranth (In Latin *Amarantus*) is a pseudo-cereal superfood. Rajgira or Amaranth has been grown in India, South America and what is now Mexico for over 8000 years for its nutritional benefits. It was the main staple of The Aztecs (modern Mexico) till the arrival of Spanish invaders. When the Spanish arrived with Cortez, they not only converted the natives into Christianity but also outlawed many native traditions and cultural practices. Rajgira was banned because the Aztecs built statues of their deity using Amaranth grains and honey. After the worship, they would break these statues and distribute for eating. Even though the Spanish tried their best to destroy Amaranth by burning fields and punishing farmers, it survived in remote areas as a wild plant. It was rediscovered and introduced in the US in the 1970s as a wonder grain. While it was almost dead in the west, Rajgira continued to thrive in India. It was named Rajgira (Royal grain) or Ramdana (God's grain) because of its powerful health benefits (Becker *et al.,* 1981; Kauffman and Weber 1990; Andini *et al.,* 2013). According to modern researches, Rajgira is a powerhouse that can help prevent several chronic health conditions such as heart disease, cancer and stroke (Martirosyan *et al.,* 2007). It's also helpful in treating inflammatory issues like arthritis. It's hard to believe that these tiny seeds are packed with so much nutrition. It provides most of the nutrition we need for a healthy life. Even NASA astronauts used amaranth as part of their diet (Tucker 1986; Pederson *et al.,* 1987; Vora *et al.,* 2014).

Grain amaranth has the potential to contribute to improvement in nutrition of populations, especially in developing countries, because of its unique agricultural, nutritional, and functional properties. It is fast-growing, high-yielding, stress-resistant, nutritious, and has nutraceutical properties. The color of the grain varies from milky white to yellow, gold, red, brown and black, which depends on the content of the betalaine pigments.

Grain amaranth is rich in proteins, lipids, energy, and fiber. Grain amaranth protein is of superior amino acid profile compared to proteins found in most other plants. Amaranth grains contain 2x the level of Ca as in milk, 5x the level of iron in wheat, higher sodium, potassium, and vitamins A, E, C, and folic acid than cereal grains. Amaranth grain contains 6 to 10 per cent unsaturated oil that is good for your health. Rajgira has about 13 per cent protein. This translates into 26 grams/cup of raw Rajgira, which is much higher than most other grains. The protein in Rajgira is called "complete protein" because it contains Lysine, an amino acid that helps in building body muscles. Lysine is missing from or is present in negligible quantity in other grains. Lysine is a building block for protein. It is also an essential amino acid that plays a role in calcium absorption. Since our bodies can't produce essential amino acids, we need to obtain them from food. Lysine plays an important role in supporting our immune system (Becker *et al.,* 1981).

Figures 8.1A-D. (A) Rajgira plant; (B) and C) Rajgira Grains; (D) Poped Rajgira Dishes/Chikki.

Soluble fiber found in Amaranth grain helps in improving digestion. The dietary fiber can eliminate constipation, bloating, cramping, and more serious conditions, like colon cancer and gastric ulcers. Fiber attaches to cholesterol in the digestive system and takes it out of the body. It benefits heart as well as helps in digestion. Dietary fibre also helps body in absorbing nutrients from the food. Rajgira grain may have probiotic properties. Probiotics help improve digestion by supporting the growth of beneficial gut bacteria. Rajgira can increase insulin levels in the body and bring down sugar levels in the blood. Surprisingly the anti-diabetic and antioxidative properties on Rajgira help keep blood sugar levels in check even though its GI is high (López *et al.,* 2011). But more research is needed to prove this beyond conclusively.

8.2 Chemical Constituents of Rajgira Grains

The total carbohydrate content of the amaranth grain is slightly lower than that of wheat grain. Starch represents the main carbohydrate component, constituting 48-69 per cent of dry grain matter. The content of amylose is very low, from 0.1 to 11.1 per cent. The starch content is also subject to large variations. The content of dietary fibre in amaranth grain is slightly lower than in wheat, and is generally similar to quinoa and common cereals - about 20 per cent of dry grain matter. As per reports of researchers content of total dietary fibers, soluble and insoluble, is between 9.8 and 14.5 per cent, depending on the variety. Only small quantities of mono- and disaccharides were detected in amaranth grain. Sucrose is dominant, but glucose, galactose, fructose, maltose, raffinose, stachyose and inositol were also detected (Pastor and Akanski 2018; Caselato-Sousa and Amaya-Farfan 2012).

Proteins are the second most common component of amaranth grains. It is believed that the amino acid composition of the amaranth protein is close to the ideal protein. All varieties of amaranth are characterized by high protein content, from 13 to 18 per cent dry matter which is generally higher than in cereals, and other pseudocereals. Unlike cereals, amaranth proteins consist of albumins (about 40 per cent), 5 (25-30 per cent) and globulins (20 per cent), and contain very small amounts of prolamins (2-3 per cent), which are the main group of proteins in cereals. Amaranth does not contain gluten, therefore it is therefore suitable for the diet of celiac patients (Alvarez-Jubete *et al.,* 2009; 2010). Albumins and globulins contain less glutamic acid and proline than prolamins, and are richer in essential amino acids, such as lysine, which makes amaranth proteins closer to the composition of animal proteins. Due to their favorable chemical content and the application in developing functional food products, the use of pseudocereal species in human nutrition is constantly increasing (Tagwira 2006).

Amaranth is one of the three pseudo-cereals, next to buckwheat and quinoa, most frequently used in modern formulations of functional bakery, pastry and confectionary food products. The content of lipids in amaranth grain is higher than in common cereals, 6-9 per cent. The content of lipids in amaranth is 2 to 3x higher than in buckwheat grains and common cereals. Amaranth lipids consist of fatty acids, triglycerides, sterols, phospholipids, glycolipids, tocopherols, and hydrocarbons. Amaranth lipids are characterized by high content of unsaturated fatty acids, with a particularly high content of linoleic acid. Linoleic acid makes up over 50 per cent of total fatty acids and therefore represents the most common fatty acid. Phospholipids make up about 5 per cent of the lipid fraction; the most common are lecithin, cefalin and phosphoinositol.

Generally, amaranth is not an important source of vitamins. However, it does contain more riboflavin (vitamin B2), folic acid (vitamin B9) and ascorbic acid (vitamin C) as compared to common cereals. The content of total tocopherols (tocopherols and tocotrienols) in amaranth grains may vary significantly. The most common tocopherol is α-tocopherol. Some report signify appropriate concentrations of β- and γ-tocotrienols. The mineral content is about two times higher than in common cereals. The content of Ca, Mg, Zn and Fe is particularly high. The Ca/P ratio is very suitable, being in the range 1: 1.9-2.7. Recommended ratio is 1: 1.5. The high content of Ca in amaranth grain is particularly significant in patients suffering from celiac disease, due to the onset of osteopenia and osteoporosis in these individuals.

Gluten is a mixture of proteins found in wheat, and rye, barley and triticale. Gluten helps foods maintain their shape, acting as a glue that holds food together. The gluten protein complex accounts for 75 to 85 per cent of the total protein in bread wheat. It is conjoined with starch in the **endosperm** of various grass- related grains. Gluten proteins are termed prolamins. It is a protein that is attached to starch in the endosperm of various grass-related grains. Wheat prolamins are called gliadins and glutenins, barley prolamins are hordeins, rye prolamins are secalins and oats prolamins are avenins. Prolamins are a group of plant storage proteins found in the seeds of cereal grains: wheat (**gliadin**) [Structure Shown], barley (hordein), rye (secalin), corn (zein), sorghum (kafirin) and as a minor protein, avenin in oats. They are characterised by a high glutamine and proline content and are generally soluble only in strong alcohol solutions. Some prolamins, notably gliadin, may induce celiac disease in genetically predisposed individuals.

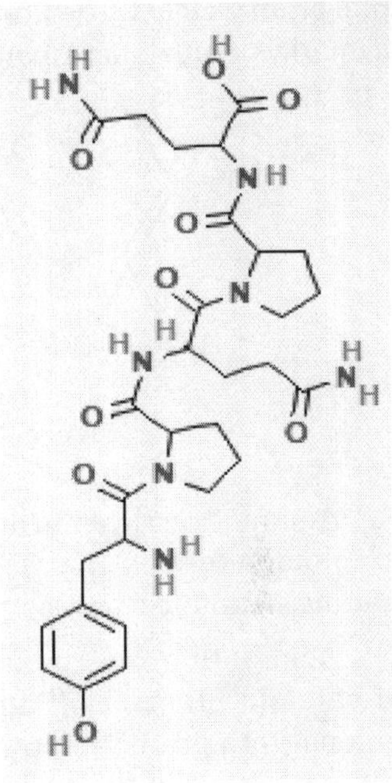

Figure 7.2. Structure of Wheat Gliadin.

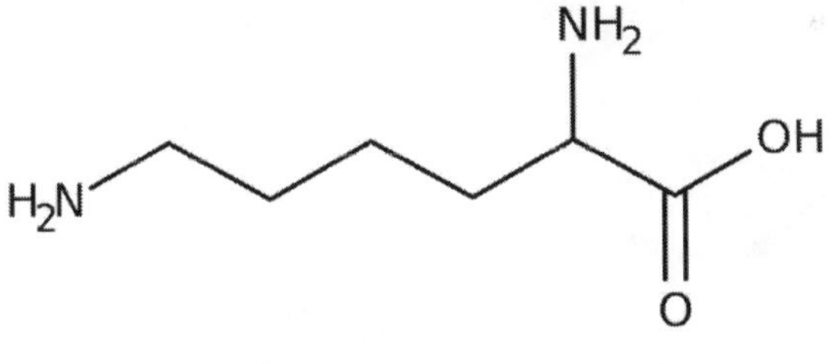

Figure 7.3. Chemical Structure of Lysine.

Amaranth grain contains protocatechuic, hydroxybenzoic, caffeic, and ferulic acid. Polyphenolic substances, such as rutine, nicotiflorin and isoquercetin, were also detected. It is known that routine has an effect on preventing the ageing process, quercetin inhibits oxidation processes, and nictoflorin participates in the preservation of memory functions (Meyers *et al.*, 2008). Squalene (Figure 7.4) is a biosynthetic precursor of all steroids in plants and animals (Petras and Venskutonis 2013). It is a highly unsaturated open chain triterpene, which is widely used in the cosmetics industry (in skin care products) and as an industrial lubricant resistant to oxidation. The richest source of squalene is a small shark liver (*Squalus acanthus*), from which comes its name, as well as some other marine species. It is also present in some vegetable oils, such as olive oil (0.1-0.5 per cent), but particularly high concentrations are detected in amaranth oil (3-8 per cent and 0.3-0.4 per cent of total grain weight). The non-saponifiable lipid fraction of amaranth grain consists primarily of squalene. Amaranth is the richest source of squalene in the plant world.

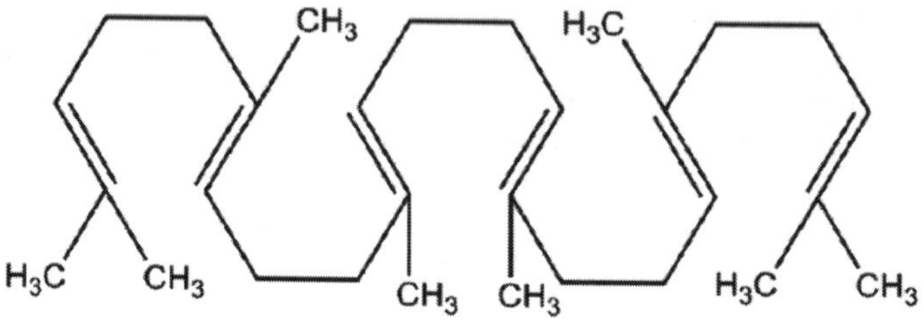

Figure 7.4. Chemical Structure of Squalene.

Amaranth is richer in proteins and lipids, but contains less starch, as compared to common cereals. It is rich in phytochemicals (secondary metabolites), such as routine, nictoflorin and isoquercetin, which have a positive effect on human health.

Amaranth grain contains pigments – betacyanins, belonging to the group of betalain pigments (Figure 7.5). Two types of these pigments are present in amaranth grain: amarantin and iso-amarantin. Tannins are present in high concentrations in the grain hull, so with grain peeling they are mostly removed. They can form complexes with proteins and digestive enzymes, thereby inhibiting protein digestion. Also, depending on the chemical composition, they can inhibit starch digestion.

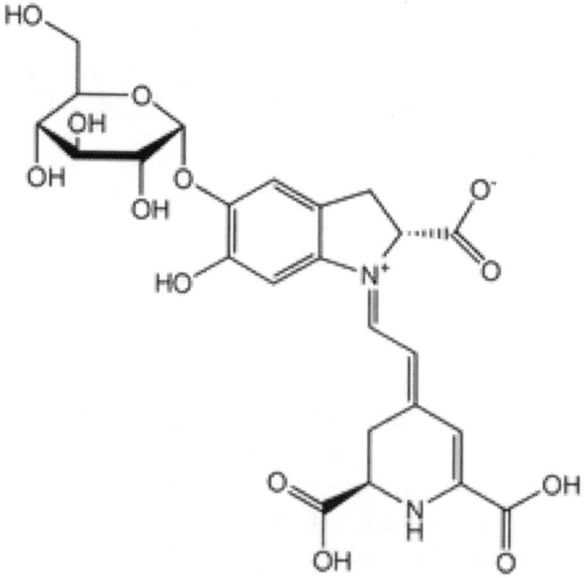

Figure 7.5. Chemical Structure of Betalain Pigment.

The phytic acid is distributed uniformly in the grain, so its content cannot be reduced by removal of the external grain layers or by water extraction. It is already known that the phytic acid has the inhibitory effect on starch digestion, and the positive effect on lowering cholesterol in the blood. However, phytic acid can produce complexes with proteins and mono- and divalent cations. Amaranth grains contain smaller amounts of protease inhibitors (trypsin and chymotrypsin) compared to common cereals. However, protease inhibitors have been found to have anti-cancer, antioxidant, and anti-inflammatory effects, as well as affecting the regulation of blood glucose levels.

Grain amaranth has been suggested as an alternative to marine animals as a natural source of squalene. Oil contents, squalene contents, and fatty acid profiles were determined in 11 genotypes of four grain amaranth species. Although the oil contents of grain amaranth were low (from 5.1 per cent in *Amaranthus tricolor* to 7.7 per cent in *A. cruentus*) as compared to other oil-containing grains, high concentrations of squalene were found in total lipids, ranging from 3.6 per cent in *A. hypochondriacus* to 6.1 per cent in *A. tricolor*. The major fatty acids in Amaranthus oil consisted of palmitic acid (19.1-23.4 per cent), oleic acid (18.7-38.9 per cent), and linoleic acid (36.7-55.9 per cent). A high degree of unsaturation was observed in Amaranthus oils, with S/U ratios of 0.26-0.32 (Chernenko and Glushenkova 1998; Gamel *et al.*, 2007; Bozorov *et al.*, 2018). A method to isolate and purify the squalene from Amaranthus oil was developed. After the saponification of K112, the squalene content increased from 4.2 per cent in the crude oil to 43.3 per cent in the unsaponifiables by the removal of the saponifiables. The unsaponifiables were fractionated by silica gel column chromatography to get highly purified squalene. The squalene purity in certain fractions was as high as 98 per cent. Combining the fractions rich in squalene gave a 94 per cent squalene concentrate, with a yield of 90 per cent. The structure of squalene in the purified sample was confirmed by comparison of its ultraviolet spectrum with a standard and from its NMR spectra (He *et al.*, 2002).

Inflammation can cause many common diseases, including heart problems. Many studies have found that Rajgira may have anti-inflammatory benefits. It may even help in the treatment of arthritis. Amaranth grain contains phytochemical compounds like rutin and nicotiflorin and peptides with the ability to help lower hypertension and incidences of cancer. Rajgira oil contains Squalene, unsaturated fatty acids and Vitamin E. These are essential nutrients for healthy skin and hair. Lysine has been linked to a stronger, healthier hair, better roots, and a reduction in hair loss. Rajgira oil is often used in beauty products to protect, moisturize, and nourish the skin and hair.

8.3 Nutritional Value of Rajgira

Amaranth is one of the three pseudo-cereals, next to buckwheat and quinoa, most frequently used in modern formulations of functional bakery, pastry and confectionary food products (Figure 7.1D). Amaranth flour can be used as a gluten-free flour option to thicken soups and stews, sauces, and more. It can also be used with other gluten-free flours and gums in baking. Rajgira is one of the best sources of vegetable protein in the world. This whole grain is full of fiber, ensuring that one stay satiated for a longer duration. It slows absorption of sugar and even aids digestion. Moreover, the amino acid – lysine also metabolizes fatty acids. Gluten free is another added advantage of this ingredient, making it fit for consumption for almost everyone. Calcium content in Rajgira is more than 4x that of other grains. In fact, it has more calcium than milk. Rajgira is also good food for Calcium for lactose intolerants (meaning one can't digest milk). Amaranth flour can be used in the production of various types of breads, such as chapatti and tortilla, but not yeast bread. There is no gluten, so there is no ability to retain gases. In order to produce bread and pasta, it can be added to wheat flour up to 20 per cent, without a product quality change. Amaranth grain can be popped, similarly to the grain of corn. It can be used in the production of soy sauce, which is commonly used in Southeast Asia. In addition to the common application of barley malt, beer can also be produced from amaranth. Since amaranth grains contain a relatively large starch content, strong alcoholic beverages can also be produced. Popped Rajgira seeds is a great option for a healthy and quick breakfast cereal. The popped Rajgira seeds sprinkled on soup or salad will make it crunchy, add slight savoury flavour to it and load it with all the good nutrients. Rajgira or Ramdana laddus made with jaggery are quite popular and tasty. Rajgira parathas or Puris are a favourite meal during fasting. In addition to eating Rajgira grains, leaves that come in vibrant colors like red, orange, and purple are eaten.

8.4 Conclusion

Amaranth has a good amount of lysine which helps the body absorb calcium, build muscle, and produce energy. Vegetables and grains are often lacking in this essential amino acid. Amaranth is a very rich source of minerals like Ca, Cu, Zn, Mg K and P. It is also a good source of zinc, potassium, and phosphorus. These build strong bones and muscles, aid hydration, boost energy, and are vital in thousands of processes throughout the body. Amaranth is also a good source of many essential vitamins, too, including vitamins A, C, E, K, B5, B6, folate, niacin, and riboflavin. These act as antioxidants, raise energy levels, control hormones, and do much more. Amaranth helps prevent premature graying, mainly due to the minerals this grain has in abundance. The fiber and phytonutrients in amaranth can lower BP. This grain tackles cholesterol, and inflammation, making it an all-around good food for heart health.

REFERENCES

Alvarez-Jubete L., Arendt EK., Gallagher E. (2009). Nutritive value and chemical composition of pseudocereals as gluten-free ingredients. *Int. J. Food Sci. Nutr.*,60(Suppl 1): 240–257.

Alvarez-Jubete L., Arendt EK., Gallagher E. (2010). Nutritive value of pseudocereals and their increasing use as functional gluten-free ingredients. *Trends Food Sci. Technol.,*21: 106.

Andini R., Yoshida S., Ohsawa R. (2013). Variation in Protein Content and Amino Acids in the Leaves of Grain, Vegetable and Weedy Types of Amaranths. *Agron.,*3: 391–403.

Becker R., Wheeler EL., Lorenz K., Stafford AE.,Grosjean OK., Betschart AA., Saundersrm RM. (1981). A compositional study of amaranth grain. *J. Food Sci.,* 46: 1175.

Bozorov SS., Berdiev NS., Ishimov UJ., Olimjonov SS., Ziyavitdinov JF., Asrorov AM.,Salikhov SI. (2018). Chemical composition and biological activity of grainoil of amaranth varieties.*Nova Biotechnol.Chimica.,* 17(1): 66-73. doi: 10.2478/nbec-2018-0007

Caselato-Sousa VM., Amaya-Farfan J. (2012).Concise Reviews in Food Science State of Knowledge on Amaranth Grain: A Comprehensive Review. J Food Sci. 2012; 77(4): 93– 104.

ChernenkoTV.,Glushenkova AI. (1998).Composition of the triacylglycerols of *Amaranthus caudatus* seeds. *Chem. Nat. Compd.* 34: 99-100.

Gamel TH., Mesallam AS., DamirAA.,Shekib LA., Linssen JP. (2007). Characterization of Amaranth grain oils. *J. Food Lipids,* 14: 323-334.

He H P., Cai Y., Sun M., Corke H. (2002). Extraction and purification of squalene from amaranthus grain J. Agric. Food Chem., 16; 50(2): 368-72. doi: 10.1021/jf010918p.

Kauffman CS., Weber LE. (1990). Grain amaranth. In: J Janick and JE Simon, editors. Advances in new crops. Timber Press, Portland, OR; 127–139.

López VR.,Razzeto GS., Gimenéz MS., Escudero NL. (2011) Antioxidant properties of *Amaranthus hypochondriacus* seeds and their effect on the liver of alcohol-treated rats. *Plant Foods Hum. Nutr.,*66: 157–162.

Meyers KJ, Rudolf JL, Mitchell AE. (2008). Influence of dietary quercetin on glutathione redox status in mice. J Agric Food Chem. 56(3): 830–836.

Martirosyan DM.,Miroshnichenko LA., Kulakova SN., Pogojeva AV., Zoloedov VI. (2007). Amaranth oil application for coronary heart disease and hypertension. Lipids Health Dis.,6: 1–12.

Pastor K, Akanski M. (2018).The chemistry behind amaranth grains. *J.Nutr. Health Food Eng.*, 8(5): 358-360. doi: 10.15406/jnhfe.2018.08.00295

Petras PK., Venskutonis R.(2013). Supercritical carbon dioxide extraction of squalene and tocopherols from amaranth and assessment of extracts antioxidant activity. *The J. Supercrit. Fluids.,* 80: 78-85. https: //doi.org/10.1016/j.supflu.2013.04.005

Pedersen B., KalinowskiLS.,Eggum BO. (1987). The nutritive value of amaranth grain (*Amaranthus caudatus*). Plant Foods Human Nutri.,36: 309–324.

Tucker JB. (1986). Amaranth: The Once and Future Crop. *BioScience.,* 36(1): 9–13.

Tagwira M., Tagwira F., Dugger R., Okum B. (2006). Using grain amaranth to fight malnutrition. *RuforumWorking Doc.,* 1: 201–206.

Vora JD., Jadhav P., Rane A. (2014). Biochemical, Anti Microbial and Organoleptic Studies on Rajgira (*Amaranthus caudatus*).*IOSR* J. Pharma. Biol. Sci., 9(2) Ver. VI: 01-05.

Chapter 9

Health Advantages of Ash Gourd [*Benincasa hispida* (Thunb.) Cogn.]

The Ash Gourd is recognized for its therapeutic properties. Ash Gourd is not just the vegetable, which is useful, but its seeds and seed oil is equally beneficial in a different condition such as pregnancy, brain health, and skin. Phytochemical analysis showed that the major constituents of Benincasa hispida vegetable are volatile oils, flavonoids, glycosides, sacchrides, proteins, carotenes, vitamins (niacin, thiamine, vitamin-B and riboflavin), minerals (P, Ca, K, Fe,), β-sitosterin and uronic acid. This plant was is medicinally in various complains such as gastrointestinal problems, respiratory disease, heart diseases, diabetes mellitus and urinary diseases. Traditionally it is used as a laxative, diuretic, tonic, aphrodisiac, cardiotonic, urinary calculi, blood disease, insanity, epilepsy, schizophrenia and other psychologic disorders, jaundice, dyspepsia, fever, and menstrual disorders. The pharmacological studies revealed that the plant exerted many pharmacological activities, including central nervous effects (anxiolytic, muscle relaxant, antidepressant, in the treatment of Alzheimer's disease and to minimize opiates withdrawal signs), antioxidant, anti-inflammatory, analgesic, antiasthmatic, diuretic, nephroprotective, antidiabetic, hypolipidemic and antimicrobial effects. The present review designed to highlight the culinary uses, chemical constituents, traditional medicinal value and proven pharmacological activities of ash gourd.

Keywords: *Anti-inflammatory, Anti-angiogenic, Detoxificant, Curvative effects, Cucurbitacin, Flavonoids, Ash gourd.*

9.1 Introduction

One of the locally available vegetable, Ash gourd [*Benincasa hispida* (Thunb.) Cogn.] belongs to the family Cucurbitaceae, found to be originated in wild form in the sea shore of Java. It is also widely distributed throughout the tropical and subtropical Asia (Dhiman *et al.*, 2012; Selvakumar *et al.*, 2017). Ash gourd naturally grows in the wilderness in South-East Asian countries such as India, Sri Lanka, China, Nepal and Indonesia, as well as in the warmer southern regions in Australia. Ash gourd (Figures 9.1A-D) goes by several vernacular names such as "Boodida Gummadi" in Telugu, "Neer Pooshnikkai" in Tamil and "Kumbalam" in Malayalam. This green vegetable has been well-regarded for its significant medicinal values since ancient times and is widely documented in Ayurvedic texts. Today, it continues to be heralded for its immense health benefits and is widely incorporated, in popular local cuisine across India, as well as in alleviating illnesses of the stomach, liver and skin, to name a few (Gupta *et al.*, 2019).

Figures 9.1A-D. Various Size and Shapes of Ash Gourds Attached to Plant.

Safed Petha is another name for this vegetable as a white gourd like winter melon and fuzzy melon. The therapeutic and remedial traits offered by the vegetable, as well as ash gourd seeds (Figures 9.2A-B) and leaves are extensive. Moreover, the roots and juice of ash gourd also have applications in skin and hair care. This remarkable natural wonder, which belongs to the cucumber and squash family, like snake gourd, also provides valuable curative properties for conditions such as fever, jaundice and diabetes (Pagare *et al.,* 2011). This is owing to its noteworthy antioxidant and anti-inflammatory qualities, due to the presence of beneficial plant compounds like flavonoids and carotenoids. In addition, the juice of ash gourd also heals hair and scalp disorders like dandruff and bald patches. Ash gourd is being consumed to a great extent nowadays and not just in its native milieus in Asia and Australia. The crop is, in fact, naturalized and propagated all over the world in tropical environments, so that people can procure its marvellous merits, for complete wellness. Ash gourd fruits are admired for their medicinal properties such antipyretic nootropic, antidepressant, anorectic, anxiolytic, antinflammatry, analgesic, diuretic, antioxidant antihistamine, antidiarrheal, angiontens in-converting enzyme inhibition and gastro protective activities.

Figures 9.2A-B: Cut Fruit and Seeds of Ash Gourd.

9.2 Nutritional Importance

More frequently ash gourd is being sought to be more than just a source of nourishment; consumers are looking for functional and nutritionally active foods. Normally the entire ash gourd plant, including fruit peel, flower, seed, and leaves are used. Biochemical activity of the fruit includes anti-oxidative, anti-inflammatory, anti-angiogenic, detoxificant, and curvative effects in treating various ailments. Ash gourd supplies plenty of nutrition, being inherently high on water content providing a cooling effect on the body, having zero cholesterol thereby augmenting heart health and abounding in a plethora of vitamins and minerals to facilitate key metabolic functions in the body (Grover and Rathi 1994). It has many amazing healing properties and is highly nutritious. This vegetable is rich in many essential vitamins and minerals and is a good source of P, Ca Fe, niacin, thiamine, vitamin-c and riboflavin.

This is ideal for weight loss because it has high water content. Ash Gourd is basic in nature, which helps in combating the formation of ulcers on the membranes stomach and intestine. This is very useful and helpful for Combat acidity due to spicy foods or fasting for a long time (Lee *et al.,* 2005). Also, Ash gourd helps us clear away all the harmful bacteria in the stomach and intestine by acting as an anti-microbial agent.

Ash Gourd has a lot of nutrients and 96 per cent has water so this is a very good conscious for weight. In order to lose weight, the best alternative to calorie-rich foods is Ash gourd juice or stew this ash gourd. Ash gourd is a natural anti-Coagulant and thereby helps control bleeding. This vegetable works wonders for internal bleeding. It is taken regularly it helps stop nose bleeds that occur commonly in the summer season. It also helps to with that condition's like haematuria, bleeding throw ulcers, piles, and other such internal bleeds that can be stopped by consuming Ash gourd juice.

Ash gourd helps deal with mental illnesses as it acts as a natural sedative. It has a calming effect on the nerves and the brain. And therefore, it is used to alleviate conditions related to the nervous system such as epilepsy, anxiety, insomnia, neurosis, and paranoia. Ash Gourds juice in a day can help deal with all the stress that the day has to give us and to keep us calm (Ali Esmail 2013). Ash gourd is very useful for our skin and for our hairs. We wouldn't expect it of Ash gourd, but it has Cosmetic use as well. When the juice is applied onto the skin along with lemon juice or as a part of any other face pack, it is also known for removing dark spots on the skin and bring a natural glow to the skin. Ash Gourd juice is also useful for hairs. This is very beneficial for our hairs, Ash gourd juice is one of the best and lasting remedies for dandruff. If we use Ash gourd with coconut oil for our hair, add ground seeds of best results for smooth to and long healthy hairs (Majumdar *et al.,* 2009).

9.3 Culinary Uses

Ash gourd is a unique melon eaten mostly in India and China. It is often cubed and added to soups in Asian cooking. However, the Yogis of India have long regarded ash gourd as one of the most naturally energizing foods due to its high quotient of what Yogic science refers to as "prana", or vital life energy. Packed with myriad vital nutrients and powerful antioxidants that confer excellent incentives for all-round healthcare, ash gourd is truly a nourishing veggie to include in the routine diet (Majumdar *et al.,* 2010). We can make a juice of Ash gourd. Ash gourd has more quantity of water so we can cut the ash gourd and mix salt and sugar Both are giving the good test in a juice, now blender in a mixture jar and we can use to do drink like juice.

Ash gourd raita is too good for us. Great Ash gourd and mix with plain yoghurt add lemon or lime juice, salt, black pepper and a sprinkling of roasted cumin Seeds and now ready to eat Raita of Ash gourd. Ash gourds halwa is also very tasteful and it is very beneficial for us. This halwa is a very sticky treat containing the positive pyranic and healthful ash gourd (Majumdar 2012). Daily Consumption of Ash gourd greatly enhances one's intellectual capabilities. Ash gourd is used in cooking a variety of staple Indian dishes such as kootu, curry, sabzi and dal, besides sweets and candies called petha (Figures 9.3A-B).

Figures 9.3A-B. Sweet Ash Gourd Candies (Petha).

9.4 Phytochemical and Pharmacological Aspects

Benincasa hispida seeds are mainly composed of saponin, urea, citrulline, linoleic acid, oleic acid and fatty acids. The seed also contains minute amounts of a triterpenoid known as isomultiflorenol, proteins such as trigonellin (Figure 9.4), coffearin, and osmotin, steroids such as betasitosterol and stigmast-5-ene-3-beta-ol, alkaloids such as 5-methylcytosine, and triterpenoids such as cucurbitacin. The major constituents of this fruits are triterpenoids, flavanoids, glycosides, saccharides, carotenes, vitamins, β sitosterin, and uronic acid (Figure 9.6). Also shows the presence of triterpenes: alnusenol, multiflorenol, isomultiflorenol; flavone: iso-vitexin; and sterols: lupeol, lupeol acetate (Figure 9.7), and beta-sitosterol (Figure 9.8). The pharmacological studies revealed that the plant exerted many pharmacological activities, including central nervous effects (anxiolytic, muscle relaxant, antidepressant, in the treatment of Alzheimer's disease and to minimize opiates withdrawal signs), antioxidant, antiinflammatory, analgesic, antiasthmatic, diuretic, nephroprotective, antidiabetic, hypolipidemic and antimicrobial effects (Shih *et al.,* 2001).

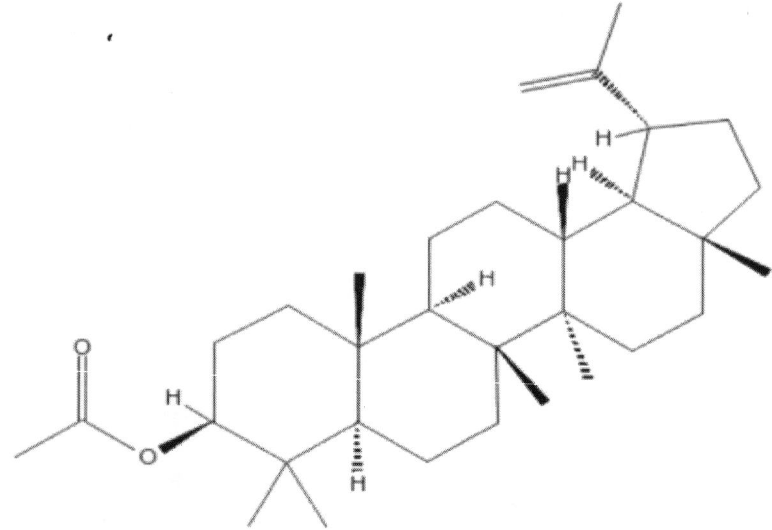

Figure 9.4. Chemical Structure of Trigonellin.

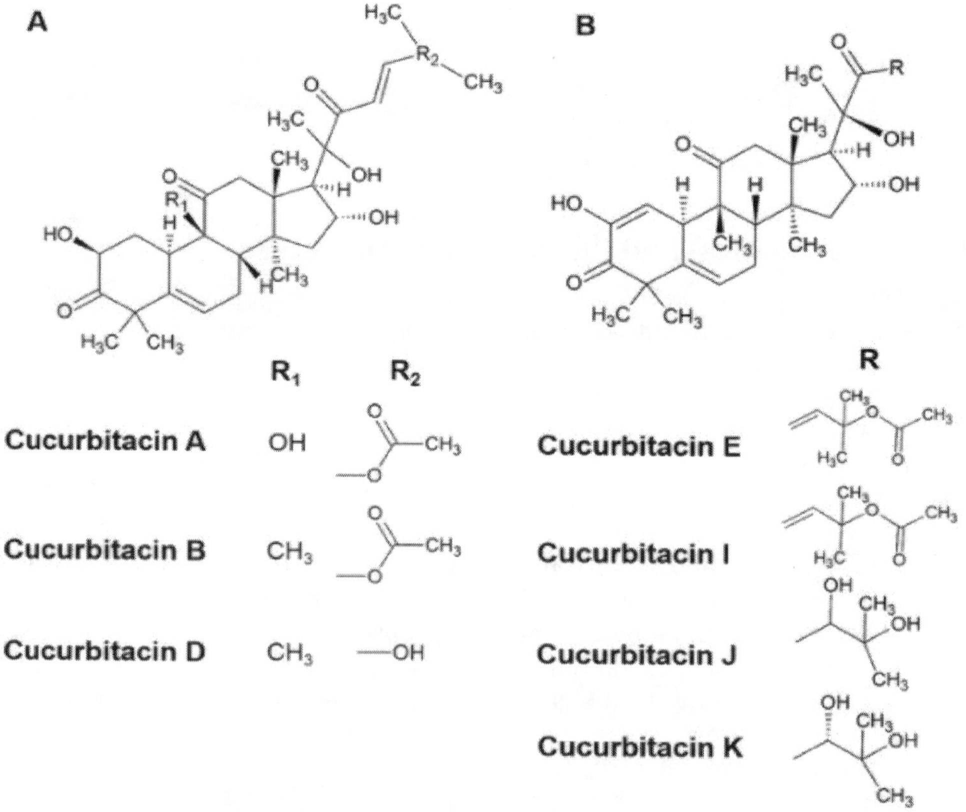

A

	R_1	R_2
Cucurbitacin A	OH	
Cucurbitacin B	CH_3	
Cucurbitacin D	CH_3	—OH

B

	R
Cucurbitacin E	
Cucurbitacin I	
Cucurbitacin J	
Cucurbitacin K	

Figure 9.5. The Structure of Cucurbitacins. (A) The skeleton of cucurbitacin A, B and D. (B) The skeleton of cucurbitacin E, I, J and K.

Figure 9.6. Chemical Structure of Uronic Acid.

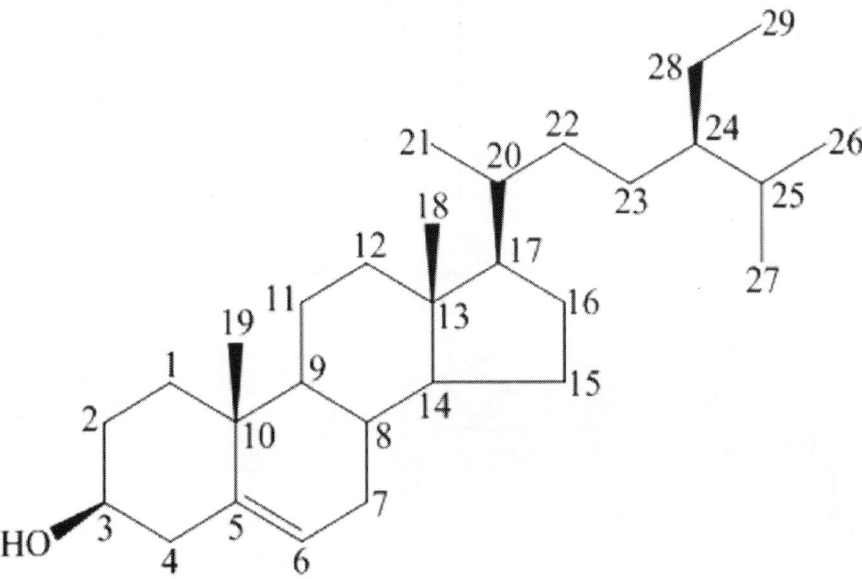

Figure 9.7. Chemical Structure of Lupeol Acetate.

Figure 9.8. Chemical Structure of β-Sterol.

9.5 Nutritional Contents in Ash Gourd

Like most flora of the gourd family, the ash gourd vegetables, seeds, leaves and juice extracts are enriched with a multitude of crucial fundamental dietary components like carbohydrates, fats, proteins and fibers, vital trace compounds such as vitamins and minerals and a host of plant substances including phenolics and cucurbitacins.

Table 6.1: The Ash Gourd Nutritional Factors for a 100 g Serving

Calories	86.2 kcal		
	Macronutrients		
Total Fat	3.9 g	Saturated Fat	0.5 g
Total Carbohydrate	12.5 g		
Dietary Fiber	0.6 g	Protein	2.0 g
Sodium	33.0 mg	Potassium	359.1 mg
	Micronutrients		
Vitamins:			
Vitamin A	9.8 per cent	Vitamin B6	11.3 per cent
Vitamin C	30.5 per cent	Vitamin E	1.1 per cent
Minerals:			
Calcium	5.1 per cent	Magnesium	6.7 per cent
Phosphorus	5.0 per cent	Zinc	7.2 per cent
Iron	5.7 per cent	Manganese	12.5 per cent
Iodine	5.9 per cent		

9.6 Health Benefits of Ash Gourd

Ash gourd, being low on calories and high in essential nutrients, can be taken regularly by those people who are strictly following a diet regime to lose weight, especially in the case of those with diabetes. Ash gourd also provides dietary fibers that can be processed easily in the stomach, keeping one feeling full for longer, reducing cravings and assisting in burning fat at a quicker pace. Having negligible cholesterol content, ash gourd can be safely consumed in the diet regularly for promoting heart health. The boiled vegetable can be effortlessly added to many standard homemade Indian dishes, as it improves the circulation of blood to and from the heart, ensuring optimal functioning of cardiac muscles (Singh *et al.*, 2014). Ash gourd stimulates the normal elimination of body wastes via the excretory system in the body. It boosts the secretion of fluids within the kidneys, promptly getting rid of accumulated toxins and at the same time, guaranteeing proper hydration of the internal organs in the body. Ash gourd juice supports the routine tasks of kidneys and bladder.

Ash gourd has a significant fiber content, which helps to prevent instances of constipation, bloating and cramping of the stomach, upon ingestion of a heavy meal. Moreover, its laxative nature regulates bowel movements, thereby easing any discomfort experienced in the gut. Ash gourd has an intrinsic expectorant quality, which implies that it can readily loosen any excess phlegm or mucus secretions and remove them from the respiratory tract (Wu *et al.,* 1987). This immensely benefits lung function and also prevents any allergies and breathing difficulties. Non-starchy vegetables are a constant addition to all keto diets. In this regard, ash gourd, being naturally low in carbohydrates and sugars, can be an ideal component of a keto diet, which focuses on lowering calorie intake by cutting down on carbs. A simple quick-fix recipe of boiling cut ash gourd vegetable and seasoning with salt and pepper can be incorporated on a regular basis for lunch in the ketogenic diet.

Ash gourd has innate content of the smoothening or emollient Vitamin E, that also has antioxidant properties. The gel extract from the vegetable, when applied on sunburns and rashes, pacifies the aggravated and dried up regions of the skin, leaving it soft and completely moisturized. The residue derived from ash gourd leaves has astringent traits. This helps in neutralizing the highly inflamed spots on the skin. It also efficiently reduces any abscesses, boils, pus or carbuncles on the areas of the skin affected by allergies, fungal infections, environmental pollutants and sun rays. Ash gourd has a multitude of vitamins and minerals which provide nourishment and vigor to the hair strands. Moreover, when applied as a gel, it penetrates deep into the layers of the scalp and protects follicles, thereby maintaining the thickness and stability of hair. Ash gourd is an ideal option if you are looking to gain long and strong hair (Zaini *et al.,* 2011).

Ash gourd has potent chemicals that can reduce the intensity of flakiness and dandruff on the scalp of hair. It can also shelter the roots of the hair strands, known as follicles, from the dirt and fungus particles that trigger dandruff. Ash gourd gel, when applied routinely to the itchy and peeling scalp and dry hair, can significantly revamp the appearance of dull hair, giving it an incredible sheen. Ash gourd possesses moisturizing and anti-fungal properties, which acts as a balm, soothing the irritated scalp and eliminating dandruff particles. When used once a month, this simple yet highly effective homemade ash gourd hair gel prevents flaking of the scalp and ensures nourished and lustrous tresses. Since time immemorial, ash gourd has been utilized in the vegetable, leaf, root and juice forms, as these are imbued with potent antioxidant and anti-inflammatory plant elements, to prepare ayurvedic concoctions and tonics, to remedy sicknesses such as fevers, jaundice, heart ailments and bone disorders (Palamthodi *et al.,* 2018).

The phytonutrients or plant compounds in ash gourd have an inherent temperature reducing potential. The leaves of ash gourd, when rubbed on a person suffering from high fever, provides immediate relief, reducing body temperature and symptoms of fatigue. Moreover, since normal metabolism is affected during fevers, ash gourd leaves also expel excess water and salts from the body, in order to help maintain ideal electrolyte balance. The leaves of ash gourd contain substances called cucurbitacins, which play a crucial role in boosting the defense system and liver function in the body. In addition, ash gourd leaves also contain remarkable amounts of Vitamin C, which adds to the defense function and antioxidant capacity in those suffering from jaundice. The Ayurvedic remedy comprises crushing the leaves of ash gourd along with coriander seeds and consuming it twice a day, to treat jaundice.

Ash gourd extract is considered as one of the best remedies for cardiac ailments such as palpitations, irregular heartbeats, chest pain, high blood pressure and coronary heart disease. In traditional Indian medicine, a dose of two cups of ash gourd extract is given to those suffering from heart problems, to promote blood circulation and ameliorate the difficulties in performing normal day to day activities. Ash gourd gel, when prepared and applied in instances of severe hair fall, invigorated the blood circulation and nerve function in the scalp, promoting rapid hair growth. Alopecia is characterized by prominent bald spots and excessive hair loss, and the high carotene content in ash gourd gel extract counters these factors, to reduce constant hair fall and enhance the strength and smoothness of hair (AlSnafi 2013).

Being an effective anti-inflammatory, ash gourd juice actually reduces bone and muscle pain and heals joint disorders such as arthritis, osteoporosis, gout and fractures. In addition, it is packed with the three-essential bone-fortifying minerals namely calcium, magnesium and phosphorous, which in turn increases bone mass and helps regain flexible movement in muscles and joints. Being a powerhouse of vitamin C and a host of flavonoids and carotenes, ash gourd is a potent agent to boost immune system function in situations of illnesses. Since the function of organs is below the optimum during disease, consuming ash gourd vegetable instils vitamin C to blood cells, which is subsequently transported to other organs to recover their peak functioning levels. It also helps to recuperate from tiredness.

The thyroid hormone levels tend to fluctuate and rise above the normal range in some people, causing hyperthyroidism. Ash gourd is profuse in iodine content, which is vital to reduce elevated thyroid hormone levels, as well as zinc, which plays a central role in facilitating enzyme function for optimizing thyroid concentrations. The prominent vitamin B6 or pyridoxine content in ash gourd juice is very beneficial in monitoring brain functions and allowing the unobstructed conduction of nerve impulses. Hence, during instances of insomnia or severe lack of sleep, having a glass of ash gourd juice can lessen the activity of neurotransmitters and promote sleep.

Ash gourd abounds in water content and immensely cools the heated body, thereby balancing the internal tissues of the liver and kidneys. It is also useful in treating urinary tract infection (UTI) and eliminates kidney stones, rectifies dysuria, constipation, to regulate renal processes and ensure smooth bowel, bladder movements. Comprising copious amounts of the potent antioxidants - cucurbitacins, ash gourd gets rid of toxic free radicals in the intestines and is very beneficial for gut wellness. Moreover, being an intrinsically alkaline vegetable, it neutralizes excess acidity in the stomach, providing instant relief from heartburn, indigestion, bloating and promoting digestion. Possessing vast reserves of soluble dietary fibers, in addition to high water content, ash gourd speeds up weight loss, helping lose extra body fat and averting obesity. It is an ideal vegetable to incorporate in the daily diet as the high water quantities aid in easy break down of food to energy, better absorption of essential nutrients, to augment metabolism (Prachi and Premavalli 2010; Pradhan *et al.,* 2020).

9.7 Conclusion

Ash gourd is composed primarily of water about 96 per cent. It also boasts a variety of beneficial minerals and vitamins including vitamin C, B-complex niacin, thiamine and riboflavin. The gourd also provides a good amount of protein, carbohydrates and dietary fiber. Ash gourd also has phytochemicals like tannins, phenols, glycosides, alkaloids, and flavonoids. Low calories, high dietary fibre and water content of ash gourd make it excellent for people aiming to shed some extra pounds. Its fibre content provides fullness and satiety. It is high in soluble fibre which makes it excellent for our gut health. Being rich in soluble fibre, promotes the growth of good bacteria, alleviates indigestion causing constipation, haemorrhoids, *etc.*, thereby keeping the colon healthy. It can be included easily in the diet to enjoy the health benefits. Some studies also support that ash gourd may also help diabetic patients in maintaining blood sugar levels of the body. It acts as a natural anticoagulant in the body, promoting healthy hair and skin.

REFERENCES

Al-Snafi AE. (2013). The pharmacological importance of Benincasa hispida. *Int. J. Pharm. Sci.Res.,* 4(12): 165–170.

Dhiman KA., Gupta DK., Sharma NS., Gill., Goyal, A. (2012). A review on the medicinally important plants of the family cucurbitaceae. *Asian J. Clin. Nut.,* 4(1): 16-26.

Gupta P., Chikkala S., Kundu P. (2019). Ash gourd and its applications in the food, pharmacological and biomedical industries. Int. J. Veg. Sci.,27(12): 1-10 doi: 10.1080/19315260.2019.1699222

Grover JK., Rathi SS. (1994). Anti-inflammatory activity of fresh juice of Benincasa hispida. Ind. J. Pharma., 13: 26-66.

Lee KH., Choi HR., Kim CH. (2005). Anti-angiogenic effect of the seed extract of *Benincasa hispida* Cogniaux. J. Ethnopharmacol., 97(3): 509-513.

Majumdar TK., Vasudhish CR., Premavalli KS., Bawa AS. (2009). Preservation of functional vegetable juices: ashgourd juice and ashgourd-pudina blended juice. Bev. Food World, 36(4): 48-50.

Majumdar TK., Vasudish CR., Premavalli KS., Bawa AS. (2010). Studies on processing and storage stability of ashgourd-mint leaves juice. J. Food Process. Preserv., 34(2): 549-556.

Majumdar TK., Wadikar DD., Bawa AS. (2012). Development of storage stability of aseptically processed ashgourd-mint leaves juice. Int. Food Res. J., 19(3): 823-828.

Pagare MS., Patil L., Kadam VJ. (2011). Benincasa hispida: A Natural medicine. Res. J. Pharm. Tech., 4(12): 1941-1944.

Palamthodi S., Kadam D., Lele SS. (2018). Physicochemical and functional properties of ash gourd/bottle gourd beverages blended with jamun J. Food Sci Technol. 56(1): 473– 482. doi: 10.1007/s13197-018-3509-z

Prachi G., Premavalli KS. (2010). Effect of particle size reduction on physicochemical properties of ashgourd (*Benincasa hispida*) and radish (*Raphanus sativus*) fibre. Int. J. Food Sci. Nutri., 61(1): 18-28.

Pradhan K., Nandi A., Jena D., Rout S., Tripathy B. (2020). Sensory Evaluation of Value Added Products and Quantification of Ascorbic Acid of Ash Gourd (*Benincasa hispida*, Thumb.) Cong. Germplasm by Volumetric Method. Biosci. Biotech. Res. Comm., 13 (2) : 964-968.

Selvakumar G., Shathirapathiy G., Jainraj R., Paul PY. (2017). Immediate effect of bitter gourd, ash gourd, Knol-khol juices on blood sugar levels of patients with type 2 diabetes mellitus: A pilot study. J. Trad. compl. med., *7*(4), 526– 531. https: //doi.org/10.1016/j.jtcme.2017.01.009

Shih CT., Wu J., Jia S., Khan A.A., Ting KH., Shih DS. (2001). Purification of an osmotin-like protein from the seeds of *Benincasa hispida* and cloning of the gene encoding this protein. Plant Sci., 16(5): 817-826.

Singh S., Kumar U., Rai A. (2014). Process optimization for the manufacture of angoori petha. J. Food Sci. Technol., 51(5): 892–899.

Wu CM., Liou SE., Chang YH. Chiang W. (1987). Volatile compounds of the wax gourd (*Benincasa hispida,* Cong.) and a wax gourd beverages. J. Food Sci., 52: 132-134.

Zaini NAM., Anwar F., Abdul Hamid A., Saari N. (2011). Kundur [*Benincasa hispida* (Thunb.) Cogn.]: a potential source for valuable nutrients and functional foods. Food Res. Int., 44(7): 2368-2376.

Chapter 10

Nutritional and Health Benefits of Taro: Arbi [*Colocasia esculenta* (L.) Schott]

Taro (Colocasia esculenta (L.) Schott) is one of the staple tuber crop grown for various purposes and considered as a rich source of carbohydrates, proteins, minerals and vitamins and accredited to have medicinal values and are used to reduce tuberculosis, ulcers, pulmonary congestion, fungal infection and reducing body temperature in a feverish patient and others. Besides its importance as a food source, taro contains high levels of gum, which has also been shown to play a role in the reduction of high BP, in hypercholesterolemia and in the management of diabetes. Taro tubers provide a numbers of desirable nutritional constituents viz. phenolic acid, pelargonidin 3 glucoside, cyanidin 3 rhamnoside, cyaniding 3 glucoside, and sucrose. It also has high starch content and antinutrients like calcium oxalate $[Ca^{2+}(C_2O_4)^{2-}]$ *and mucilage. The corms of taro are used as vegetable and consider as a rich source of carbohydrates, proteins, minerals, and vitamins viz. riboflavin and niacin,. Taro tubers contain 70 to 80 per cent of starch in them. It can control high blood sugar, improve digestive system, prevent cancer, delay the aging process, improve the skin, reduce fatigue, improve blood circulation and heart's health, improve vision, lose weight, and boost immunity. Taro corn possess various nutraceutical activities such as antioxidant, anticancer, antifungal, curing of various diseases like cancer, cardiovascular disease, coronary heart disease etc. In this review there is imperative assets about, nutritional importance and the pharmacological attributes of taro corms and leaves.*

Keywords: *Calcium oxalate, Mucilage, Phytoconstituents, Nutraceuticals, Pharmacological properties, Arbi, Taro.*

10.1 Introduction

Taro (*Colocasia esculenta* Linn.) is a vegetative propagated tropical crop belonging to the monocotyledonous family Araceae, having its origin from South-east Asia. It occupies 9th position among world food crops with its cultivation spread across Africa. *C. esculenta* also known as Chinese Potato, Cocoyam, Colocasia, Curcas, Dalo, Dasheen is thought to have originated from tropical Asia. Taro tubers are important sources of carbohydrates as an energy source and are used as staple foods in tropical and subtropical countries (Rao *et al.,* 1970; Ahmed and Khan 2013). The high levels of vitamin A, vitamin C, and various other phenolic antioxidants found in taro root boost human immune system and help eliminate dangerous free radicals (Temesgen and Ratta 2015). Taro flour is a good substitute for wheat flour in bread making. (Ammar *et al.,* 2009; Alam and Hasnain 2009; Aleantara *et al.,* 2013).The corm of taro is relatively low in protein and fat and this is similar to many other tuber crops. It is a good source of starch, fibre, and ash. Taro is (Figures 10.1A-B) also a good source of thiamine, riboflavin, iron, phosphorus, and zinc and a very good source of vitamin B6, vitamin C, niacin, potassium, copper, and manganese. Taro contains both dietary and non-dietary fiber. Crude fiber has many desirable functional properties. These include facilitating alimentary functions, helping in micro-component delivery and glucose metabolism and also slowing down the process of re-absorption of undesirable dietary components such as cholesterol, decrease intestinal transit time, reduce total and LDL cholesterol in the blood, decrease post prandial blood glucose and insulin level, buffer excessive acid in the stomach, prevent constipation, increase water holding capacity of food, increase stability of food by modifying structure and density of food, textures the food, gel formation in the food, thickening capacity in the food.Vitamin C and vitamin B complex (niacin, riboflavin and thiamin) which are important constituents of human diet, are present in appreciable quantity in corms and leaves of taro (Quach *et al.,* 2003; Prajapati *et al.,* 2011). Like other roots and tubers are deficient in most other vitamins but contain significant amounts of dietary fiber. Cooked leaf of Taro contains beta carotene, iron and folic acid which protects against anemia (Rashmi *et al.,* 2018; Shekade *et al.,* 2018).

Taros have high amount of β-carotene in the corm and will impart vitamin A and antioxidant property in the body. B-carotene differs only very slightly in terms of structure. They are very common carotenoids, and are antioxidants, as well as having other potential health benefits. As mentioned earlier, both can be converted into vitamin A by the body, though β-carotene has about twice the provitamin A activity as α- carotene. Yellow-fleshed cultivar of taro is associated with a high level of total phenolic compounds. The main antinutrients that exist in taro are: mucilage, oxalic acid, tannins, cyanide, lectins, alpha–amylase inhibitors, protease (trypsin and chymotrypsin) and inhibitors (Chair *et al.,* 2016).

Figures 10.1A-B. Potted Plant and Corm of Taro Root (Arbi).

10.2 Nutritional Value of Taro (Arbi)

Taro contains more than twice the carbohydrate content of potatoes and yield 112 kcals per 100 g. Taro contains about 11 per cent protein on a dry weight basis. It contains 85-87 per cent starch on dry matter basis and other nutrients such as minerals, Vitamin C, thiamin, riboflavin and niacin (Mandal *et al.,* 2013; Macharia *et al.,* 2014; Mengane 2015). Protein rich leaves favourably complements the high carbohydrate content of the tubers. In other parts of the world, the leaves of *C. esculenta* have been reported to be rich in nutrients, including minerals such as calcium, phosphorus, iron, and vitamins like vitamin C, thiamine, riboflavin and niacin. High levels of dietary fibre in taro are also advantageous for their active role in the regulation of intestinal transit, increasing dietary bulk and faeces consistency due to their ability to absorb water. Dietary fiber has the potential to maintain the blood sugar level by controlling the insulin production in the body. Dietary fiber plays a significant role in the human diet, as it cleans the digestive system. Taro root give benefits to those person suffering form digestive issue, since taro root is an excellent source of dietary fiber, as it adds bulk to the stool. Subsequently, stool can pass quickly through the digestive system (Tari and Singhal 2002; Sanful 2011). Healthy gut signifies a longer life span, hence, higher the Taro root intake in diet lesser are the chances of indigestion, bloating, cramping, and constipation (Jane *et al.,* 1992; Karthikeyan *et al.,* 2018).

Taro is rich in potassium that helps in controlling the heart pulse rate and relieves stress in the arteries. Also, potassium helps to regulate the blood pressure as it decreases the sodium's effect on the body. By making amendments to the diet and adding taro roots in lifestyle, people can significantly improve their heart's health. Taro roots contain vitamins like folate (22 µg/100 g), Niacin (0.600 mg/100 g), pantothenic acid (0.303 mg/100g), Pyridoxine (0.283 mg/per 100 g), Riboflavin (0.025 mg/100g), Thiamine (0.095 mg/100 g), Vitamin A (76 IU/100 g), Vitamin C (4.5 mg/100 g), Vitamin E (2.38 mg/100 g), and Vitamin K (1 µg/per 100 g). Additionally, Taro root is a rich source of antioxidants like cryptoxanthin (1) and β-carotene (2). Also, other minerals that are present in taro roots are copper (0.172 mg/100 g), Mn (0.383 mg/100 g), Se (0.7µ/per 100 g), Zn (0.23 mg/100 g), and Mg (33 mg/100g), Fe (0.55mg/100g), and Ca (43 mg/100 g). The electrolytes present in taro roots are K (591 mg/100 g) and Na (11 mg/100 g). Taro's tuber contains complex carbohydrates (26.46 g/100 g). Hence, it is an excellent energy booster. It is also rich in dietary fiber.It is a good source of starch (70–80 g/100 g dry taro), fiber (0.8 per cent), and ash (1.2 per cent). Taro starch contains amylose/ amylopectin (Fig. 3 and 4) in in the ratio 1:7. Taro tubers provide a numbers of desirable nutritional constituents *viz.* phenolic acid,pelargonidin 3-glucoside (5), cyanidin 3 rhamnoside(6) and cyanidin 3 glucoside (7). It is the most important sugar in taro is sucrose (8), but fructose, maltose, glucose and raffinose (9) are also present. Organic acid *viz.* Malic acid, citric acid (10) and oxalic acid (11)also occur in Taro leaves. The protein fraction is rich in essential amino acids of threonine (12), leucine, arginine (13), valine and phenylalanine (14). Among the essential amino acids methionine, lycine, cystine, phnylalanine and leucine are relatively abundant in the leaf than the corm (Nip *et al.,* 1989; Iwashina *et al.,* 1999; Ndabikunze *et al.,* 2011).

Table 10.1. Nutritional Value of Taro

Nutritional Value per 100g			
Principle/Nutrients	*Nutritional Value*	*Principle/Nutrients*	*Nutritional Value*
Energy	112K cal	Carbohydrates	26.46g
Protein	1.50g	Total fat	0.20g
Cholesterol	-	Dietary fibres	4.1g
Vitamins			
Folates	22µg	Niacin	0.600mg
Pantothenic acid	0.303mg	Pyridoxine	0.283mg
Riboflavin	0.025mg	Thiamin	0.095
Vit. A	76IU	Vit. C	4.5mg
Vit. E	2.38mg	Vit. K	1µg
Minerals			
Na (Sodium)	11mg	K (Potassium)	591mg
Ca (Calcium)	43mg	Cu (Copper)	0.172mg
Fe (Iron)	0.55mg	Mg (Magnesium)	33mg
Mn (Manganese)	0.383	Se (Selenium)	0.7µg
Zn (Zinc)	0,23mg		

Source: USDA National Nutrient Data Base.

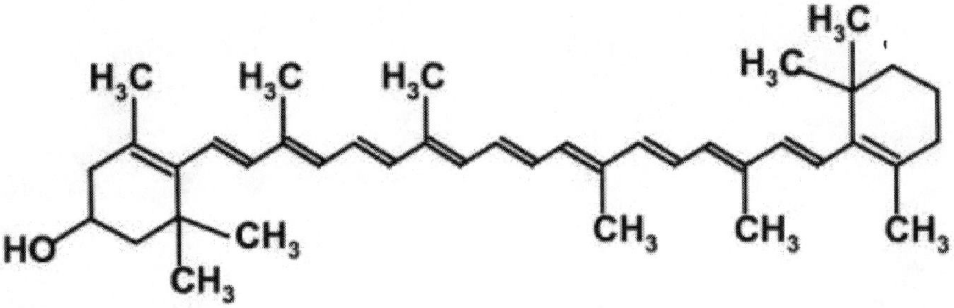
H3C
H3C
H3C
H3C
H3C
HO
CH3
CH3
CH3
CH3
CH3

1

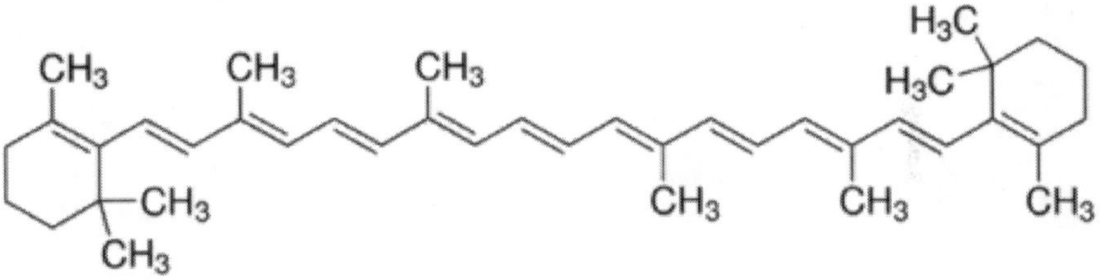
CH3
CH3
CH3
H3C
H3C
CH3
CH3
CH3
CH3
CH3

2

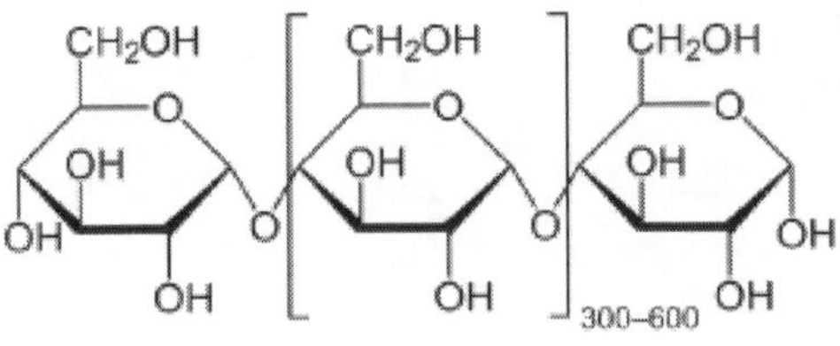
CH2OH
CH2OH
CH2OH
OH
OH
OH
300–600

3

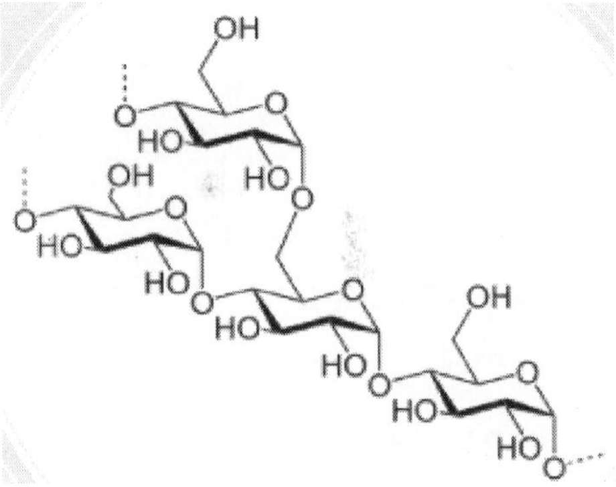
OH
HO
HO

4

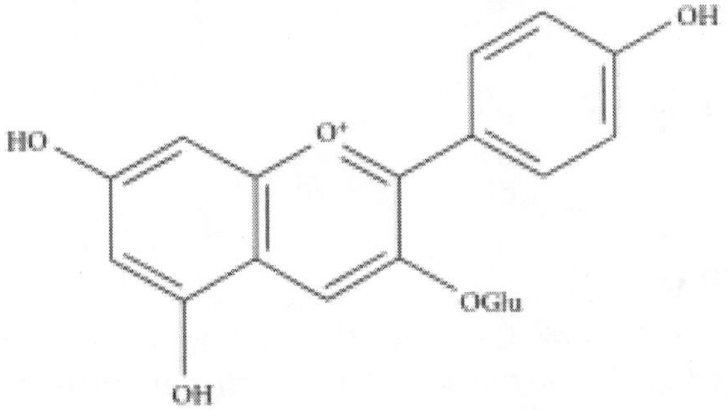
OH
HO
O+
OGlu
OH

5

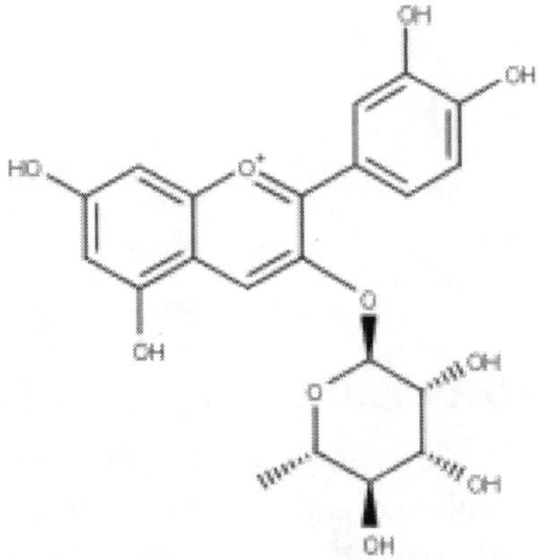
OH
OH
HO
O+
OH
O
OH
OH
OH

6

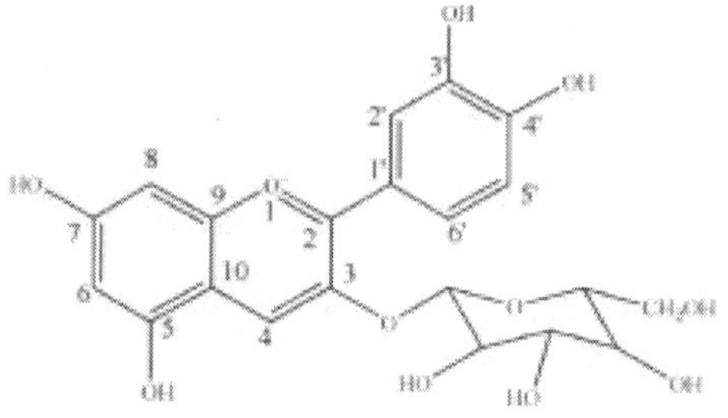

7 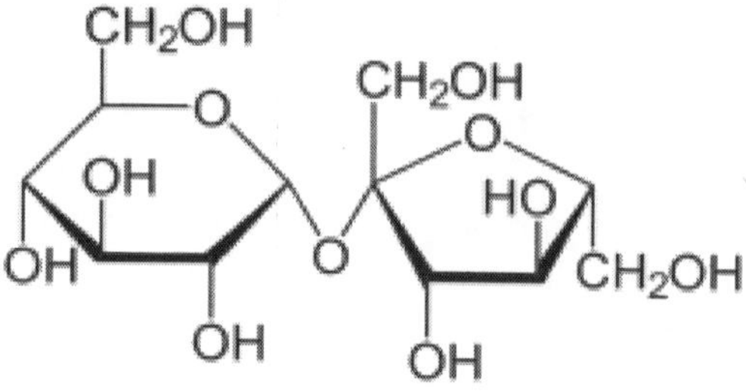 8

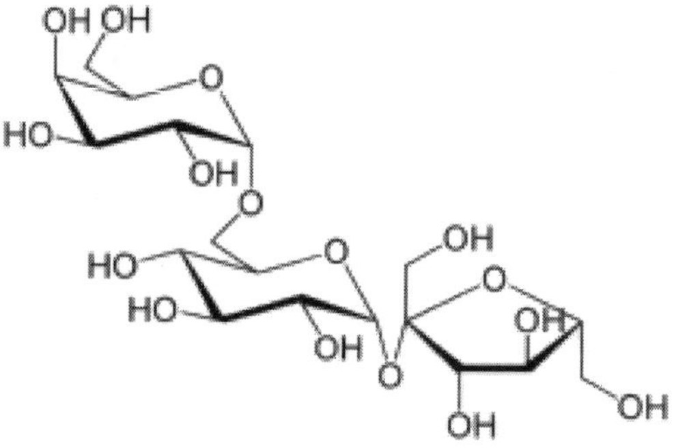

9 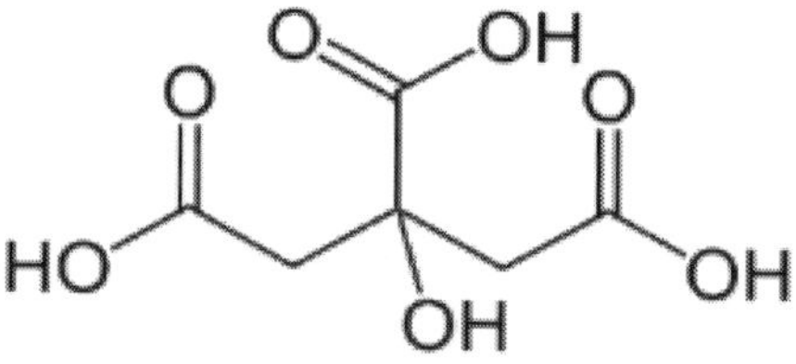 10

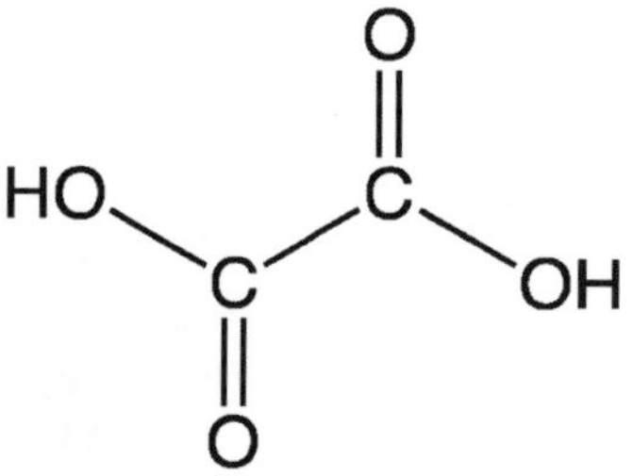

11 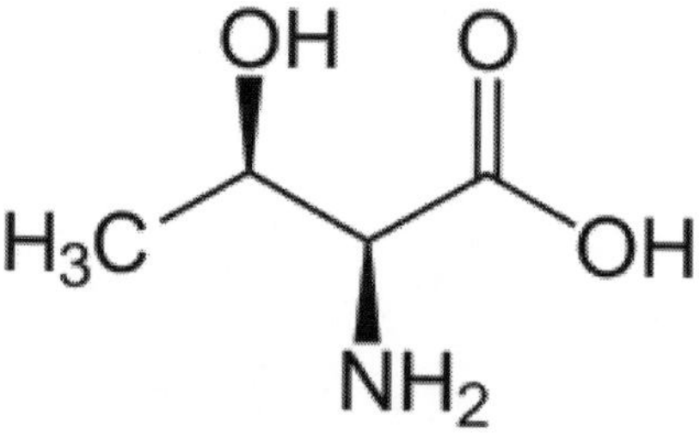 12

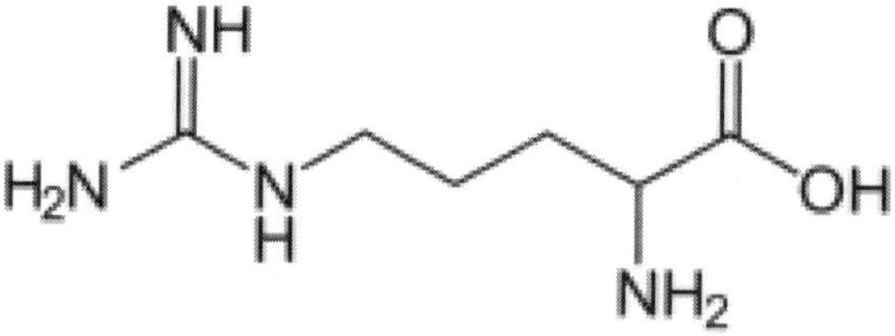

13 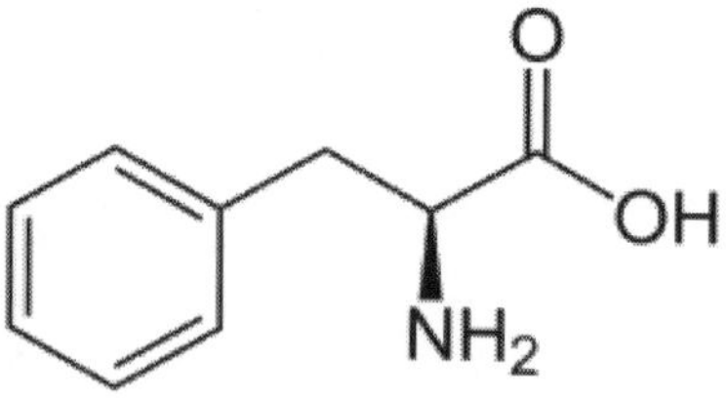 14

10.3 Health Benefits of Taro (Arbi)

Taro root is an incredible source of Vitamin C - an excellent immunity booster. Regular consumption of Taro root will improve the levels of Vitamin C in the body and strengthen our body' immunity against illnesses. Antioxidants like beta-carotene and cryptoxanthin present in taro root strengthens eyesight and promotes general eye health. These antioxidants slow down the ageing of cells present in the eyes that may cause macular degeneration and cataract. For healthy eyes, regularly consuming taro root is wise. The powerful combination of iron and copper in taro root does wonders for the blood's health. Iron helps to prevent anemia, and promotes blood circulation. Further, the taro root helps to fight the consequences of anemia like fatigue, lack of concentration, and headache. Vitamins A and E present in the Taro root have a great impact on the skin. These vitamins boost the formation of new and healthy skin cells and improve the life of degenerating skin cells. By regularly consuming Taro root, you can reduce the appearance of blemishes and wrinkles on the skin. Taro root promotes the skin healing process, and hence this food that does wonders for your skin. Another health benefit of taro root is that research indicates that since taro root contains a low glycemic index, it is beneficial for athletes. Taro plant root enables the athletes in keeping the energy level high for a longer time. Taro root also has the right amount of carbohydrate that boosts energy and reduces fatigue. Losing weight seems a difficult task due to poor diet and stressful lifestyle, today. Hence for an effective weight loss, people must consume foods that are high in nutritional value but low on calories. Taro root used for weight loss, as one cup of cooked taro root contains some amazing nutrients and only 187 calories. Hence, one should consider adding taro root when trying to lose weight. Antioxidants in taro root have certain properties that help in fighting against various diseases and delaying the aging process. This root is cholesterol-free, low in sodium, gluten-free and comprises of protein. Magnesium and vitamin E are responsible for maintaining the health of the muscles. Taro corms are rich in magnesium and vitamin E, and thus should be regularly consumed for healthy muscles.

Taro corms plays an important part for the antioxidant activity in our body. High levels of vitamin A, vitamin C, and various other phenolic antioxidants found in taro root helps to boost immune system and help eliminate dangerous free radicals from our system. Taro roots are extremely beneficial for overcoming rheumatoid arthritis because low levels of vitamin B6 are related with increased signs of rheumatoid arthritis, like more severe pain. Taro root s consists of considerable amount of potassium which is considered other essential minerals that are essential to remain healthy and efficient. It not only enables healthy fluid transfers between membranes and tissues throughout the body, but also helps to relieve stress and pressure on blood vessels and arteries. Taro roots play an important role in the immune system. Vitamin C is found in taro roots which help to encourage immune system to create more white blood cells which help to defend the body from foreign pathogens and agents. Taro roots consist of considerable amount of potassium 615 mg which is 13.09 per cent of the daily recommended value. Muscle cramps are one of the common side effects of low potassium levels. Taro is good for people allergic to milk or cereals and can be consumed by children who are sensitive to milk. It acts as a probiotic because it contains the predominant lactic acid bacteria. Studies have been reported its possible beneficial role in the treatment of diarrhoea, gastroenteritis, irritable bowel syndrome and inflammatory bowel disease, cancer, depressed immune function and inadequate lactase digestion (Shah *et al.,* 2007). Some tribes use taro as an antidote to the stings of wasps and other biting insects, heated tubers were applied to painful parts in rheumatism.

10.4 Culinary Uses of Taro

Apart from sweet vanilla, floral and potato-y taste taro is a good source of nutrients, vitamins and minerals. It can also be used for entrapment of flavoring compounds. Taro corms are highly perishable owing to their high moisture content. The incorporation of taro flour into wheat-based products has been reported to increase their keeping quality. Given these characteristics, taro flour has much potential for use in food formulations. Indeed, taro has been shown to be an important ingredient in the production of beverages and for partial replacement of wheat flour in bread, cookies, taro-based desserts and miscellaneous taro-based products. Taro can also be used for entrapment of flavoring compounds.

Taro's tuber and leaves are full with flavor (Njintang *et al.,* 2008). The taro tubers have a nutty flavor, while its leaves taste like a cabbage. Subsequently, they are used for preparing tasty food items like Taro chips, crispy taro pancakes, taro cheesecakes, taro fries, and taro buns. Taro root is a great energy booster; therefore, heated taro root is used in many warm compresses and plasters. Alopecia disease can be treated by regularly applying the juice of the taro roots at the hair roots. The leaves of taro are consumed as sauces, purees, stews and soups. In some parts of the world taro is made into ice cream and drinks. The traditional ways to cook taro are roasting on stones or baking in a ground oven. More modern ways are boiling and steaming, or baking in an oven. Taro retains its food value if cooked whole and in its skin. It must be cooked thoroughly to prevent mouth and throat itching. Taro corms and leaves are usually consumed by humans after heat treatments, such as boiling, blanching, steaming, stewing, frying and pressure cooking. These methods are found to be effective in improving digestibility, increasing nutrient bioavailability and also minimizing anti-nutritional factors and food-borne diseases. Processing taro corm affects its proximate composition, mineral content, phytochemical components and antinutrient (oxalate and phytate) contents. When taro corms processed into powder and further decrease will occur when processed into taro noodles and cookies. Therefore the combination of cooking time temperature program is necessary to preserves the nutrients and deactivates the anti-nutritional factors. On other hand cooking increases antioxidant activity, crude fat, crude protein and crude fibre contents cooking substantially can used in the management non communicable illness such as obesity, heart disease, blood pressure, diabetes, cancer and gastrointestinal disorders because of the high fibre content. The taro flour can then be used in other food formulations such as taro bread, taro cookies, baby food, pasta, or other products (Del Rosario and Lorenz 1999; Njintang and Mbofung 2003). The corm's flour is also a good source of carbohydrate for diabetics and production of weaning food for infants and for those with gastrointestinal disorders. Taro chips made from the bottom of the corm are rated better in appearance than those made from the top part (Reyad-ul-Ferdous *et al.,* 2015).

10.5 Conclusion

Taro leaves are a rich source of protein, ascorbic acid, dietary fiber, and other important minerals including Fe, P, Zn, Cu, Mn and vitamins *viz.* vitamin B6, vitamin C, thiamin, riboflavin, and niacin. Studies have suggested that they have analgesic, antiinflammatory properties, and antifungal properties. As it has been previously mentioned, taro root is loaded with several antioxidants, like β-carotene and cryptoxanthin. Taro roots consist of P which is essential for both bone health as well as maintaining teeth and gum health. Apart from proper bowel movement dietary fiber can also help lower the chances of developing diabetes because it helps to regulate the release of insulin and glucose in the body. A taro root consists of significant amount of vitamin E which is a powerful antioxidant that helps to decrease environmental damage to your hair. It also promotes circulation to the scalp. Taro roots consist of several mineral contents. But the presence of Fe and Cu in taro root make it an important food to prevent anemia and boost circulation throughout the body. Cu present in taro roots plays an important role in slow down bone loss and osteoporosis in elderly persons when taken in combination with other important minerals like Zn, Ca and Mn. It consists of sufficient amount of vitamin A and vitamin E. Each of these vitamin helps to eliminate skin problems and boost overall cellular health. When raw taro corms are cut in to pieces, the exposed surfaces often exude droplets of a slimy substance called mucilage. The pieces or slices of corm are washed or placed in portable water; a great quantity of mucilage is quickly released or avoided. The crude mucilage is a complex mixture composed mainly of neutral polysaccharides, with small quantities of fiber and protein. Oxalates are one major limiting factor in the utilization of taro. It is the presence of oxalates which impart acrid taste or cause irritation when raw or unprocessed foods from them are eaten. In large quantities, oxalic acid is poisonous to humans and can also reduce the nutritional value of a food by binding with Ca.

REFERENCES

Ahmed A., Khan F. (2013). Extraction of Starch from Taro (*Colocasia esculenta*) and Evaluating it and further using Taro Starch as Disintegrating Agent in Tablet Formulation with Over All Evaluation Inventi Rapid: Nov. Excepie.,2: 1-5.

Alam, F., Hasnain A. (2009). Studies on and solubility of modified starch from taro (*Colocasia esculenta*) effect of pH and temperature.Agri.Conspectu. Scintificus,74(1): 45-50.

Alcantara RM., Hurtada WA., Dizon EI. (2013). The Nutritional Value and Phytochemical Components of Taro [Colocasiaesculenta (L.) Schott] Powder and its Selected Processed Foods. J. Nut. Food Sci., 3: 207-211.

Ammar MS., Hegazy AE., Bedeir SH. (2009). Using of taro flour as partial substitute of wheat flour in bread making. World J. Dairy Food Sci., 4: 94-99.

Chaïr H., Traore RE., Duval MF., Rivallan R., Mukherjee A. (2016). Genetic diversification and dispersal of Taro (*Colocasia esculenta* (L.) Schott). PLoS ONE,11: 1-19.

Del Rosario AV., Lorenz K. (1999). Pasta products containing taro (*Colocasia esculenta* L. Schott) and chaya (*Cnidoscolus chayamansa* L. Mcvaugh). J. Food ProcessPreserv.,23: 1-20 44.

Hang DT., Preston TR. (2009). Taro (*Colocacia esculenta*) leaves as a protein source in Central Vietnam. Livest.Res. Rural Dev., 21: 222–234.

Huang AS., Tanudjaja L. (1992). Application of anion-exchange high performance liquid chromatography in determining oxalates in taro (*Colocasia esculenta*) corms. J. Agri. Food Chem.,40: 2123–2126.

Iwashina T., Konishi T., Takayama A., Fukada M., Ootani S. (1999.) Isolation and identification of the.flavanoids in the leaves of taro.*Ann. Tsukuba Bot. Gard.,* 18: 71–74.

Jane J, Shen L, Lim S, Kasemsuwantt T, Nip K. (1992) Physical and chemical studies of taro starches and flours. Cereal Chemistry. 69: 528-535.

Karthikeyan M., Gnanasekaran A., Rashmi DR.,GopenathTS.,PalanisamyT.S. Chandrashekrappa GK., Kanthesh M Basalingappa KM., Kanthesh BM. (2018). Taro (*Colocasia esculenta*): An overview. J. Med. Plants Studies, 6(4): 156-161.

Macharia WM., Nuro MS., Muchugi AN., Palapala V (2014). Genetic structure and diversity of East African Taro *Colocasia esculenta* L. Afr. J.Biotechnol., 139: 2950-2955.

Mandal R., Mukherjee A., Mandal N., Tarafdar J., Mukherjee A. (2013). Assessment of genetic diversity in Taro using morphometrics. Curr.Agri. Res. J., 1: 79-85.

Mengane SK. (2015). Antifungal activity of the crude extracts of *Colocasia esculenta* leaves in vitro on plant pathogenic fungi. *Int. Res. J. Pharma.*, 6(10): 713–714.

Ndabikunze BK., TalwanaHAL.,Mongi RJ., Issa-Zacharia A., Serem AK. (2011). Proximate and mineral composition of cocoyam (*Colocasia esculenta* L. and *Xanthosoma sagittifolium* L.) grown along the Lake Victoria Basin in Tanzania and Uganda. Afr. J.Food Sci.,5: 248-254.

Nip WK., Muchille J., Cai T., Moy JH.(1989). Nutritive and non-nutritive constituents in taro (*Colocasia esculenta* (L.) Schott) from American Samoa. J. Hawaii Pac, Agri., 2: 1-5.

Njintang,YN., Mbofung CMF (2003) Development of taro (*Colocasia esculenta* L. Schoff) flour as ingredient in food processing: Effect of gelatinization and drying temperature on the dehydration kinetics and color of flour. J. Food Engg.,58: 259-265.

Njintang YN., Scher J., Mbofung CMF (2008). Physicochemical, thermal properties and microstructure of six varieties of taro (*Colocasia esculenta* L. Schott) flours and starches. J. Food Engg., 86(2): 294-299.

Prajapati RM., Kalariya R., Umbarkar S., Parmar N. (2011) *Colocasia esculenta*: a potent indigenous plant. Int.Nut.Pharmacol. Neurol. Dis.,1(2): 90–96.

Quach ML, Melton LD, Harris PJ, Burdon JN, Smith BG. Cell wall compositions of raw and cooked corms of taro (*Colocasia esculenta*). J.Sci. Food Agri.,2003; 81: 311-318.

Rashmi DR., Raghu N., Gopenath TS., Palanisamy P., Bakthavatchalam P., Karthikeyan M., Gnanasekaran A., Ranjith MS., Chandrashekrappa GK., Kanthesh M Basalingappa KM. (2018). Taro (*Colocasia esculenta*): An overview. J. Med. Plants Stud., 6(4): 156-161.

Rao N., Shurpalekar M., SundaravalliK. (1970). Purification and properties of an amylase inhibitor from colocasia (*Colocasia esculenta*) tubers. Ind. J. Biochem., 7, 241-243.

Reyad-ul-Ferdous M., Arman MSI., Tanvir MMI., Sumi S., Siddique KMMR., Md. Mustahsan Billah., Islam MS. (2015) Biologically potential for pharmacological and phytochemicals of medicinal plants of *Colocasia esculenta*: a comprehensive review. Am. J.Clin. Exp. Med. 3(5–1): 7–12.

Shah BN., NayakBS., Bhatt SP., Jalalpure SS., Sheth AK. (2007). The antiinflammatory activity of the leaves of *Calocasia esculenta*. Sau.Pharma. J. 15: 3–4.

Shekade DP., Patil PD., Gurunath V Mote GV3 Akashya K Sahoo AK. (2018).Potential Use of Dragon Fruit and Taro leaves as functional food: a Review Europ. J.Engg. Sci. Technol., 1 (1): 10-20.

Sanful RS. (2011). Organoleptic and Nutritional Analysis of Taro and Wheat Flour Composite Bread. World J. Dairy Food Sci.,6(2): 175-179.

Tari AT., Singhal RS. (2002).Starch based spherical aggregates: Stability of a model flavoring compound, vanillin entrapped therein. Carbo.Poly., 50: 417-421.

Temesgen M., Ratta N. (2015). Nutritional Potential, Health and Food Security Benefits of Taro *Colocasia esculenta* (L.): A Review. Food Sci. Qual. Manag., 36: 23-30.

Chapter 11

A Reservoir of Nutritional Components: Sesame Seeds (*Sesamum indicum* L.)

Sesame seed is a reservoir of nutritional components with numerous beneficial effects along with health promotion in humans. The bioactive components present in the seed include vital minerals, vitamins, phytosterols, polyunsaturated fatty acids (PUFA), tocopherols and unique class of lignans such as sesamin and sesamolin. The presence of phenylpropanoid compounds namely lignans along with tocopherols and phytosterols provide defense mechanism against reactive oxygen species and increases keeping quality of oil by preventing oxidative rancidity. Black sesame seeds are rich in magnesium that helps to prevent hypertension. Polyunsaturated fats and the compound sesamin present in sesame oil are known to keep blood pressure levels in check. The sesamin helps to protect the liver against the damage caused by free radicals in the body. Also, the seeds are rich in fiber, lignans (cell reinforcements) and phytosterol (phytochemicals), which can protect you against the development of colon cancer. In this article, we have delineated precisely the nutraceutical, pharmacological, traditional and industrial applications of sesame seeds with respect to bioactive components that hold high antioxidant value.

Keywords: *Sesamin, Lignans, Vitamin B, PUFA, Omega-6, Bioactive components, Nutraceuticals, Phytosterols, Tocopherols, Phytate, Til.*

11.1 Introduction

Sesame seeds, popularly known as '**Til**' in Hindi, '**Nuvvulu**' in Telugu, '**Ellu**' in Tamil, Malayalam, Kannada, '**Teel**' in Marathi and '**Til**' in Bengali are the oldest condiment known to the mankind. They are nutty and fragrant seed variants. They add a lovely crunch to many Asian and Middle Eastern dishes. Both these races are known for their longevity. Sesame seeds include a combination of vitamins and minerals and are one of the healthiest foods in the world. Sesame (*Sesamum indicum* L.) belonging to the order Tubiflorae, of family Pedaliaceae, is a herbaceous annual plant cultivated for its edible seeds, oil and good flavor.

It is also known as gingelly, til, benne seed and popularly as "Queen of Oilseeds" due to its high degree of resistance to oxidation and rancidity. Sesame seed contains 50-60 per cent of high quality oil which is rich in polyunsaturated fatty acids (PUFA) and natural antioxidants, sesamin, sesamolin and tocopherol homologues. These bioactive components enhance the suitability and keeping quality of sesame oil (Figure 11.1).

Figure 11.1. Black and White Coloured Seeds of Sesame and Sweets (Burfi and Laddu) Prepared from it.

Sesame seeds are considered as valuable foods since they enhance the diet with the pleasing flavor and offer nutritional and physiological benefits. Recent studies on the antioxidant and anti-carcinogenic activities of sesame seed have greatly increased its applications in health food products that assert for liver and heart protection and tumor prevention. Sesame seed is high in protein, vitamin B1, dietary fiber as well as an excellent source of phosphorous, iron, magnesium calcium, manganese, copper and zinc. In addition to these important nutrients, sesame seeds contain two unique substances, sesamin and sesamolin. Both of these substances belong to a group of special beneficial fibers called lignans and have a cholesterol lowering effect in humans and prevent high blood pressure and increase vitamin E supplies in animals (Milder *et al.*, 2005; ; Kuo *et al.*, 2011; Gouveia *et al.*, 2016; Khosravi-Boroujeni *et al.*, 2017).

11.2 Culinary Usage

Black sesame seeds and its oil are used widely across India. The seeds are often blended with warm jaggery, sugar, or palm sugar and made into balls that are eaten as a snacks. In Manipur, black sesame is used for the preparation of *Thoiding* and *Singju* (a kind of salad). Thoiding is prepared along with ginger, chilli and vegetables and is served along with the spicy *Singju* dish. In Assam, black sesame seeds are used to make *Til Pitha* and *tilor laru* (sesame seed sweet balls) during the festival of Bihu. They are nutty and fragrant seed variants. They add a lovely crunch to many Asian and Middle Eastern dishes. These seeds have the highest oil content among all seeds and are delicate, with nutty flavor that becomes more pronounced when they are roasted under low temperature for a few minutes.

These seeds can be consumed as raw, in dried form, or as roasted snacks. They are used as an ingredient in several cuisines. Dry roasted sesame seeds are ground with olive into a thin, light brown coloured paste, known as 'Tahini', which is a popular Middle Eastern dip. In Europe, they are commonly used in the manufacture of margarine. There are several varieties of sesame seeds depending on the type of cultivar such as white, black and brown seeds. White sesame seeds have higher iron content than the black ones and are mostly used as ingredients in food or in extraction of oil. Black sesame seeds are more flavorful and have a stronger aroma than white or brown sesame seeds and are preferably used in medicines. They contain 60 per cent more calcium than the white ones. White sesame seeds are hulled, whereas the black and brown ones retain their hulls. The black sesame seed can help in curing constipation due to the high fiber content and unsaturated fatty acid content. The oil found in the seed can lubricate your intestine, while the fiber in the seed helps in smooth bowel movements. These seeds also help in clearing up worms in your intestinal tract and improve the digestion process (Permaul *et al.*, 2009; Ben-Shoshan *et al.*, 2010).

11.3 Health Benefits of Sesame Seeds

Charak Samhita (Saxena 2023 p. 279) mentioned use of dew collected from flowers of sesame helped in cure of eye diseases. Oil obtained from black til and seeds of Bakuchi (*Psoralea corylifolia*) are extensively used as topical treatment of skin disorder psoriasis and vitiligo. Kidney shaped seeds of Bakuchi are bitter help in cough cold removal and have cardiotonic properties as well (Saxena 2023 p. 401).

In addition to their culinary uses, the seeds possess, preventive, and curative properties, which render them usable in traditional medicines. Sesame seed oil is a rich source of phytonutrients such as omega-6 fatty acids, flavonoid, phenolics, antioxidants, vitamins and dietary fiber. Sesame seeds are a good source of dietary protein, with high-quality amino acids making up 20 per cent of the seed.

Just sprinkle them over your salads, veggies, and noodles. Sesame seeds contain magnesium and other nutrients that have been shown to combat diabetes. The usage of sesame seed oil as the sole edible oil has been found to be effective in lowering the blood pressure and plasma glucose in hypersensitive diabetics. Sesame seeds, particularly the black ones, are rich in iron. Hence, they are highly recommended for those suffering from anemia and weakness. Sesame seed oil prevents atherosclerotic lesions and hence, is beneficial for the heart. They contain an antioxidant and anti-inflammatory compound called sesamol that also exhibits anti-atherogenic properties, thus improving the cardiovascular health. Sesame seeds are high in the monounsaturated fatty acid, oleic acid, which helps in lowering the bad cholesterol and increasing the good cholesterol in the body (Khosravi-Boroujeni *et al.,* 2017). This prevents the risk of coronary artery disease and strokes.

Sesame seeds contain magnesium which has anti-cancer properties. It also prevents asthma and other respiratory disorders by preventing airway spasms. They also contain an anti-cancer compound called phytate. Sesame seeds have been found effective in reducing the risk of colorectal tumors or cancer. This high fiber content helps in smooth functioning of the intestine, thus facilitating waste disposal and relieving constipation. Sesame seeds contain copper, a mineral that is vital for antioxidant enzyme systems, thus reducing the pain and swelling associated with arthritis. Besides, this mineral provides strength to the blood vessels, bones, and joints (Gouveia *et al.,* 2016).

Sesamol, found in sesame seeds and sesame oil, has been found to prevent the DNA from being damaged by radiation. It also prevents damage to the intestines and the spleen. Sesame seeds contain zinc that boosts the bone mineral density and the bone health. The deficiency of this mineral can cause osteoporosis in the hip and spine area. Sesame seeds are a great source of calcium, a trace mineral that is vital to bone health. Sesame seeds and sesame seed oil help in boosting the oral health by removing dental plaque and whitening your teeth (Milder *et al.,* 2005).

Sesame seeds help the liver to decompose the harmful effects of alcohol as well as other substances that generate poisoning in the body. Sesame seeds contain several nutrients that have stress-relieving properties. Minerals like magnesium and calcium act as an antispasmodic by regulating the muscle function *i.e.* contraction and relaxation. Thiamin or vitamin B1 has calming properties that aid in proper nerve functioning. The deficiency of this vitamin can lead to muscle spasms, moodiness, and depression. Tryptophan is an essential amino acid that is involved in the production of serotonin, a neurotransmitter that reduces pain and regulates the sleep pattern and mood. The inadequacy of serotonin production and transmission in the brain can result in anxiety and depression (Kuo *et al.,* 2011).

Lignans have a cholesterol lowering effect as they are rich in dietary fiber. Black sesame seeds also contain plant compounds called phytosterols which have a structure similar to that of cholesterol. Their consumption not only decreases the blood cholesterol levels, but also reduces the risk of developing certain types of cancer. Sesame seeds have the highest phytosterol content of all seeds and nuts.

According to traditional Chinese medicine, there is a strong relationship between the internal organs and external parts such as the eyes and the liver. The liver stores blood and since a certain branch of the liver channel goes to the eyes, the liver can also send blood to the eyes to support their functioning. Black sesame seeds are beneficial for the liver as they increase the liver blood, thereby nourishing the eyes. Their therapeutic effects help in treating blurred vision and tired, dry eyes. Black sesame seeds have been found to increase energy, nourish the brain and slow down aging. The regular consumption of black sesame seeds can help reduce the symptoms of backache, painful or tight and stiff joints, and weakness in the joints (Bedigian *et al.,* 1986).

Studies have indicated that using this oil can help reduce hypertension. The magnesium in this oil helps reduce blood pressure. Using black sesame seed oil, either topically or by consumption, can help reduce ailments and conditions caused by inflammation. The high amount of copper in this oil helps the users cope better with inflammation caused conditions affecting the body joints (Brar and Ahuja 1979).

Sesame is rich in powerful antioxidants and possesses antibacterial and antiviral properties. It offers several benefits to your skin by bringing blood and nourishment. The oil extracted from sesame seeds have been used as products for beautification. It is beneficial for the skin. Sesame seed oil is a natural anti-inflammatory agent and has excellent healing properties. Its antibacterial properties help to get rid of skin pathogens like *Staphylococcus* and *Streptococcus* as well as common skin fungi such as athlete's foot fungus. Sesame seed oil mixed with warm water can control vaginal yeast infections. It prevents the harmful ultraviolet rays of the sun from damaging your skin, thus preventing the appearance of wrinkles and pigmentation. The regular usage of this oil significantly reduces the risk of skin cancer and prevents the skin from the effects of chlorine in swimming pool water. The antioxidants contained in sesame seed oil help in detoxifying your skin (Cheng *et al.,* 2006).

Sesame seed oil protects the tender skin against the rashes. Applying it to the nose and ears provides protection against common skin pathogens. It also combats dryness of skin. Sesame seed oil can provide you with glowing skin. It maintains skin flexibility by keeping it soft and supple, and heals the areas of mild cuts, scrapes and abrasions. It helps tighten the facial skin, particularly the area around the nose, and controls the enlargement of pores. Sesame seeds contain essential fatty acids such as omega-3, omega-6 and omega-9 which promote hair growth. Sesame seed oil stimulates hair growth by nourishing, conditioning, and promoting a healthy scalp. The regular massage with warm sesame oil penetrates your scalp, thus increasing the blood circulation. Charak Samhita (Saxena 2023 p. 297) mentioned use of sesame seed oil along with opium externally help in cure of exzematous otitis or ear ailments.

Sesame seeds are rich in replenishing vitamins, minerals and nutrients that are vital for a healthy scalp. Massaging your scalp with sesame seed oil combats dryness, flakiness and clogged pores that cause hair thinning and hair loss. Besides, its antifungal, antibacterial and anti-inflammatory properties help in treating scalp infections and dandruff and soothing an irritated scalp. It acts as a natural sunscreen for your hair by protecting it from the damaging effects of the sun's UV rays and pollution. Sesame seed oil acts as a deep conditioning treatment for dry, damaged hair, split ends or chemically treated hair. It restores the lost moisture and strengthens the hair shaft, enabling dull and brittle hair to regain its shine, bounce, elasticity and softness. The sesame oil is known for its hair darkening qualities that make it effective for people suffering from premature graying of hair. It can be used with carrier oils like olive or almond oil to reap the maximum benefits (Pathak *et al.,* 2014).

11.4 Nutritional Facts about Sesame Seeds

All varieties of sesame seeds are extremely nutritious. They have a high oil content of 40 per cent to 60 per cent. They are a rich source of minerals like copper and manganese. They also contain magnesium, calcium, phosphorus, iron, selenium, vitamin B1 and zinc, and are rich in dietary fiber and monounsaturated fats. The nutritional value of sesame seeds is explained in Table 11.1.

Table 11.1. Sesame Seeds (*Sesamum indicum*), Whole, Dried Nutritional Value per 100 g

Principle	*Nutritive Value*	*Percentage of RDA*
Energy	573 Kcal	29 per cent
Carbohydrates	23.45 g	18 per cent
Protein	17.73 g	32 per cent
Total Fat	49.67 g	166 per cent
Cholesterol	0 mg	0 per cent
Dietary Fiber	11.8 g	31 per cent
	Vitamins	
Folates	97 µg	25 per cent
Niacin	4.515 mg	28 per cent
Pantothenic acid	0.050 mg	1 per cent
Pyridoxine	0.790 mg	61 per cent
Riboflavin	0.247 mg	19 per cent
Thiamin	0.791 mg	66 per cent
Vitamin A	9 IU	<1 per cent
Vitamin C	0	0 per cent
Vitamin E	0.25 mg	2 per cent
	Electrolytes	
Sodium	11 mg	1 per cent
Potassium	468 mg	10 per cent

Principle	Nutritive Value	Percentage of RDA
	Minerals	
Calcium	975 mg	98 per cent
Copper	4.082 mg	453 per cent
Iron	14.55 mg	182 per cent
Magnesium	351 mg	88 per cent
Manganese	2.460 mg	107 per cent
Phosphorus	629 mg	90 per cent
Selenium	34.4 µg	62.5 per cent
Zinc	7.75 mg	70 per cent
	Phytonutrients	
Carotene-β	5 µg	–
Crypto-xanthin-β	–	–
Lutein-zeaxanthin	–	–

Source: USDA National Nutrient Database.

11.5 Genomic Data

Sesamum indicum, is an important oil crop, which provides part of the daily edible oil allowance for almost half of the world's population. A limited number of co-dominant markers has been developed and applied in sesame genetic diversity and germplasm identity studies. Here we report a whole genome survey used to develop simple sequence repeat (SSR) markers and to detect the genetic diversity of sesame germplasm. From the initial assembled sesame genome, 23,438 SSRs (≥5 repeats) were identified (Wei *et al.,* 2014). The most common repeat motif was dinucleotide with a frequency of 84.24 per cent, followed by 13.53 per cent trinucleotide, 1.65 per cent tetranucleotide, 0.3 per cent pentanucleotide and 0.28 per cent hexanucleotide motifs. From 1500 designed and synthesized primer pairs, 218 polymorphic SSRs were developed and used to screen 31 sesame accessions that from 12 countries. Structure and phylogenetic analyses indicated that all sesame accessions could be divided into two groups: one mainly from China and another from other countries. Cluster analysis classified Chinese major sesame varieties into three groups. These novel SSR markers are a useful tool for genetic linkage map construction, genetic diversity detection, and marker-assisted selective sesame breeding.

Data collected from Ghana were subjected to Principal Component Analysis (PCA) and a dendrogram showing similarity between the accessions were drawn. Data on number of capsules per plant, number of seeds per capsule, and plant height at flowering were subjected to analysis of variance using Gen Stat Discovery Edition 4. Molecular genetic diversity was assessed by using thirty eight SSR markers widely distributed across sesame genome to characterize the materials. Twenty-one out of the 38 primers were polymorphic. Cluster analyses using the Euclidean similarity test and a complete link clustering method were used to make a dendrogram out of the morphological data. Analysis of variance showed that capsule number was significantly different; a range of 54.9 and 146.7 was produced. Te number of seeds per capsule varied significantly and the variation between highest and lowest accession in seed production was 33 per cent. Plant height was also significantly different ranging from 60.6 to 94.1 cm. Using morphological traits the accessions clustered into two major groups and two minor groups and variation among accessions were 10-61 per cent. On the other hand, SSR marker-based dendrogram revealed five major and two minor groups. It showed that variation among the accessions was low, 10-20 per cent. Heterozygosity was 0.52, total alleles produced were 410, and average allele per locus was 19.52. Six accessions, C3, C4, S5, W1, W3, and W5 fell in five different clusters in the SSR dendrogram and in six clusters in the morphomolecular based dendrogram. These accessions were noted for high capsule number per plant and seeds number per capsule and are recommended for consideration as potential parental lines for breeding programme for high yield.

11.6 Conclusion

Black Sesame seeds, also known as *kala til,* is one of oldest condiments known to history. They are highly valued for their oil. "Open Sesame"- the famous phrase from the Arabian Nights symbolizes the distinguishing feature of the sesame seed pod, which bursts open when it reaches maturity. It may be a small seed but no doubt a very powerful one, used for many health-promoting and anti-ageing benefits. Black sesame seeds are a good source of energy due to the high fat content. The sesamin found in seeds is found to protect the liver against the damage caused by free radicals in the body. Also, the seeds are rich in fiber, lignans (cell reinforcements) and phytosterol, which can protect us against the development of colon cancer.

REFERENCES

Ben-Shoshan M, Harrington DW, Soller L. (2010). A population-based study on peanut, tree nut, fish, shellfish, and sesame allergy prevalence in Canada. The Journal of Allergy and Clinical Immunology. 125 (6): 1327–35. doi: 10.1016/j.jaci.2010.03.015.

Bedigian D, Harlan JR. (1986) Evidence for cultivation of sesame in the ancient world. Econ Bot. 1986; 40: 137–54.

Brar G, Ahuja KL. (1979) Sesame: its culture, genetics, breeding and biochemistry. Annu Rev Plant Sci. 1979; 1: 245–313.

Cheng FC, Jinn TR, Hou RC, Tzen JT. (2006) Neuroprotective effects of sesamin and sesamolin on gerbil brain in cerebral ischemia. Int J Biomed Sci. 2006; 2: 284–8.

Gouveia Lde A, Cardoso CA, de Oliveira GM, Rosa G, Moreira AS (2016). "Effects of the Intake of Sesame Seeds (*Sesamum indicum* L.) and Derivatives on Oxidative

Stress: A Systematic Review". Journal of Medicinal Food. 19 (4): 337–345. doi: 10.1089/jmf.2015.0075.

Khosravi-Boroujeni H, Nikbakht E, Natanelov E, Khalesi S (2017). "Can sesame consumption improve blood pressure? A systematic review and meta-analysis of controlled trials". Journal of the Science of Food and Agriculture. 97 (10): 3087–3094. doi: 10.1002/jsfa.8361

Kuo PC, Lin MC, Chen GF, Yiu TJ, Tzen JT (2011). "Identification of methanol-soluble compounds in sesame and evaluation of antioxidant potential of its lignans". J Agric Food Chem. 59` (7): 3214–9. doi: 10.1021/jf104311g.

Milder, Ivon E. J.; Arts, Ilja C. W.; Betty; Venema, Dini P.; Hollman, Peter C. H. (2005). "Lignan contents of Dutch plant foods: a database including lariciresinol, pinoresinol, secoisolariciresinol and matairesinol". British Journal of Nutrition. 93: 393–402. doi: 10.1079/BJN20051371.

Pathak N, Rai A K, Kumari R. (2014) Value addition in sesame: A perspective on bioactive components for enhancing utility and profitability. Pharmacogn Rev. 2014 Jul-Dec; 8(16): 147–155. doi: 10.4103/0973-7847.134249

Permaul P, Stutius LM, Sheehan WJ, Rangsithienchai P, Walter JE, Twarog FJ, Young MC, Scott JE, Schneider LC, Phipatanakul W. (2009) Sesame allergy: role of specific IgE and skin-prick testing in predicting food challenge results. Allergy Asthma Proc. Nov-Dec; 30(6): 643-8. doi: 10.2500/aap.2009.30.3294.

Saxena O.P. (Nidar) (2023) Charak Samhita in Hindi by DPB Pub. Delhi pp. 279 and 401.

Wei X, Wang L, Zhang Y, Qi X, Wang X, Ding X, Zhang J, Zhang X. (2014) Development of simple sequence repeat (SSR) markers of sesame (*Sesamum indicum*) from a genome survey. Molecules. 2014 Apr 22; 19(4): 5150-62. doi: 10.3390/molecules19045150. PMID: 24759074; PMCID: PMC6270694.

Chapter 12

Nutritive Vegetable: Pointed Gourd (*Trichosanthes dioica* Roxb.)

Pointed gourd (parwal) scientifically known as Trichosanthes dioica is a highly nutritious cucurbitaceous vegetable. It is an elite vegetable crops which can be grown quite intensively in tropical and sub-tropical conditions. The fruit is widely consumed as a vegetable and different parts of this plant are used in the traditional medicine to treat various types of human ailments Studies have found promising antihyperglycemic, antihyperlipidemic, antitumor, cytotoxic, arsenic poisoning ameliorative, anti-inflammatory, antidiarrheal, and varieties of pharmacological activities of T. dioica. Different studies have proven that its consumption can reduce the bad cholesterol level in our body, such as LDL cholesterol and improve the good cholesterol level in our body. Pointed Gourd is highly rich in dietary fiber, which helps to improve the production of insulin in our body. It contains 20 k cal, Cu 1.1 mg (122.22 per cent), Ascorbic acid 29 mg (32.22 per cent), 4 per cent Protein 2 g, Carbohydrate 4.2 g (3.23 per cent), Ca 30 mg (3 per cent). Different types of bioactive compounds have been identified and isolated from T. dioica including peptides viz. trichosanthin, lectin, cucurbitacin B, euphol, α-amyrin, β-amyrin, lupeol, taraxerol, betulin, karounidiol, sterols, tannin, flavonoids, and vitamin A, saponins, alkaloids, proteins, tetra and pentacyclic triterpenes, etc. It is possible to accurately and rapidly determine the sex of T. dioica plants using two RAPD markers, which requires a small amount of leaf sample. In this review, scientific studies on therapeutic potential, its different folkloric uses, advances in ethnopharmacological studies and phytochemistry of different parts of T. dioica are presented.

***Keywords**: Cucurbitacin, Diabetes, Hepatoprotective, Therapeutic potential, Phytoconstituents.*

12.1 Introduction

Trichosanthes dioica Roxb. belonging to Family-Cucurbitaceae is a perennial herbaceous vine commonly known as pointed gourd, Putulika in Sanskrit, Parval in Hindi, and Potol in Bengali. It is mainly cultivated in Bangladesh, India, Pakistan, Myanmar, Nepal and Sri Lanka. It is cultivated in subtropical and tropical regions around the globe (Robinson and Decker-Walters 1999). It is propagated through stem cuttings taken, ideally, from mature plants to ensure the sex and fruit types. Although each fruit contains several well-developed seeds having hard seed coat, propagation using seed is not feasible primarily due to poor germination (Kumar *et al.,* 2008) and slow growth of the seed-derived plants (Singh 1989). Numerous studies have been performed to evaluate the validity of the traditional uses. *T. dioica* is used in the Ayurveda as well as by the rural and tribal folkloric practitioners in a number of human ailments. In Ayurveda, it is used alone or in combination with other herbs to treat hyperacidity. It has been described as antiparasitic, antiperiodics, febrifuge, refrigerants, anticoagulants, blood purifiers, antiseptics, and tonic (Hariti and Rathee 1995; 1996; Rai *et al.,* 2008; 2010). The plant is also used as stomachic, antitumor and antidiabetic agent. In Ayurveda leaves and fruits of *T. dioica* are used to treat alcoholism and jaundice. Fruit is used in improving appetite and digestion. Juice of leaves of *T. dioica* is used as tonic, febrifuge, in edema, alopecia, and in subacute cases of enlargement of liver. In Charaka Samhita, leaves and fruits find mention for treating alcoholism and jaundice. According to Ayurveda, leaves of the plant are used as antipyretic, diuretic, cardiotonic, laxative, antiulcer, *etc.* The various chemical constituents present in *T. dioica* are vitamins A, C, tannins, saponins, alkaloids, proteins tetra and pentacyclic triterpenes, *etc.* (Toshihiro *et al.,* 1997). The unripe fruit is cooked either alone or with other vegetables. It is also used as antipyretic, diuretic, cardiotonic, and laxative. The vegetable is rated as one of the nutritious foods and is offering vitamins like vitamin A, vitamin B1, vitamin B2 and vitamin C. Pointed gourd provides a few calories as low as 20 kcal in 100g and ample amount of other nutrients. The seeds contain a significant amount of peptides. Pointed gourd is laden with mammoth amount of fiber, which helps to encourage proper digestion. It is useful in gastrointestinal and liver disorders. It also cures cough, eczema, blood-impurities, excessive phlegm and flatulence. It helps to clean our blood, tissues and thus offering skin care as well. It helps to fight with aging factors. Parwal has the capacity to boost your immunity. It is used like a medicine to treat flu, throat problems and high temperatures (Alom *et al.,* 2013; Roy and Mani 2020).

12.2 Phytochemical Constituents

Pointed gourd is rich in vitamins and contains 9.0 mg Mg, 2.6 mg Na, 83.0 mg K, 1.1 mg Cu, and 17.0 mg S per 100 g edible part. The seeds of *T. dioica* contain a large amount of peptides (Mehta and Sharma 2012).The seed peptides have the unique property of being resistant to the action of silver nitrate, a sensitive reagent commonly used to stain proteins. The various chemical constituents present in *T. dioica* are vitamin A, vitamin C, tannins, and saponin (Saxena and Dave 1995). Phytochemical evaluations of aqueous and ethanolic extracts have shown the presence of saponins and tannins. Two main phytosterols present in

T. dioica are, namely, 24α-ethylcholest-7-enol and 24β-ethylcholest-7-enol. Total phenolic content of *T. dioica* leaves has been about 2x more than that obtained from the fruits and seeds of *M. olifera* and *E. officinalis*, respectively. Fruits contain cucurbitane glycosides *viz.* khekadaengosides, cucurbitacin 2-O-bglucopyranoside and cucurbitacin 2-O-b-glucopyranoside), a hexanorcucurbitane glucoside (khekadaengoside) and octanorcucurbitane (khekadaengoside) along with 25-O-acetyl-cucurbitacin. Seeds of *T. dioica* also contain lectin, a carbohydrate (specifically galactose) binding protein. The seed extract of *T. dioica* contains 7-oxidihydrokarounidol-3-benzoate as the most predominant component in the highly polar fraction of the non-saponifiable lipid (Kumar *et al.,* 2012b).

Majority of the glandular trichomes are with 4 to 5 celled uniseriate stalk and unicellular head very few are short and with uni- to bi-cellular stalk and uni- to multi-cellular head especially from that of petiole region. Simple covering multicellular uniseriate thick walled trichomes are of various sizes and usually of the cells of both simple and glandular trichomes are often embedded with cystolith. Phytochemical studies of the powdered leaves revealed the presences of alkaloids, resins, glycosides, flavonides and some carbohydrates. The pharmacognostic profile of the leaves will assist in standardization for quality, purity and sample identification.

Singh *et al.* (2010) during study of male leaves of pointed gourd found that majority of the glandular trichomes were with 4 to 5 celled uniseriate stalk. The unicellular heads very few in number are short and with uni- to bi-cellular stalk and uni- to multi-cellular head especially from that of petiolar region. Simple covering multicellular uniseriate thick walled trichomes are of various sizes and usually of the cells of both simple and glandular trichomes are often embedded with cystolith. Phytochemical studies of the powdered leaves revealed the presences of alkaloids, resins, glycosides, flavonides and some carbohydrates. The aqueous extract of fruits of *T. dioica* possess antioxidant properties due to the presence of phenolic acids, polyphenols and flavonoids scavenge free radicals such as peroxide, hydroperoxide or lipid peroxyl and thus inhibit the oxidative mechanisms that lead to degenerative diseases (Shivhare *et al.,* 2010a).

12.3 Traditional Uses

Trichosanthes dioica is rated as one of the nutritious foods and is offering vitamins like vitamin A, B1, B2 and C. In Ayurveda, pointed gourd is used to treat the gastric problems and is also used as a natural aphrodisiac that can improve sex life (Singh and Singh 1985). It is used like a medicine to treat flu, throat problems and high temperatures. Green colored parwal is loaded with a huge amount of fiber, which helps to promote proper digestion. It also helps to solve gastrointestinal problems and some problems related to the liver. Parwal consists of good amount of antioxidants, vitamin A and C that helps to fight with the free radicals which are encouraging the aging signs (Shivhare *et al.,* 2010b). Seeds present in the parwal

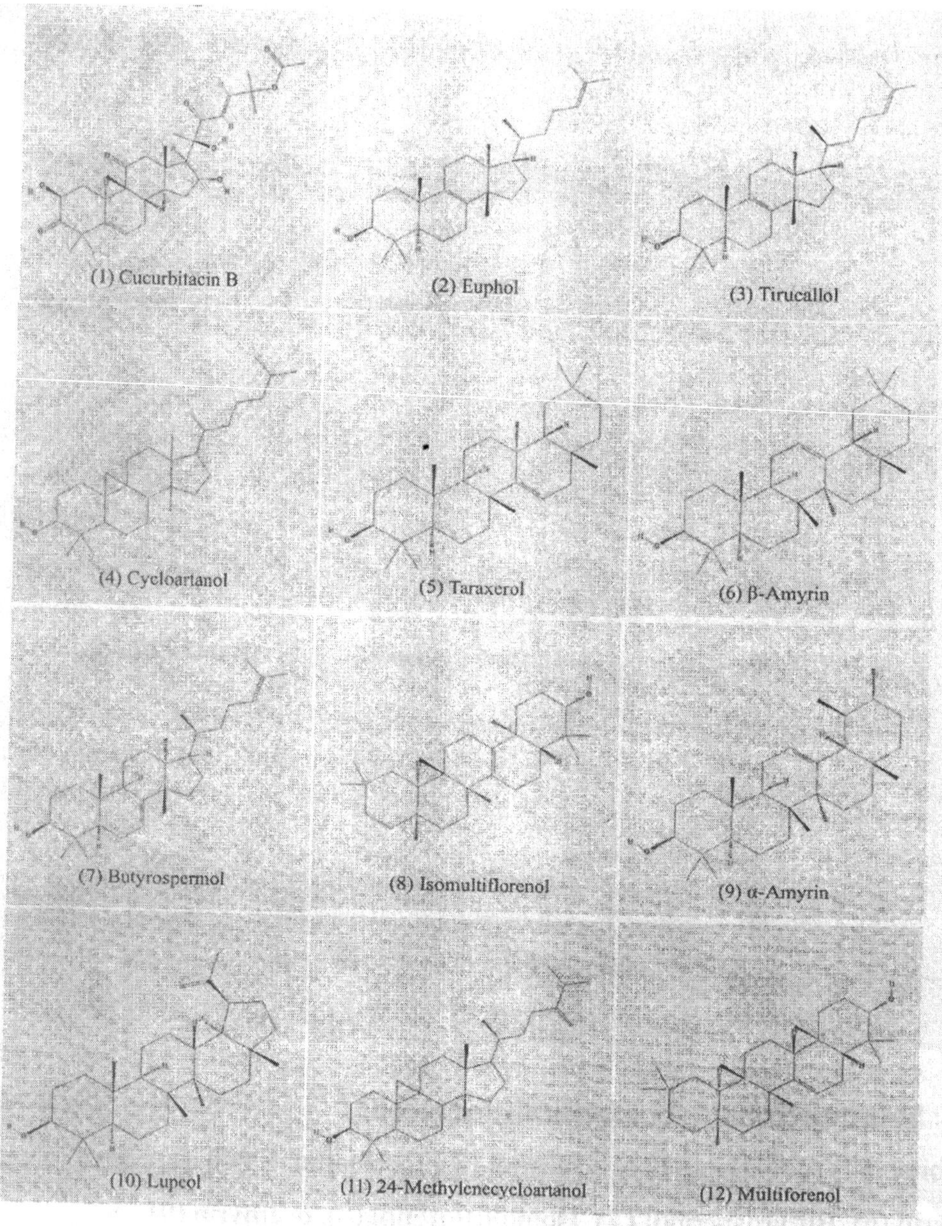

Figure 12.1A. Structural Formula of Chemical Compounds

Cucurbitacin B (1), Euphol (2), Tirucallol (3), Cycloartanol (4), Taraxerol (5), β-Amyrin, (6), Butyrospermol (7), Isomultiflorenol (8), α-amyrin (9), Lupeol (10), 24-Methylenecycloartanol (11), Multiflorenol (12).

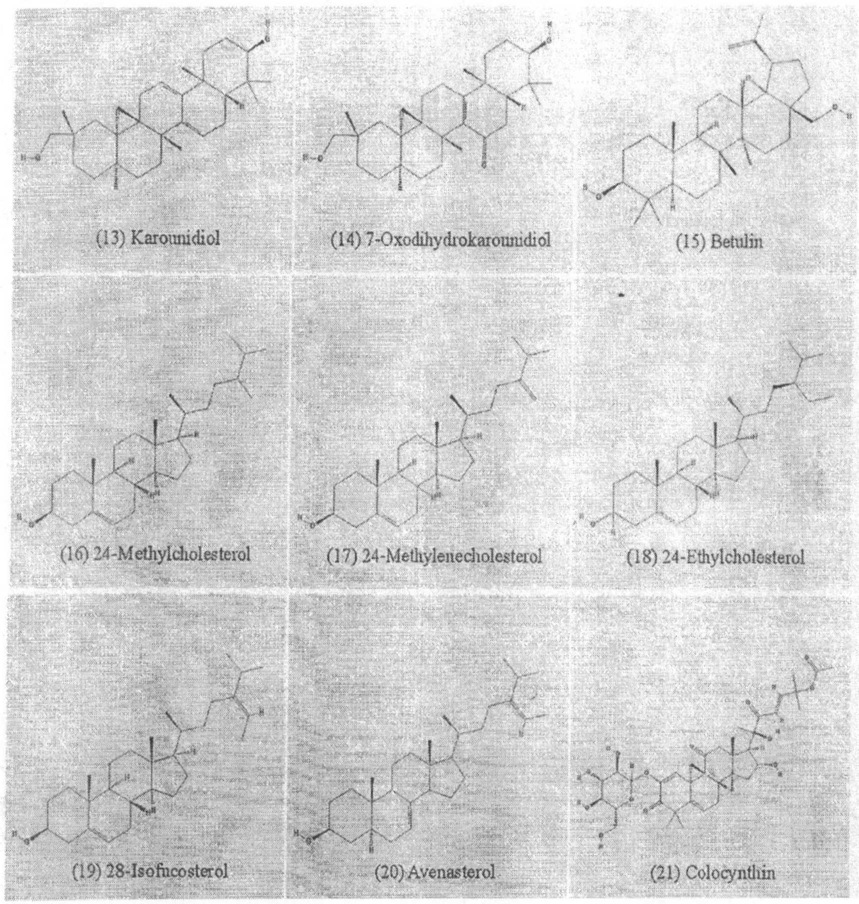

Figure 12.1B. Structural Formula of Chemical Compounds

Karounidiol (13), 7-Oxodihydrokarounidiol (14), Betulin (15), 24-Methylcholesterol (16), 24-Methylenecholesterol (17), 24-Ethylcholesterol (18), 28-Isofucosterol (19), Avenasterol (20), Colocynthin (21).

vegetables help to ease stool and is recommended to reduce constipation problem. It is useful in gastrointestinal and liver disorders (Kabir 2000).

It is used in jaundice, viral infections, and skin diseases. It helps in throwing out worms from the stomach and cures constipation. It gives one with strength and sexual powers. It also cures cough, eczema, blood-impurities, excessive phlegm and flatulence. The leaves juice of plant is very effective in treatment of febrifuge, edema, alopecia and in sub-acute cases of enlargement of liver. Leaves are used in treatment of enlarged liver and spleen, hemorrhoids, fistula in ano, fevers, leprosy, intrinsic hemorrhage, erysipelas, alopecia, diseases of mouth, inflammations and wounds. Leaves of plant are boiled in water and this decoction is taken internally for hyper acidity and biliousness. Paste of leaves is applied topically for boils,

Figure 12.2. A Plant of Pointed Gourd and some Preparations.

wounds (Shivhare *et al.,* 2010b). Decoction of tender shoots with sugar improves digestion. Leaf juice is rubbed over the chest in liver congestion and over the whole body in intermittent fevers (Mehta and Sharma 2012). Fresh juice of the unripe fruit is often used as a cooling and laxative adjunct to some alternative medicines. Fruit is used as a remedy for spermatorrhoea.

12.4 Culinary Uses

Pointed gourd is prepared in various dishes, such as stir fry, cooked in coconut milk, or steamed in Indonesia. Cooked parwal is helpful in improving appetite and digestion. It forms a staple of many dishes and is used as an ingredient for soup, stew, curry, sweet, or eaten fried. Parwal can also be pickled in vinegar with an assortment of herbs and spices, as a side dish or a salad. It can be stuffed, or cut into cubes and sautéed, or made into soups. Parwal soup is traditionally prepared for sick people in Nepal. It is a very nutritious vegetable which also has potential for industrial use in making different types of jam, jelly, and pickles.

12.5 Sex Expression-Associated RAPD Markers

Sex a genetic phenomenon, assures the continuous generation of new gene combinations needed for genetic improvement through breeding. In addition, sex usually forms the largest single division within a species and causes differences in physiology, behavior and economic usefulness of different individuals of a species. Of the estimated 2,500,000 angiosperm species, only about 6 per cent (14,620 species) are dioecious (Kumar *et al.,* 2008), while dioecy is prevalent in animals, where it is controlled by sex chromosomes. In contrast, sex determination in dioecious plants may often be genetic or environmental, and only a tiny fraction of them has evolved sex chromosomes. Genetic sex determination may be due to a single locus, multiple tightly linked loci on autosomes, multiple unlinked loci on autosomes, or several genes located on heteromorphic chromosomes (Parasnis *et al.,* 2000). Dioecy exists in a number of cucurbitaceous genera, including *Trichosanthes,* but the mechanism of sex determination has been established in only few of them, *e.g.* XY system in *Coccinia*.

In *T. dioica*, male and female plants strictly maintain their respective sexual phenotypes but we are not aware of any report of sex-specific peculiarities in chromosome morphology in this species. In case of dioecious species, sex identification at the juvenile plants is of considerable importance to cultivation practices (Yakubov *et al.,* 2005). A reliable method for determining the gender of plants before flowering would facilitate economy of the various resources, including the time and effort of the breeders. In addition, this would allow the identification and even isolation of the gene(s) involved in the process. Random amplified polymorphic DNA (RAPD) has been used to identify sexassociated markers in plants like *Pistacia* (Yakubov *et al.,* 2005), papaya (Parasnis *et al.,* 2000), and *Asparagus* (Bracale *et al.,* 1991).

The two RAPD markers reliably identified nine unrelated male and nine female *T. dioica* plants. The reproducibility of RAPD markers has often been questioned (Lessels and Mateman 1998) and currently the trend is to develop a more reliable and simple to use sequence characterized amplified region (SCAR; Paran and Michelmore 1993) markers from sequenced RAPD fragment and designing a pair of longer (17 bases or more) primers from the sequence (Paran and Michelmore 1993). This has been achieved in the case of *Mercurialis annua* (Khadka *et al.,* 2002), *C. papaya* (Urasaki *et al.,* 2002), and *Cannabis sativa* (Mondolino *et al.,* 1999). The results indicated that it is possible to accurately and rapidly determine the sex of *T. dioica* plants using two RAPD markers, which requires a small amount of leaf sample (Kumar *et al.,* 2008). Tagging and cloning of sex determination genes will provide information that is crucial to decipher the sex-determination system. Cloning of sex determination genes is quite a challenging task because of genetic recombination that make map-based cloning approaches ineffective. The generation of sex reversal mutants might be a good alternative approach towards identifying the sex determining genes (Kumar *et al.,* 2012a).

Nanda *et al.* (2013), used inter simple sequence repeats (ISSRs) primers to isolate a suitable marker, TdSTSM to identify sex at the seedling stage in pointed gourd. The significant achievement of this study is that the male specific STS marker (TdSTSM) was found to be highly reliable and reproducible for effective distinction of male and female *T. dioica*. The result is particularly convincing because southern blot analysis showed DNA fragment in male but not in female plant. This suggests that TdMSM, might be located in a chromosome region specific to male genome, indicating that the two sexes in *T. dioica* might be having active chromosomal differentiation. Further, the TdSTSM marker could also be used as a starting material towards sequence characterization of sex linked genes for better understanding the developmental as well as evolutionary pathways in sexual dimorphism.

12.6 Conclusion

The Pointed Gourd is also known as green potato is one of the wonderful vegetable cultivated many regions throughout the world with numerous health advantages. Parwal fruits contain free amino acids, nicotinic acid, riboflavin, vitamin C, thiamine, 5-hydroxytryptamine; Root contains Colocynthin, trichosanthin, hentriacontane; Plant and root contain Cucurbita-5, 24-dienol. The edible part of fresh Parwal contains about: moisture 92 per cent, protein 2 per cent, fat 0.3 per cent, fiber 3 per cent, carbohydrate 2.2 per cent, vitamin C 29mg/100g, carotene 153 mcg/100g, Mg 9mg/100g, Na 2.6mg/100g, K 83mg/100g, Cu 1.1mg/100g, and S 17, mg/100g.Due to very low in calories and high in dietary fiber content found in bitter gourd, it is very beneficial in weight management. 100 g of bitter gourd provides only 20 calories of energy. Some of the vitamins that Pointed Gourd has Vitamin C, Vitamin A, Folates, Niacin, Pantothenic acid, Pyridoxine, Thiamin and Riboflavin. Apart from this Fiber helps to promote the bowel movement.

REFERENCES

Alom M, Nag B, Islam M, Ahmed F, Akther S (2013) Performance of different crop species with pointed gourd (*Trichosanthes dioica* Roxb.) Bangla. J. Agric. Res. 38: 523-529.

Bracale M, Caporali E, Galli MG, Longo C, Marziani-Longo G, Tassi F (1991) Sex expression and determination in *Asparagus officinalis.* Plant Sci 80: 67-77

Hariti M, Rathee P S (1995) Antibacterial activity of the unsaponifiable fraction of the fixed oil of *Trichosanthes* seeds. Asian J. Chem.,7: 909–911.

Hariti M, Rathee P S (1996) Antifungal activity of the unsaponifiable fraction of the fixed oil of *Trichosanthes* seeds. Asian J. Chem. 8: 180–182.

Kabir S (2000) The novel peptide composition of the seeds of *Trichosanthes dioica* Roxb. Cytobios. 103: 121–31.

Khadka D K, Nejidat A, Tal M, Golan-Goldhirsh A (2002) DNA markers for sex: Molecular evidence for gender dimorphism in dioecious *Mercurialis annua* L. Mol Breed 9: 251-257

Kumar S, Singh B D, Sinha D P, Rai M (2008) Sex expression-associated RAPD markers in pointed gourd (*Trichosanthes dioica*) Proceedings of the IXth EUCARPIA meeting on genetics and breeding of Cucurbitaceae (Pitrat M, ed), INRA, Avignon (France), May 21-24th, 2008

Kumar S, Singh B D, Sinha D P (2012a) RAPD markers for identification of sex in pointed gourd (*Trichosanthes dioica* Roxb.). Indian Journal of Biotechnology 11: 251-256

Kumar N, Singh S, Manvi, Gupta R (2012b). *Trichosanthes dioica* Roxb.: An overview. Pharmacog. Rev. 6(11): 61–67.doi: 10.4103/0973-7847.95886

Lessels C, Mateman A (1998) Sexing birds using random amplified polymorphic DNA (RAPD) markers, Mol Ecol, 7 (1998) 1071-1076.

Mandolino G, Carboni A, Forapani S, Faeti V, Ranali P (1999) Identification of DNA markers linked to male sex in dioecious hemp (Cannabis sativa L.). Theor Appl Genet 98: 86-92

Mehta D, Sharma A (2012) *Trichosanthes dioica* (roxb.): A review on pharmacological update. Invent. Impact – Planta Activa. 12 (10): 187-194

Nanda S, Kar B, Nayaka S, Jha S, Joshi R K (2013) Development of an ISSR based STS marker for sex identification in pointed gourd (*Trichosanthes dioica* Roxb.). Scientia Horticulturae 150: 11–15 Proceedings of the IXEUCARPIA meeting on genetics and breeding of Cucurbitaceae (Pitrat M, ed), INRA, Avignon (France), May 21-24th, 2008

Paran I M, Michelmore R W (1993) Development of reliable PCR-based markers linked to downy mildew resistance genes in lettuce, Theor Appl Genet, 85 (1993) 985-993.

Parasnis A S, Gupta V S, Tamhankar S A, Ranjekar P K (2000) A highly reliable sex diagnostic PCR assay for mass screening of papaya seedlings. Mol Breed 6: 337-344

Rai P K, Mehta S, Gupta R K, Watal G (2010) A novel antimicrobial agents *Trichosanthes dioica.* Int. J. Pharma. Bio. Sci.1: 1–9.

Rai P K, Jaiswal D, Singh R K, Gupta R K,Watal G (2008) Glycemic Properties of *Trichosanthes dioica* Leaves. Pharmaceut. Biol. 46 (12): 894-899. doi: 10.1080/13880200802370167

Robinson R W, Decker-Walter D S (1999) Cucurbits. CAB International Publishing, Oxon (GB)

Roy S, Mani A (2020) Nutritional importance and medicinal properties of pointed gourd. 10.13140/RG.2.2.36051.32808.

Saxena V, Dave R (1995) A new steroidal saponin from *Trichosanthes dioica* (Roxb.) Asian J. Chem. 7: 490-495.

Shivhare Y, Singh P, Rajak H, Patil U K, Pawar R S (2010a) Antioxidant potential of *Trichosanthes dioica* Roxb. (fruits). Pharmacognol. J,2: 107–111.

Shivhare Y, Singh P, Patil U K (2010b) Healing potential of *Trichosanthes dioica* Roxb. on burn wounds. Res. J. Pharmacol. Pharmacodyn. 2: 168–171.

Singh K (1989) Pointed gourd (*Trichosanthes dioica* Roxb.) Ind. Hort. 33: 35–8.

Singh K P, Singh R H (1985) Clinical investigation of importance of *T. dioica*. J. Res. Ayur. Sidh. 6: 132–140.

Singh S, Machawal L, Chauhan M G (2010) Pharmacognostic study of male leaves of *Trichosanthes dioica* Roxb. with special emphasis on microscopic technique. J. Pharmacog. Phytother. 2: 71–75.

Toshihiro A, Yumico K, Yoshimasa K, Kunio K, Thakur S, Tamura T (1997)7-oxodihydrokarounidiol-3-benzoate and other triterpenes from the seeds of Cucurbitaceae. Phytochem. 6: 1261–1266.

Urasaki N, Tokumoto M, Torora K, Ban Y, Kayano T, Tanaka N, Oku H, Chinen I, Terauchi R (2002) A male and hermaphrodite specific RAPD marker for papaya (*Carica papaya* L.) Theor Appl Genet 104: 281-285

Yakubov B, Barazani O, Golan-Goldfish A (2005) Combination of SCAR primers and touch down PCR for sex determination in *Pistacia vera* L. Scientia Hort 103: 473-478

Chapter 13

Nutritional and Therapeutic Benefits of Tamarind (*Tamarindus indica* L.)

Tamarind is a fruit with a distinctive sweet and sour taste. Used in cuisines from around the world, this delicious pod is a nutritional powerhouse with an abundance of health benefits. The health benefits of tamarind have been well-studied and they include its ability to reduce inflammation throughout the body, improve vision, boost respiratory health, heal skin conditions, and improve the digestive system. Tamarind also relieves pain, boosts the immune system, reduces fever, lowers cholesterol to improve cardiovascular health, treats piles, prevents cancer, and protects against parasites and worms. The edible pulp of ripe fruit is used as a flavouring agent in cooking, soups, jams chutneys, sauces, and juices and the preparation of beverage. Studies on tamarind have shown it to be effective in reducing blood pressure and cholesterol. The fiber content in tamarind certainly has something to do with the reduction in cholesterol, since it is known to scrap excess LDL cholesterol from the veins and arteries. The potassium in tamarind may be responsible for a reduction in blood pressure since it is known as a vasodilator that reduces the stress on the cardiovascular system. The impressive level of vitamin C in tamarind also have something to do with it as well. Vitamin C is an antioxidant that can reduce the impact of free radicals, those pesky by-products of cellular metabolism that have been linked to heart diseases and a number of other health conditions.

Keywords: *Pectin, Cellulose, Cholesterol, Polygalacturonic acid, Rhamnose, Tamarind Tamarindus indica.*

13.1 Introduction

Tamarindus indica L. commonly known as tamarind a member of the family Fabaceae. It is native to dry Savanna of tropical Africa, and is an important woody perennial fruit species known for its adaptability to variable climatic and edaphic conditions. In Sudan, the tamarind grows wild on sandy soils and near Khors (water courses) in short grasses Savanna in Kordofan, Darfur, Blue Nile, Bahr ElGhazal. India is the main producer and consumer of tamarind in the world. Pectin, which is chemically similar to cellulose, may be thought of as the "glue" that holds the cellulose together in the cell walls of plants.

Figure 13.1. A Branch of Tamarind Tree with Fruits.

Pectin and cellulose are both polysaccharides, but pectin is primarily a-linked polygalacturonic acid (partly esterified with methyl ester) in which rhamnose may be found, and cellulose is essentially a polymer chain made up of β-1,4 linked sucrose units with no esterification. Not all plant materials are rich in pectin. The oldest pectin source is apple pomace left over from juicing. The most common source of pectin nowadays is from citrus peel, primarily lemon–but lime, orange and grapefruit may also be used. Novel sources of pectin include sugar beets and sunflower heads, but these are not, at the moment, commercially significant. In order to qualify as pectin, the anhydro-galacturonic acid must make up at least 65 per cent of the ash-free dry matter in pectin sold as a commercial product (Bhattacharya *et al.,* 1994). Tamarind is a fruit with a distinctive sweet and sour taste (Figure 13.1). Used in cuisines from around the world, this delicious pod-like fruit is a nutritional powerhouse with an abundance of health benefits. Tamarind trees are cultivated in abundance in Africa and Asia and South America. Although raw tamarind is edible, it has an unusual sour taste which makes you change your facial expression every time you take a bite. Ripe tamarind pods or mature pods when cracked open have a brown or brownish-black colored fruit which contain hard black seeds (Brand]Williams 1995). The brown flesh of the fruit is juicy and acidic and has a sweet-tangy flavor. Aside from being used in world cuisine, Tamarind fruit is hugely popular because of its health benefits and medicinal benefits. Leaves, bark, seeds, and flowers of tamarind too have medicinal uses. Tamarind trees grow up to 40-80 meters depending on soil condition and weather condition. A mature tree is capable of producing between 175- 250 kg tamarind fruit per year. Flowers and leaves of tamarind are edible and have a delicious flavor. Thailand is known for its supreme variety of sweet tamarind and boasts of some of the fleshiest and sweetest ripe pods. This fruit is loaded with minerals and vitamins required in daily diet. Tamarind offers excellent nutrition value which is why this is the one of the most popular condiment in the world (Haghju and Almasi 2015).

Figure 13.2. Tamarind Fruits and Chutney.

Figure 13.3. Tamarind Fruit Juice.

13.2 Health Benefits of Tamarind

It's not just the fruit of tamarind that offers health benefits. Combinations of herbs and spices mixed with tamarind powders are used by herbalists and Ayurvedic practitioners for various natural treatments. Fruit, leaves, seeds, bark, flowers and pulp of tamarind have been used in several ancient cultures and by medieval herbalists. Not only the leaves of the Tamarind Tree and Pulp of the fruit are articles of diet but alo the seeds, they are powdered and boiled into thick Conjee which the poor Natives of the Hill Countries chiefly eat (Robinson 2008). Listed below are the numerous health benefits and medicinal uses of tamarind.

Pulp of tamarind acts as a laxative and is an excellent remedy for sluggish bowel movement. The presence of dietary fiber in tamarind pulp plays an important role on binding bile salts. Two teaspoons of tamarind paste is very good for bowel movement (Figure 13.2). Tamarind has the ability to lower levels of bad cholesterol (LDL) thereby promoting healthy cardiovascular health, this because of the presence of phenols, antioxidants beneficial for levels of HDL. Leaves of tamarind are used in herbal tea decoctions. Medicinal teas that contain extracts of tamarind leaves are given to individuals suffering from fever and chills. This fruit pod is loaded with minerals and vitamins essential for red blood cell production and healthy cardiovascular function. Scoops of thick tamarind paste are diluted in water and given to individuals with intestinal parasites. Tamarind is an excellent home remedy for persons with bile problems. In Ayurvedic medicine, tamarind is a major ingredient used to treat digestive and gastric problems. Tamarind fruit is rich in dietary fiber such as tannins, mucilage and pectin, a combination that plays a vital role in preventing constipation. 13 per cent to 15 per cent per cent dietary fiber is provided by 100 grams of tamarind paste. Pulp of tamarind is known to protect against Vitamin C deficiency.

Diluted tamarind paste is an excellent home remedy for sore throat. A person with throat infection can gargle diluted lukewarm tamarind water for relief. Tamarind pulp with a mix of crushed herbs such as coriander and mint is an excellent treatment for mouth ulcers. Tamarind has a cooling effect on ulcers caused by hot and spicy foods. In several cultures flowering leaves of tamarind are crushed to extract juice. This juice is a home remedy for piles.

Powdered tamarind seeds mixed with turmeric paste are used as a treatment for inflammation and sores in many Asian cultures. Tamarind juice is used as a traditional medicine for jaundice sufferers.

This fruit has powerful antioxidants that protect against harmful chemicals that cause cancer. fruit are good for bone health. Paste of tamarind seeds acts as an antidote and is used as a home remedy for insect bites and stings. Paste made from powdered tamarind seeds or paste made from bark powder is used as a home remedy for healing open sores. Individuals that suffer from thyroid disorders should include tamarind in their diet. Fruit extract of tamarind is consumed because it acts as a natural blood purifier(Figure 13.3)

This fruit is essential for healthy muscle function beneficial for good health. A cup of tamarind water or two tablespoons of thick tamarind paste is an excellent home remedy for an individual that has suffered mild sunstroke. Paste of tamarind is excellent for diabetics. It is used as traditional medicine along with jamun and herbs for diabetes control. Pulp of this fruit helps lower glucose levels that tend to rise after meals. Tamarind juice is a natural antiseptic that is used with herbal ingredients to cure Scurvy that arises from deficiency in Vitamin C. Paste or powder of tamarind seeds is used as a home remedy for relief from styes and burns. Extracts of leaves and flowers are used as a treatment for hemorrhoids, dysentery and erysipelas. Flower, fruit and leaf juice of tamarind are used as an antiseptic and vermifuge. Juice of tamarind is given to individuals with a severe hangover resulting from alcohol consumption. Intoxication from various substances can be treated because tamarind plays the role of antidote. Paste made from leaves of tamarind can be applied on areas of the body to reduce swelling of joints. This paste provides much-needed relief for arthritis sufferers that are not able to move their hands, legs or shoulders because of joint pain. Paste of seeds can also be used to treat boils and sprains. Five grams of tamarind pulp consumed twice a day for 3 to 4 weeks can be used to delay fluorosis progression. Root and bark infusions of tamarind are used as an alternative treatment for leprosy. In many Asian countries tamarind juice is used to clean brass, copper and other metals. This is done to restore shine on metal. The juice when scrubbed on metal gets rid of grime, tarnish and formation of green patina. Tamarind wood is hard and extremely durable. Natural bright-red tamarind wood is used to make furniture and floor panels with natural hues. Tamarind trees are used in landscape designs and ornamental gardening in several Asian countries (Soong and Barlow 2004; Hooda and Jood 2005; Hough *et al.,* 2006; Siddhuraju 2007; Oluseyi and Temitayo 2015).

13.3 Nutritional Properties

Tamarind nutrition facts and properties state that this herb is rich in various minerals, volatile chemical compounds, vitamins as well as dietary fiber. One of the health benefits of tamarind is that the dietary fiber obtained from this herb is beneficial to naturally treat digestive problems such as loose motions and also for the prevention of constipation as the fiber tends to add bulk and which is required for healthy bowel movements. This fiber also helps in binding together any toxins that may be present in the food and thus protects the mucus membrane of the colon from chemicals that cause cancer. The nutritional value of the tamarind herb is also tremendous as the dietary fiber obtained from tamarind bind the bile salts that are produced from cholesterol and thus minimize their re-absorption into the colon. The tamarind herb is therefore beneficial in lowering the levels of bad cholesterol in the body. The health benefits of tamarind also include it being effective on regulating the blood pressure and the heart rate on account of the presence of potassium in tamarind. Tamarind is also rich in other minerals such as calcium, iron, copper, zinc and magnesium. The iron present in tamarind helps in the production of red blood cells. Another one of the health benefits of tamarind is that it is used as a digestive, laxative and also as a natural treatment for bile disorders and biliousness (Stratil 2006; Soultani 2014). The medicinal uses of tamarind also include preparing a poultice of the pulp, flowers and leaves and then applying this poultice on the joints in case of swelling.

Health benefits and therapeutic uses of tamarind also include the juice of the tamarind herb flowers being used in the natural treatment of bleeding piles. For those who are prone to suffering from dry eye syndrome, one of the medicinal uses of tamarind is to use eye drops that are prepared from the extracts of the tamarind seeds. Tamarind health benefits also include it being used to naturally treat medical ailments such as constipation and fever. In fact the tamarind herb is also proven to be very effective in dealing with sunstroke, intoxication and *Datura* poisoning. One of the tamarind leaves uses is that it is used as a poultice for treating boils, swollen joints and sprains. On account of the numerous health benefits of tamarind, it is also used in various lotions and extracts that are used in the treatment of jaundice.

The bark of the tamarind tree is also used to treat open sores whereas the seeds of this plant are used in the form of a paste to draw out boils (Sudha *et al.,* 2007).

13.4 Prominent Health Benefits

Charak Samhita mentioned use of leaves of tamarind and turmeric powder mixed with alum tied in a cloth bag, when applied with water on eyes relieves from conjuctivitis and eye problems (Saxena 2023). The health benefits of tamarind have been well-studied and they include its ability to reduce inflammation throughout the body, improve vision, boost respiratory health, heal skin conditions, and improve the digestive system. Tamarind also relieves pain, boosts the immune system, reduces fever, lowers cholesterol to improve cardiovascular health, treats piles, prevents cancer, and protects against parasites and worms. The tamarind tree is indigenous to tropical regions of Africa, particularly the Sudan, but its cultivation has spread to almost all tropical areas of the world since it is such a useful and desirable source of nutrition and unique flavors. It has been cultivated for thousands of years and likely made its way to Asia about 5,000 years ago. It wasn't until 500 years ago that it made its way to the Americas via Spanish explorers. South America and Mexico remain the largest consumers and producers of tamarind in the world (Sudjaroen *et al.,* 2005).

It can be consumed as a raw fruit, added to desserts once it is fully ripe or can be dried and ground into a spice. It is often used in jams and sauces and is even dried and processed into candies in some parts of the world. It is a normal ingredient in soups and dishes throughout Asia and South America, owing to the health benefits this little seed pod contains. Let's find out more about what components make tamarind such a powerful and essential part of your diet! Tamarind is a very valuable commodity in the world because of its nutritional components. These include a significant level of vitamin C, E, and B, calcium, iron, phosphorous, potassium, manganese, and dietary fiber. There are also a number of organic compounds that make tamarind a powerful antioxidant and anti-inflammatory agent (Sun *et al.,* 1998).

Tamarind has long been considered a natural laxative and its dietary fiber content is a major reason for this property. Eating tamarind as a fruit or as a spice can increase the efficiency of your digestive system, while the fiber can bulk up your stool, making it move through the smooth muscles of the intestinal tract easily. Tamarind is also a bilious substance, meaning that it stimulates the activity of bile, which can help dissolve food faster, and the fiber stimulates gastric juices to speed up digestion. All of this together means that things run through your digestive tract faster, making it a powerful laxative if you are suffering from chronic constipation. Oddly enough, the fiber can also reduce loose stools, and studies have shown tamarind to be effective against chronic diarrhea as well! Studies on tamarind have shown it to be effective in reducing blood pressure and cholesterol. The fiber content in tamarind certainly has something to do with the reduction in cholesterol, since it is known to scrap excess LDL cholesterol from the veins and arteries. The potassium in tamarind may be responsible for a reduction in blood pressure since it is known as a vasodilator that reduces the stress on the cardiovascular system. The impressive level of vitamin C in tamarind also have something to do with it as well. Vitamin C is an antioxidant that can reduce the impact of free radicals, those pesky by-products of cellular metabolism that have been linked to heart diseases and a number of other health conditions (Tril *et al.,* 2014).

Tamarind is a very good source of iron and a single serving can provide more than 10 per cent of your daily requirement. A healthy supply of iron in the body guarantees proper red blood cell count in the body, which can ensure appropriate oxygenation of different muscles and organs to function properly. Also, iron deficiency results in anemia, which is characterized by weakness, fatigue, headaches, cognitive disorders, and stomach issues. So, eat plenty of tamarinds to keep anemia at bay! One of the most significant vitamins in tamarind is the B complex. Thiamine, an important part of the vitamin B family, is found in high quantities within tamarind. Thiamine is responsible for improving nerve function, as well as muscle development, which can help you remain active, maintain your reflexes, and stay strong. One of the unique compounds that can be extracted from tamarind or gained as a benefit from it when used as a spice is called hydroxycitric acid (HCA). HCA is connected to weight loss because it has been shown to inhibit an enzyme in the body that specifically helps to store fat. Furthermore, tamarind has been known to suppress the appetite by increasing the serotonin neurotransmitter. Research is still ongoing in these respective areas, but it shows promising signs as a weight loss supplement (Christine 2011).

Along with its ability to stop weight gain, inhibiting the enzyme alpha-amylase mainly stops carbohydrates from being absorbed. A carbohydrate-heavy diet can increase the chances of uncontrolled glucose and insulin levels, which is the biggest problem for people suffering from diabetes. Tamarind can help to monitor and control these fluctuations.

The essential oil of tamarind has been connected to a number of anti-inflammatory abilities, including the reduction of joint pain and inflammation, arthritis, rheumatic conditions, and gout. It also reduces eye irritation. One of the most common forms of this is conjunctivitis, also known as pink eye. Tamarind has shown a definite soothing and anti-inflammatory ability and is, therefore, used in many herbal remedies for inflammation. High levels of vitamin C, as well as other antioxidant effects in the essential oils, make tamarind a wonderful way to boost your immune system and ensure long-term health from microbial and fungal infections. It also reduces the occurrence of parasites in the body due to its antiseptic and antimicrobial effects. It has specifically been linked to eliminating stomach worms in children in tropical areas where tamarind in cultivated.

13.5 Industrial Importance

Currently, industries are interested in developing value]added products from the waste]by products generated by the food and agricultural processing industries. The waste products including seeds, peels, stalks, stems, and leaves of plants contain substantial amount of phenolics and thus can be used as cheap sources of natural antioxidants for pharmaceutical, cosmetic, and food application. Fruits and vegetable waste products including seeds have been reported to have higher content of bioactive phytochemicals than the edible portions (Sheilla *et al.*, 2016). Fruit seeds contain a variety of biologically active phytochemical compounds, especially phenolic constituents, flavonoids, anthocyaninins, vitamin C, and carotenoids. These phytochemicals positively influence human health and indicate high antioxidant activity hence, it is considered crucial to increase the antioxidant intake in the human diet and one way of achieving this is can be through enriching food products with seeds which are rich in phytochemicals.

Tamarind seed is a by]product in the tamarind pulp industry. Recently, a large amount of the seed waste is discarded from the tamarind industry. It is a rich source of phytochemicals which consists of phenolic antioxidants, such as 2]hydroxy]32 ,42]dihydoxyacetophenone, methyl 3, 4]dihydroxybenzoate, 3,4]dihydroxyphenyl acetate and epicatechin. Tamarind seed extracts exhibit antioxidant potential by reducing lipid peroxidation *in vitro* (Tsuda *et al.*, 1994; Brand]Williams *et al.*, 1995) and anti]microbial activity. Tamarind seed, therefore, has the potential of providing low cost nutritional and nutraceutical value.

In this study, tamarind seed powder (TSP) was evaluated as a source of antioxidants for inclusion in two commonly consumed products; cookies and mango juice. Cookies and mango juice are regularly consumed by nearly all age groups in developing countries. Cookies are popular compared to other processed foods because of their low cost, diverse taste, myriad shapes, and long shelf life reported that cookies were selected by consumers as a good vehicle for antioxidants. Cookies also contain fat which possess' sensory characteristics that are ideal to mask the astringent flavor and taste associated with tamarind seed which is undesirable for most consumers. Mango (*Mangifera indica*) is the most vital and widely cultivated among the tropical and subtropical fruits. Thus, mango juice was selected because of the abundance, seasonal availability, popularity, and distinctive flavor and taste. Mangoes also have a strong flavor and substantial levels of pectin which can mask the astringency of the tamarind seed without affecting antioxidant activity (Sheilla *et al.*, 2016).

While tamarind seeds are known to contain substantial levels of bioactive phyto-compounds, it was not clear if they could be used to enhance nutraceutical properties of widely consumed processed foods while retaining product acceptability. This study was therefore undertaken to assess the effect of tamarind seed powder on the physico-chemical properties, phytonutrient content, antioxidant, and sensory properties of the enriched selected food products (Christine *et al.*, 2011; Tril *et al.*, 2014).

13.6 Other Uses

Besides the medicinal uses, tamarind is also used as an emulsifying agent in different pharmaceutical products such as decoctions and syrups. Tamarind also has a variety of culinary uses in jams, chutneys, and pickles. In fact tamarind is also an ingredient in the preparation of barbecue and Worcestershire sauce and is used in many meat preparations. Similarly the leaves of the tamarind plant are used to feed silkworms and as fodder for livestock. The pulp of the fruit of tamarind is also used to coagulate rubber latex and as a dyeing fixative. The tamarind pulp when mixed with butter and some other key ingredients results in the preparation of an ointment which is then used on pets and livestock to get rid of parasites such as fleas.

Epidemiological studies have consistently demonstrated that consumption of plant derived foods rich in bioactive phytochemicals has a protective effect against oxidation stress. Oxidative stress is strongly associated with mutagenesis, carcinogenesis, aging, and atherosclerosis in humans. The bioactive phytochemical compounds thus decrease the risk of chronic diseases,cardiovascular disease, cancer, and degenerative diseases of the aging.Jam is one of the important fruit products with considerable economic and dietary importance in Sudan. The pectin used in jam making is imported as there is no pectin industry yet in Sudan [so more attention should be directed towards utilization of pectin obtained from local substitutes. From this point of view, the agricultural by-product like tamarind seed kernel can be used as a cheap source for functional food to increase the added value of tamarind seeds. Most researches, carried out in Sudan on tamarind, were on the pulp and for medicinal purposes the seed was found to be rich in protein, sugar and potassium, thus can suffice the human needs and was consumed as famine food (Christine *et al.,* 2011).

Research concerning the chemical, technological and usage of tamarind kernel in Sudan are scarce of this study was to extract polysaccharides from tamarind kernel and study its physicochemical and functional properties compared to commercial pectin (Tsuda *et al.,* 2004; Soong and Barlow 2004). Seeds are flattened more or less rounded. Seed powder is pinkish pale brown in colour. Palisade cells of outer testa can be seen in seed powder. Tannin as gallic acid epiactechinm flavonoidsm phenolsm sterol, proteins and polysaccharides present (Chauhan and Kevalia 2014). seed powder can be used to cure diabetes.

13.7 Conclusion

Tamarind is a very well known tree having many uses such as culinary, medicinal, ornamental uses and as an economic dye. Tamarind is included in daily diet as it has many health benefits that are sure to keep you healthy or improve you overall health. Tamarind is a delicious, sweet, and sour fruit that has a wide variety of uses, both for medicinal and culinary purposes. Tamarind is used as an Ayurvedic medicine for gastric problem, digestion problems and cardio protective activity. Tamarind can be used as a mild laxative. The pulp which comes from the pods of the tamarind tree is a gentle laxative that improves general sluggishness of the bowels. One to two tablespoons of the pulp in evening can improve bowel movements. Tamarind is very useful to treat bilious disorders. Being acidic it excites the bile and other juices in the body. Tamarind leaves are used in herbal tea for reducing malaria fever.

REFERENCES

Bhattacharya S, Bal S, Mukherjee R. K., and Bhattacharya S. (1994) Functional and nutritional properties of tamarind (*Tamarindus indica*) kernel protein. Food Chem. 49: 1–9.

Brand]Williams W, Cuvelier M. E., and Berset C. (1995) Use of a free radical method to evaluate antioxidant activity. Food Sci. Technol. – LEB 28: 25–30.

Chauhan M.G. and Kevalia J.K. (2014) Microscopic profile of drugs used in Indian systems of medicine. Vol. 3 Seed drugs Part II, Gujarat Ayurvedic University, Jamnagar. 299pp.

Christine Van der Stege, Sarah Prehsler, Anna Hartl and Christian Reinhard Vogl (2011) Tamarind (*Tamarindus indica* L.) in the traditional West African diet: not just a famine food 2011; 66(3): 171-185https: //doi.org/10.1051/fruits/2011025

Haghju S, and Almasi H. (2015) Antioxidant, antibacterial and therapeutic properties of some endemic medicinal plants of Iran: a review. Adv. Plants Agric. Res. 2: 00053.

Hooda S, and Jood S. (2005) Organoleptic and nutritional evaluation of wheat biscuits supplemented with untreated and treated fenugreek flour. Food Chem. 90: 427–435.

Hough G, Wakeling I, Mucci A, Chambers E., Gallardo I M, and Alves L R (2006) Number of consumers necessary for sensory acceptability tests. Food Qual. Pref. 17: 522–526.

Oluseyi E. O., and Temitayo O. M. (2015) Chemical and functional properties of fermented, roasted and germinated tamarind (*Tamarindus indica*) seed flours. J. Nutr. Food Sci. 45: 97–111.

Robinson T (2008) William Roxburgh: The founding father of Indian Botany. Pub. By Phillimore and Company England. Pp 286.

Saxena O.P. (Nidar) (2023) Charak Samhita in Hindi by DPB Pub. Delhi pp.265

Sheilla N. J. Muyonga, H. and Mukisa I. M. (2016) Effect of tamarind (*Tamarindus indica* L.) seed on antioxidant activity, phytocompounds, physicochemical characteristics, and sensory acceptability of enriched cookies and mango juice. Food Sci Nutr. 4(4): 494– 507. doi: 10.1002/fsn3.3111.

Siddhuraju P. (2007) Antioxidant activity of polyphenolic compounds extracted from defatted raw and dry heated *Tamarindus indica* seed coat. Food Sci. Technol. – LEB 40: 982–990.

Soong Y. Y., and Barlow P. J. (2004) Antioxidant activity and phenolic content of selected fruit seeds. Food Chem. 88: 411–417.

Soultani G., Evageliou V., Koutelidakis A. E., Kapsokefalou M., and Komaitis M. (2014) The effect of pectin and other constituents on the antioxidant activity of tea. Food Hydrocoll. 35: 727–732.

Stratil P., Klejdus B., and Kuban V. (2006) Determination of total content of phenolic compounds and their antioxidant activity in vegetables evaluation of spectrophotometric methods. J. Agr. Food Chem. 54: 607–616.

Sudha M., Baskaran V., and Leelavathi K. (2007) Apple pomace as a source of dietary fiber and polyphenols and its effect on the rheological characteristics and cake making. Food Chem. 104: 686–692.

Sudjaroen Y., Haubner R., Würtele G., Hull W. E., Erben G., Spiegelhalder B. (2005) Isolation and structure elucidation of phenolic antioxidants from Tamarind (*Tamarindus indica* L.) seeds and pericarp. Food Chem. Toxicol. 43: 1673–1682.

Sun B., Ricardo]da]Silva J. M., and Spranger I. (1998) Critical factors of vanillin assay for catechins and proanthocyanidins. J. Agr. Food Chem. 46: 4267–4274.

Tril U., Fernández]López J., Álvarez J. Á. P., and Viuda]Martos M. (2014) Chemical, physicochemical, technological, antibacterial and antioxidant properties of rich]fibre powder extract obtained from tamarind (*Tamarindus indica* L.). Ind. Crop Prod. 55: 155–162.

Tsuda T, Watanabe M, Ohshima K, Yamamoto A, Kawakishi S and Osawa T (1994) Antioxidative components isolated from the seed of tamarind (*Tamarindus indica* L.), J. Agric. Food Chem., 42(12): 2671–4.

Chapter 14

Nutritious Fruit in Indian Deserts: Cactus pear [*Opuntia ficus-indica* (L.) Mill.]

Prickly pear (Opuntia ficus-indica (L.) Mill) is a multipurpose crop, not only providing food and feed but also being a source of health-promoting substances that can be used as natural plant-based drugs to prevent diseases. The beneficial properties of this plant, already known by ancient civilizations, are generating a growing interest among the scientific. According to FAO (2013), the prickly pear is widely distributed around the world and the plant is found either cultivated or wild. Concerning production data, this information is scarce, but Mexico cultivated around 50,000 ha, Brazil 40,000 ha, and Peru 35,000 ha. In North Africa countries, in Tunisia, approximately 500,000 ha are cultivated (FAO 2013). The prickly pear fruit is highly appreciated due to its flavor, which is sweet and slightly acid. From prickly pear, it is possible to prepare various traditional foods such as jams, gels, jellies, juices, and nectar, dehydrated fruits, juice concentrates, syrups, liquors, wines, vinegar, canned fruit, and frozen fruit and pulp.

***Keywords**: Nutritious fruit, Indian deserts, Cactus pear, Opuntia ficus-indica, Cochineal dye.*

14.1 Introduction

Opuntia ficus-indica (L.) Mill. is commonly known as prickly pear, cactus pear, Barbary fig, cactus fruit, Indian fig, or Nopal cactus (Figures 14.1 and 14.2) and belongs to the dicotyledonous angiosperm Cactaceae family. It originates from the arid and semiarid regions of Mexico since it has special adaptive mechanisms and high biomass production capacity, which allows it to grow in adverse conditions, such as high temperatures and nutritionally poor soils subject to erosion. It is a shrub or arborea plant that can reach up to 5 m in height. The root system is fleshy, very branched, and develops horizontally and can spread laterally up to 10 to 15 m from the base of the plant. In general, fruits consist of a thick peel, with small prickles, and the pulp contains a consistent number of seeds, which most of the time are not discarded from the pulp and are also eaten. Although, some variations in shape, size, and color are frequently observed. Prickly pear cacti are native to the Western Hemisphere. Several are cultivated, especially the Indian fig (*O. ficus-indica*), which is an important food for many peoples in tropical and subtropical countries. Some prickly pear species are cultivated as ornamental and are valued for their large flowers. They are easily propagated from stem segments. Two of the best-known species, Engelmann prickly pear (*O. engelmannii*) and the beaver tail cactus (*O. basilaris*), commonly occur in the southwestern United States. Some species have become invasive in regions outside their native ranges. When prickly pears were first introduced to Australia and southern Africa by early explorers, they prospered, and, having left behind their natural parasites and competitors, they eventually became pests.

Figure 14.1. Plants of Prickly Pear Showing Attached Pink Fruits.

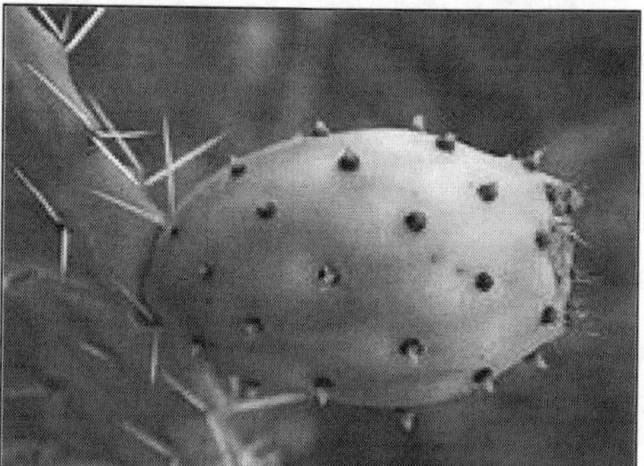

Figure 14.2. Yellow Colored Fruit.

The Indian fig is bushy to treelike, growing to a height of 5.5 m. It bears large yellow flowers, 7.5 to 10 cm across, followed by white, yellow, or reddish purple fruits. It is widely grown in warmer areas for the fruit and edible paddles and as a forage crop. The hard seeds are used to produce an oil. Because of their high water content, the stems, especially of spineless varieties, are used as emergency stock feed during drought.

14.2 Habit of *Opuntia ficus-indica*

Opuntia ficus-indica is a succulent shrub or tree ranging 1.5-3 m but it can grow up to 5 m. It develops a sturdy trunk with age. The branches (Phylloclads) are flattened, grey to grey-green. The branches are 30-60 cm long and 6-15 cm wide. Its younger branches arch upwards. The leaves of *O. ficus-indica* are minute or spiny and are shed early. The flowers of *O. ficus-indica* are conspicuous, bright yellow or orange, red in colour. It bears succulent edible berries which are reddish when ripe, about 8 cm long and covered with clusters of minute spines. Reproduces vegetatively and through seeds.

The plant flowers in three distinct colours: white, yellow, and red. The flowers first appear in early May through the early summer in the Northern Hemisphere, and the fruits ripen from August through October. The fruits are typically eaten, minus the thick outer skin, after chilling in a refrigerator for a few hours. They have a taste similar to sweet watermelon. The bright red/purple or white/yellowish flesh contains many tiny hard seeds that are usually swallowed but should be avoided by those who have problems digesting seeds

14.3 Origin of Sweet Prickly Pear

Opuntia ficus-indica (L.) Mill. is a crop species (Figure 14.1) that figures prominently in the modern folklore of ethnobotany. The narrative of Luther Burbank's development and promotion of certain clones of *O. ficus-indica* in the early 20th century is vividly chronicled (with sharp wit) by Benson (1982). Opuntioid cacti are recognized as ideal crops for arid regimes (Felger 1979; Russell and Felker 1987) because *O. ficus-indica* is extremely efficient at converting water into biomass (Kluge and Ting 1978). *O. ficus-indica*, one of several long-domesticated cactus species (Casas *et al.,* 1997; Casas and Barbera 2002), is the most widespread and economically important of these cactus crops (Nobel 2002; Nobel *et al.,* 2002), as important as corn and tequila agave in the agricultural economy of modern Mexico (Barrientos 1966). The facile introgression of *Opuntia* is very well documented; this genus is among the most interspecifically promiscuous plants, perhaps rivaled only by *Quercus* L. in this regard (Benson and Walkington 1965; Grant and Grant 1971b, 1979; Felger 1979; Baker and Pinkava 1987, 1999; Griffith 2001a, b, 2003; Pinkava 2002). The relative ease of vegetative propagation of *Opuntia* is demonstrated by its occasional clonal dominance of certain areas (Grant and Grant 1971a; Mandujano *et al.,* 1996). This aspect of *Opuntia* marks it as a noxious weed in some places (Dodd 1936; Nobel 1994). This ease of clonal propagation was probably not lost on the very early human population of the New World. Evidence exists for the use of *Opuntia* as human food at least 9000 yr before the present (Kiesling 1998) or even as early as 12 000 yr ago (Callen 1967), probably before cultivation (Ostolaza 1994).

Although most recent authors concede that this species is not native to the Old World, the geographic and evolutionary origins of *O. ficus-indica* remain clouded in obscurity (Britton and Rose 1919; Bravo-Hollis 1978; Benson 1982; Anderson 2001). This mystery arises partly from the widespread cultivated distribution of the plants, both within and away from the New World (Anderson 2001), but also from the phenotypic variability and artificial selection of the plants (Benson 1982). The specific epithet (Miller 1768) may reflect the 16th -century European opinion that the plants are native to the West Indies (Barbera *et al.,* 1992; Casas and Barbera 2002). The name "tuna" is Caribbean in origin (Bravo-Hollis 1978;

Kiesling 1998). Other early work proposed that *O. ficus-indica* was a spineless cultivar derived from *O. megacantha*, a species of central Mexico (Griffiths 1914). A recent, important paper explored this relationship with amplified fragment length polymorphism (AFLP) data and reported that *O. ficus-indica* had the closest affinity with *O. megacantha* (Labra *et al.*, 2003), corroborating the Griffiths (1914) hypothesis. Other authors, however, consider *O. megacantha* also to be a cultivated taxon (Benson 1982) or a name applied to multiple ruderal reversions to spineness from escaped, cultivated *O. ficus-indica* (Kiesling 1998) and treat *O. megacantha* as a later synonym (Benson 1982). Uphof (1968) and Zeven and Zhukovsky (1975 p. 164) place the origin of this cactus within "Mexico" or the "Central American and Mexican centre" of domestication, but without reference to a specific region. Some recent cytological work proposes that octoploid, cultivated *O. ficus-indica* is derived from a diploid Mexican (but unnamed) progenitor (Kiesling 1998). Benson (1982) states that it is native to mild tropical upland habitat (not deserts), but does not state where. Given the conflicting, sometimes confusing, and often unclear hypotheses as to the origins of this cactus crop, a wide molecular phylogenetic analysis has potential to elucidate the biogeography of this species. The current study addresses the biogeographic and evolutionary origins of *O. ficus-indica* through modern molecular means. To this end, DNA sequence data have been analyzed from a number of accessions of *O. ficus-indica* representing plantings throughout the world.

14.4 Different Species of *Opuntia*

Opuntia ficus-indica, the Indian fig opuntia, or prickly pear, is a species of cactus that has long been a domesticated crop plant grown in agricultural economies throughout arid and semiarid parts of the world. *O. ficus-indica* is the most widespread and most commercially important cactus. It is grown primarily as a fruit crop, and also for the vegetable nopales and other uses. Cacti are good crops for dry areas because they efficiently convert water into biomass. *O. ficus-indica*, as the most widespread of the long-domesticated cactuses, is as economically important as maize and blue agave in Mexico. *O. ficus-indica* is polyploid, hermaphroditic, and autogamous. As *Opuntia* species grow in semiarid environments, the main limiting factor in their environment is water. They have developed several adaptations to dry conditions, notably succulence

Opuntia species hybridize easily, but the wild origin of *O. ficus-indica* is likely to have been in central Mexico, where its closest genetic relatives are found *Opuntia ficus-indica* is invasive in parts of Kenya and Tanzania (A.B.R. Witt Pers.). It is present in Uganda. *Opuntia ficus-indica* is very similar to *Opuntia monacantha* (drooping prickly pear) and *Opuntia stricta* (common prickly pear). These species can be distinguished by the following differences:

- ☆ ***O. monacantha*** is a relatively tall shrub or tree-like plant (usually 2-5 m tall) with flattened and elongated (oblong or obovate - egg-shaped in outline but with the narrower end at the base) cladodes. These cladodes are hairless and have one or two large spines on most of the areoles on their surfaces. It has yellow flowers and reddish-purple fruit, and some of its cladodes droop towards the ground during fruiting.

- ✰ ***O. ficus-indica*** is a relatively tall shrubby or tree-like plant (usually 1.5-3 m tall) with very large flattened and elongated (oblong, elliptic or obovate) cladodes. These cladodes are hairless and do not have any spines on the areoles on their surfaces. It has yellow flowers and reddish coloured fruit.
- ✰ ***O. stricta*** is a low-growing plant (usually 50-100 cm tall) with relatively large flattened and elongated (elliptic or obovate) cladodes. These cladodes are hairless and generally do not have any spines (sometimes one or two large spines are present) on the areoles on their surfaces. The flowers are bright yellow and the fruit reddish-purple.

14.5 Different Uses

Opuntia ficus-indica is cultivated as an ornamental or as live hedge. It can be used as a medicinal plant, and the fruit is edible. In some parts of Kenya, syrup is made from the fruit to consume in a variety of ways, such as use as a "mixer" in fancy cocktails served at tourist lodges. Prickly pear cactus is a plant native to Mexico and its fruit and stems are commonly eaten and used as medicine in Mexican cultures. Prickly pear might lower blood sugar and cholesterol by reducing absorption of food. People most commonly use prickly pear cactus for diabetes. It is also used for enlarged prostrate and hangover, high cholesterol, and many other purposes, but there are no clear evidence to support these uses.

14.6 Phytochemicals Present in *Opuntia* and their Effects

O. ficus-indica for human and animal consumption is valuable for its water content in an arid environment, containing about 85 per cent water as a water source for wildlife. The seeds contain 3–10 per cent protein and 6–13 per cent of fatty acids, mainly linoleic acid. However, the seeds contained in the fruits can be unpleasant to chew because of their hardness and can lead to constipation.28 For this reason, some agronomic studies in Italy and Mexico have focused on decreasing the seed content of *O. ficus-indica* fruits.

As the fruit contains 25–30 mg per 100g vitamin C, it was once used to mitigate scurvy. *Opuntia* contains selenium. The red color of the fruit and juice is due to betalains, (betanin and indicaxanthin). The plant also contains flavonoids, such as quercetin, isorhamnetin, and kaempferol.

14.6.1 Antioxidant Effects

The antioxidant effects of the prickly pear may be due to one or several components but also due to their synergic effect. Butera *et al.* (2002) have shown that prickly pear extracts have a higher antioxidant activity than other fruits like pears, apples, grapes, oranges, grapefruits, and tomato, according to an ABTS assay which was consistent with results from ORAC assays previously presented by

Wang *et al.* (2000). These results may be due to several compounds which can be found in prickly pear fruits, such as quercetin, kaempferol, isorhamnetin, vitamin C, and carotenoids (Kuti, 2004). When compared to other species of Opuntia sp., the prickly pear has the lowest polyphenols and flavonoids content but higher antioxidant activity.

Flavonoids are especially efficient as antioxidants due to their inhibitory actions in prooxidative effects on DNA, proteins, and lipids, as well as in preventing the generation of stable radicals (Azqueta and Collins, 2016). An experimental study revealed benefits from the administration of prickly pear cladodes extract in oxidative lithium-induced damage. The results have shown that supplementation with the extract increased hepatic catalase, superoxide dismutase, and gluthatione peroxydase activities (Saad *et al.,* 2017). The authors suggested that this effect is due to the presence of flavonoids, like rutin, isorhamnetin, quercetin, kaempferol, as well as six phenolic acids (gallic acid, catechin, caffeic acid, epicatechin, vanillic acid, and coumarin) determined by HPLC. More recently, Smida *et al.* (2017) have also shown the protective effect of the extract against the organophosphate pesticide chlorpyrifos. There is a clear lack of clinical data on this subject, but Lopez-Romero and colleagues have found that the prickly pear extract improves antioxidant status in healthy adults and type 2 diabetics (Lo´pez-Romero *et al.,* 2014). Results from Tesoriere *et al.,* have shown that supplementation with prickly pear decreases lipid peroxidation and improves antioxidant status in healthy individuals (Tesoriere *et al.,* 2004). Multiple mechanisms can be involved in the antioxidant actions of prickly pear compounds. Butera *et al.,* have shown a radical scavenging ability of prickly pear extracts which have proven to be especially efficient in reducing lipid peroxidation (Butera *et al.,* 2002). This was in part also confirmed by another study where results included reduced levels of lipid oxidative damage markers (*e.g.,* isoprostanes and malondialdehyde) as well as an improved low-density lipoproteins oxidative status (Tesoriere *et al.,* 2004). In the Brahimi *et al* research, supplementation with the prickly pear promoted a normalization of innate antioxidant enzymes, mainly catalase and superoxide dismutase, which was in part confirmed by a recent study (Smida *et al.,* 2017). This subject deserves further debate in future research, mainly regarding the clinical studies, considering that there is known data on in vitro antioxidant activity and the actions of several compounds in antioxidant systems, but there is a lack of information on the behavior of the same substances in the human body in different conditions.

14.6.2 Neuroprotective Effects

The brain, due to its complexity, remains a quite unavailable place or at least a place where the manipulation through correction surgery to remove tumors can probably cause irreversible collateral damage. The damage can include amnesia which can affect the most recent or past memory but can also be a cause of paralysis. Considering this, several components have been proposed as possible neuroprotective elements, which means that their administration can prevent or even revert some of the damage caused. The mechanisms of action can be diverse because some protect against neuronal cell degeneration, while others protect neurons from dopamine and retina. The clear majority have antioxidant effects and mostly originate from plant extracts. These agents can be a valuable resource because the brain is very susceptible to damage caused by oxidative processes due to its elevated content of unsaturated fatty acids and transition metals like iron, which elevates the peroxidation risk. In addition to these features, the antioxidant mechanisms are not very efficient (Gandhi and Abramov 2012) and the active free radicals have been previously implicated in the pathogenesis of several neurologic diseases including brain trauma, ischemia, and neurodegenerative diseases. Thus, with prickly pear being a very good source of bioactive compounds, prickly pear can act as a protectant in the prevention and treatment of neurologic diseases. Dok-Go *et al.* identified three flavonoids (quercetin, dihydroquercetin, and quercetin-3-methyl-ether) isolated from prickly pear fruits and stems as the main antioxidant compounds (Dok-Go *et al.*, 2003). These inhibited the neuronal damage induced by hydrogen peroxide or xanthine oxidase in rat cortical cells. When compared, the quercetin-3-methyl-ether was shown to be six to nine times more potent than quercetin, and 13 25 times more potent than dihydroquercetin in terms of inhibiting H_2O_2 or xanthine oxidase damage, which can be due to its hydrophobic nature. In addition to its neuroprotective action as a scavenger, the mechanism that can better reflect its action is the ability to inhibit lipid peroxidation.

14.7 Culinary Uses

Most culinary references to the "prickly pear" refer to this species. The Spanish name *tuna* is also used for the fruit of this cactus and for *Opuntia* in general; according to Alexander von Humboldt, it was a word of Taino origin taken into the Spanish language around 1500. Common English names for the plant and its fruit are *Indian fig opuntia, Barbary fig, cactus pear, prickly pear,* and *spineless cactus,* among many others. In Mexican Spanish, the plant is called nopal, a name that may be used in American English as culinary terms. Peninsular Spanish mostly uses *higo chumbo* for the fruit and *chumbera* for the plant.

The fruits are commercialized in many parts of the world, eaten raw, and have one of the highest concentrations of vitamin C of any fruit. The young "leaves" (actually cladodes, which technically are stems) are cooked and eaten as a vegetable known as *nopalitos.* They are sliced into strips, skinned or unskinned, and fried with eggs and jalapeños, served as a breakfast treat. They have a texture and flavor like green beans. The fruits or leaves can be boiled, used raw, or blended with fruit juice, cooked on a frying pan, used as a side dish with chicken, or added to tacos. Jams and jellies are produced from the fruit, which resemble strawberries and figs in color and flavor. Mexicans may use *Opuntia* fruit to make an alcoholic drink called *colonche.*

In Sicily, a prickly pear-flavored liqueur called *ficodi* is produced, flavored somewhat like a medicinal aperitif. In Malta, a liqueur called *bajtra* (the Maltese name for prickly pear) is made from this fruit, which grows wild in almost every field. On the island of Saint Helena, the prickly pear also gives its name to locally distilled liqueur, *Tungi Spirit.*

The fruit must be peeled carefully to remove the small spines on the outer skin before consumption. If the outer layer is not properly removed, glochids can be ingested, causing discomfort of the throat, lips, and tongue, as the small spines are easily lodged in the skin. Native Americans like the Tequesta would roll the fruit around in a suitable medium (*e.g.* grit) to "sand" off the glochids. Alternatively, rotating the fruit in the flame of a campfire or torch has been used to remove the glochids. Today, parthenocarpic (seedless) cultivars are also available. In Mexico, prickly pears are often used to make appetizers, soups, salads, entrees, vegetable dishes, breads, desserts, beverages, candy, jelly, and drinks.

O. ficus-indica has been introduced to Europe, and flourishes in areas with a suitable climate, such as the south of France and southern Italy: In Sicily, they are referred to as *fichi d'India* (Italian literal translation of Indian fig) or *ficurinia* (Sicilian language literal translation of Indian fig). In Sardinia, they are called *figumorisca* – Moorish figs).

The prickly pear also grows widely on the islands of Malta, where it is enjoyed by the Maltese as a typical summer fruit (known as *bajtar tax-xewk*, literally 'spiny figs'), as well as being used to make the popular liqueur known as *bajtra*. The prickly pear is so commonly found in the Maltese islands, it is often used as a dividing wall between many of Malta's characteristic terraced fields in place of the usual rubble walls.

In Morocco, Tunisia, Libya, Saudi Arabia, Jordan, and other parts of North Africa and the Middle East such as Israel, prickly pears of the yellow and orange varieties are grown by the side of farms, beside railway tracks and other otherwise noncultivable land. It is sold in summer by street vendors, and is considered a refreshing fruit for that season. In Libya, it is a popular summer fruit and called by the locals *Hindi*, which literally means *Indian.Tungi* is the local St. Helenian name for cactus pears. The plants (Indian fig opuntia) were originally brought to the island by the colonial ivory traders from East Africa in the 1850s. Cactus pear is being promoted and researched by ICARDA for India, Jordan, and Pakistan especially. It is an underappreciated crop in these countries and has undergone recent expansion in cultivated area. In some particularly promising areas of India and Pakistan it has given a 30 per cent increase in milk yield/hectare.

14.8 *Opuntia* as Feed

The cattle industry of the Southwest United States has begun to cultivate *O. ficus-indica*, both as a feed source for cattle and a boundary fence. Cattle are normally fed a spineless variety of the cactus. The cactus pads are low in dry matter and crude protein but useful as a supplement in drought conditions. In addition to the food value, the moisture content adequately eliminates watering the cattle during drought. Numerous wildlife species use the prickly pear for food.

In severe drought years, the spines of wild prickly pear are sometimes burned off for emergency feed.

14.9 In Dye Production

Dactylopius coccus is a scale insect from which cochineal dye is derived. *D. coccus* itself is native to tropical and subtropical South America and Mexico. This insect, a primarily sessile parasite, lives on cacti from the genus *Opuntia*, feeding on moisture and nutrients in the cactus sap. The insect produces carminic acid, which deters predation by other insects. The carminic acid can be extracted from the insect's body and eggs to make the red dye. Various species of *Opuntia* are involved as host plant for insects to produce different lac colours (Robinson 2008).

Cochineal is used primarily as a red food colouring and for cosmetics. The cochineal dye was used by the Aztec and Maya peoples of Central and North America, and by the Inca in South America. Produced almost exclusively in Oaxaca, Mexico, by indigenous producers, cochineal became Mexico's second-most valued export after silver. The dyestuff was consumed throughout Europe, and was so highly valued, its price was regularly quoted on the London and Amsterdam Commodity Exchanges. The biggest producers of cochineal are Peru, the Canary Islands, and Chile. Current health concerns over artificial food additives have renewed the popularity of cochineal dyes, and the increased demand is making cultivation for insect farming an attractive opportunity in other regions, such as in Mexico, where cochineal production had declined again owing to the numerous natural enemies of the scale insect. Apart from cochineal, the red dye betanin can be extracted from some *Opuntia* plants themselves (Savio 1989).

14.10 As a Source of "Vegan Leather"

The thick skin of nopal cactus can be harvested as an environmental-friendly leather replacement (Ozdemir 2020).

14.11 Environmental and other Impacts

Opuntia ficus-indica is considered a landscape transformer as it displaces indigenous species and dependent animals. The plant lowers the value of pastures since it cannot be browsed and it also impedes the movement of grazing animals. This species particularly threatens large scale cattle ranching in the arid and semi-arid lands. In the fall of 1961, Cuba had its troops plant an 13-kilometre barrier of *Opuntia* cactus along the northeastern section of the 28-km fence surrounding the Guantanamo Bay Naval Base to stop Cubans from escaping Cuba to take refuge in the United States. This was dubbed the "Cactus Curtain", an allusion to Europe's Iron Curtain and the Bamboo Curtain in East Asia.

O. ficus-indica has been included in the Global Invasive Species Database (GISD 2005). It has been listed as a noxious weed in South Africa. They serve no economic purpose and possess characteristics that are harmful to humans, animals or the environment and in most Australian states. Not listed as a noxious weed by the state or governments in Kenya, Tanzania and Uganda.

14.11.1 Soil Erosion Prevention

O. ficus-indica is planted in hedges to provide a cheap but effective erosion control in the Mediterranean basin. Under those hedges and adjacent areas, soil physical properties, nitrogen, and organic matter are considerably improved. The structural stability of the soil is enhanced, runoff and erosion are reduced, and water storage capacity and permeability are enhanced. Prickly pear plantations also positively impact the plant growth of other species by improving severe environmental conditions, which facilitate the colonization and development of herbaceous species. *O. ficus-indica* is being advantageously used in Tunisia and Algeria to slow and direct sand movement and enhance the restoration of vegetative cover, thus minimizing deterioration of built terraces with its deep and strong rooting system.

14.11.2 Used for fuel

Bioethanol can be produced from some *Opuntia* species (Rosaria 2017).

14.11.3 Used for bioplastic

The prickly pear cactus has been used for centuries both as a food source and a natural fence that keeps in livestock and marks the boundaries of family lands. They are resilient and often grow back following removal (Gyekye 2019). Nopal juice can be used to produce bioplastic.

14.11.4 Management of *Opuntia* Menace

The precise management measures adopted for any plant invasion will depend upon factors such as the terrain, the cost and availability of labour, the severity of the infestation and the presence of other invasive species. The best form of invasive species management is prevention. If prevention is no longer possible, it is best to treat the weed infestations when they are small to prevent them from establishing. Consistent follow-up work is required for sustainable management. Manual control can be effective when numbers of plants are very low but must be done carefully otherwise plant fragments will resprout into new plants, thus exacerbating the infestation. Both the spines and the glochids make this a difficult and uncomfortable process. Burning of uprooted plants will help minimize this risk if there is enough dry material to ensure that the material burns. Plants can be treated by herbicide stem injections. Controlled burns have been used to control *Opuntia species.*

The moth *Cactoblastis cactorum* feeds on this plant in its larval stages and can help control *Opuntia ficus-indica* as part of an integrated control programme. It was introduced to Tanzania but never established. It is likely that some cochineal species which feed on *Opuntia* species have been introduced to East Africa. This group offers some prospects for biological control of *Opuntia* species.

REFERENCES

Azqueta, A., Collins, A. (2016) Polyphenols and DNA damage: a mixed blessing. Nutrients 8 (12), 1- 21. https: //doi.org/10.3390/nu8120785.

Baker, M. A., and Pinkava D.J (1999) A new Arizona hybrid cholla, *Opuntia* 3 *campii* (Cactaceae). Cactus and Succulent Journal (U.S.) 71: 320–322.

Baker, M. A., and Pinkava D.J. (1987) A cytological and morphometric analysis of a triploid apomict, *Opuntia* 3 *kelvinensis* (subgenus *Cylindropuntia*, Cactaceae). Brittonia 39: 387– 401.

Barrientos, P. F. (1966) El nopal y su utilizacio´n en Mexico. La Sociedad Mexicana de Historia Natural, Mexico City, Mexico.

Benson, L. H. (1982) The cacti of the United States and Canada. Stanford University Press, Stanford, California, USA.

Benson, L. H., Walkington D.L. (1965) The southern California prickly pears–invasion, adulteration, and trial-by-fire. Annals of the Missouri Botanical Garden 52: 262–273.

Butera, D., *et al.* (2002) Antioxidant activities of sicilian prickly pear (*Opuntia ficus-indica*) fruit extracts and reducing properties of its betalains: betanin and indicaxanthin. J. Agric. Food Chem. 50 (23), 6895- 6901. https: //doi.org/10.1021/jf025696p

Callen, E. O. (1967) Analysis of the *Tehuacan coprolites. In* D. S. Byers [ed.], Prehistory of the Tehuacan valley, vol. 1, Environment and subsistence, 261–289. University of Texas Press, Austin, Texas, USA.

Casas, A. J, Barbera G. (2002) Meso-american domestication and diffusion. *In* P. S. Nobel [ed.], Cacti: biology and uses, 143–162. University of California, Berkeley, California, USA.

Casas, A., Caballero J., Mapes C, and Zarate S. (1997) Manejo de la vegetacio´n, domesticacio´n de plantas y origin de la agricultura en Mesoame´rica. Boletý´n de la Sociedad Bota´nica de Me´xico 61: 31–47.

Dodd, A. P. (1936) The control and eradication of prickly pear in Australia. Bulletin of Entomological Research 27: 503–517

Dok-Go, H., *et al.* (2003) Neuroprotective effects of antioxidative flavonoids, quercetin, (1)-dihydroquercetin and quercetin 3-methyl ether, isolated from Opuntia ficus-indica var. saboten. Brain Res. 965 (12): 130-136. Available from: https: //doi.org/10.1016/S0006- 8993(02)04150-1

FAO, 2013 Food and Agriculture Organization of the United Nations, 2013. Agro-industrial utilization of cactus pear. Rome. Available at: http: // www.fao.org/3/a0534e/a0534e.pdf

Felger, R. S. (1979) Ancient crops for the twenty-first century. *In* G. A. Ritchie [ed.], New agricultural crops. American Association for the Advancement of Science Symposium, vol. 38, 5–20. Westview Press, Boulder, Colorado, USA.

Felker P, Stintzing FC, Mussig E, Leitenberge M, Carle R, Vogt T, Bunch R. (2008) Colour inheritance in cactus pear (*Opuntia ficus-indica*) fruits. Annal Appl Biol 2008; 152: 307- 318

Gandhi, S., Abramov, A.Y. (2012) Mechanism of oxidative stress in neurodegeneration. Oxid. Med. Cell. Longev. 2012. https: //doi.org/10.1155/2012/428010.

Grant, V., and Grant K.A. (1971a.) Dynamics of clonal microspecies in cholla cactus. Evolution 25: 144–155.

Grant, V., and Grant K.A. (1971b) Natural hybridization between the cholla cactus species *Opuntia spinosior* and *Opuntia versicolor*. Proceedings of the National Academy of Sciences, USA 68: 1993–1995.

Griffith M.P., Rancho S R. (2004) The origins of an important cactus crop, *Opuntia ficus-indica* (Cactaceae): new molecular evidence American Journal of Botany 91(11): 1915–1921

Griffith, M. P. (2001a.) A new Chihuahuan Desert prickly pear, Opuntia x rooneyi. Cactus and Succulent Journal (U.S.) 73: 307–310.

Griffith, M. P. (2001b.) Experimental hybridization in northern Chihuahuan desert region Opuntia. Aliso 20: 37–42.

Griffith, M. P. (2003) Using molecular data to elucidate reticulate evolution in *Opuntia*. Madron˜o 50: 162–169.

Griffith, M. P., and Porter J.M.(2003) Back to the basics: a simple method of DNA extraction for mucilaginous cacti. Bradleya 21: 126– 128.

Griffiths D. (1914) Reversion in prickly pears. Journal of Heredity 5: 222– 225.

Gyekye L. (2019). "BBC covers biodegradable bioplastics made from cactus juice". *Bio Market Insights* (*5th June 2019*). Retrieved 25 February 2020.

Huelsenbeck, J. P., and Ronquist F. (2001) Mr. Bayes version 2.01. Department of Biology, University of Rochester, Rochester, New York, USA.

Idell, A. (1957) The Bernal Diaz chronicles: the true story of the conquest of Mexico doubleday garden city New York USA

Kiesling R. (1998) Origen, domesticacio´n y distribucio´n de *Opuntia ficusindica*. Journal of the Professional Association for Cactus Development 3. Online at http: //www.jpacd.

Kluge, M., and Ting I.P. (1978) Crassulacean acid metabolism: an ecological analysis. Ecological studies series, vol. 30. Springer-Verlag, Berlin, Germany.

Kuti, J.O. (2004) Antioxidant compounds from four Opuntia cactus pear fruit varieties. Food Chem. 85 (4), 527 533. Available from: https: //doi.org/10.1016/S0308-8146(03)00184-5

Lo´pez-Romero, P., *et al.,* 2014. The effect of nopal (*Opuntia ficus indica*) on postprandial blood glucose, incretins, and antioxidant activity in Mexican patients with type 2 diabetes after consumption of two different composition breakfasts. J. Acad. Nutr. Dietetics 114 (11): 1811-1818. Available from: https: //doi.org/10.1016/j.jand.2014.06.352.

Mandujano, M. C., Montana C, Mendez I. and Eeguiarte L.(1996) Reproductive ecology and inbreeding in *Opuntia rastera* (Cactaceae) in the Chihuahuan Desert: why are sexually derived recruitments so rare? American Journal of Botany 83: 63–70.

Nobel, P. S. (2002) Cacti: biology and uses. University of California,

Nobel, P. S. (1994) Remarkable agaves and cacti. Oxford University Press, New York, New York, USA.

Nobel, P. S., Barrera, E D L, Beilman, D.W., Doherty, J.H. and Zutta. B.R. (2002) Temperature limitations for cultivation of edible cacti in California. Madron˜o 49: 228– 236. org/contents1998.htm

Ostolaza C.(1994) Cactus y etnobota´nica. Quepo 8: 79–86.

Ozdemir D. (2020). "Creating Leather From Cactus to Save Animals and the Environment". Interesting Engineering (23 June 2020).

Pinkava, D. J. (2002) On the evolution of continental North American Opuntioideae. Succulent Plant Research 6: 59–98.

Russell C.E. and Felker P.(1987) The prickly pears (*Opuntia* spp., Cactaceae): a source of human and animal food in semiarid regions. Economic Botany 41: 433–445.

Robinson T (2008) William Roxburgh: The founding father of Indian Botany. Pub. By Phillimore and Company England. Pp 286.

Rosaria C., Riccardo D., Albanese, Lorenzo A, Francesco M, Mario P. (2017). "Opuntia ficus-indica seed oil: Biorefinery and bioeconomy aspects". European Journal of Lipid Science and Technology. *119* (*8*)*: 1700013.* doi*:* 10.1002/ejlt.201700013. ISSN 1438- 7697.

Saad, A.B., *et al.* (2017) Phytochemical, antioxidant and protective effect of cactus cladodes extract against lithium-induced liver injury in rats Pharm. Biol. 55 (1), 516

Smida, A., *et al.* (2017) Immunoprotective activity and antioxidant properties of cactus (*Opuntia ficus-indica*) extract against chlorpyrifos toxicity in rats. Biomed. Pharmacother. 88, 844- 851. Available from: https: //doi.org/10.1016/j.biopha.2017.01.105

Tesoriere, L., *et al.* (2004) Supplementation with cactus pear (*Opuntia ficus-indica*) fruit decreases oxidative stress in healthy humans: a comparative study with vitamin C. Am. J. Clin. Nutr. 80 (2), 391 395. http: //www.ncbi.nlm.nih.gov/pubmed/15277160.

Wang, J.F., *et al.* (2000) A dose-response effect from chocolate consumption on plasma epicatechin and oxidative damage. J. Nutr. 130 (8Suppl), 2115S-2119SS.

Chapter 15

In vitro Mass Multiplication and *Ex situ* Conservation of *Opuntia ficus-indica* (L.) Mill. or Beles in Tigray Region of Ethiopia, Africa

Opuntia ficus-indica (L.) Mill. commonly known as prickly pear or Beles is a xerophytic, succulent, CAM (Crassulacean Acid Metabolism) plant introduced in Ethiopia (Africa), particularly to Tigray region between 1848 and 1920. It is known as a multipurpose plant since it can be used as human food (fruits and vegetables), medicine and ornamental plants, fodder, natural wind barrier, soil stabilizer, re-vegetation resource in eroded soils. Stem of this plant has been reported to treat diabetes and useful in the cure of hyperlipidemy (excess of lipids in the blood) and obesity. It is well recognized for their wound healing properties and anticancer effects. Conventionally, it is propagated by seeds but physiological limitations of the seeds, such as; low germination rate, genetic segregation, less guarantee of genetic stability, a

Shastri Prasad Shukla[1]* and Mehari G. Micheal[2]

[1]Professor (Assistant), Department of Horticulture,
[2]Head, Department of Horticulture,
College of Agriculture and Environmental Sciences, Adigrat University, Adigrat, Tigray, Ethiopia, Africa
e-mail: [1]sp_shukla2002@yahoo.co.in, [2]mehari1999@yahoo.com

long juvenile stage, less availability and low viability, and slow seedling growth rate are major constraints in the multiplication of this species in nature. In the present study we developed an efficient procedure for mass multiplication, acclimatization and ex-situ conservation of this succulent plant in Adigrat region of Ethiopia (Africa). In the present study out of six used in vitro - raised clones, maximum average height of the plantlet (13 cm.) and width (8.5 cm.) were observed in the plantlets (X6) and (X5) clones at 30.29 and 29.15°C temperatures, respectively on a modified substrate. Here, the effects of temperature and duration were significant in the increment of clones height and width. Thus, the developed efficient acclimatization techniques of this crop will ensure the supply of the in vitro-raised plantlets throughout the year to the consumers in Tigray region of Ethiopia.

Keywords: *Biodiversity, CAM plant, Conservation, Micropropagation, Opuntia ficus-indica.*

15.1 Introduction

Cactus has about 130 genera and 1,500 species within Cactaceae family (Shetty *et al.*, 2012). Of these, *Opuntia ficus-indica* (L.) Mill. commonly known as prickly pear or "Beles" is a tree-like cactus native to Mexico and it was introduced in Ethiopia, particularly to Tigray region between 1848 and 1920 (Neumann, 1997, Habtu, 2005). The region is largely covered by cactus (*O. ficus-indica*) fruits, locally known as "Beles" is the only source of food for many farmers during raining season (June-September), just before the harvesting of staple crops (Habtu, 2005). The cactus is a xerophytic, succulent, CAM (Crassulacean Acid Metabolism) plant (Bravo-Hollis, 1997). Over the last few decades, interest in Beles as food and feed has increased due to its drought resistance, high biomass yield, high palatability and tolerance to salinity (Ben *et al.*, 1996).

O. ficus-indica can attains the maximum height of about 5m tall and a multipurpose plant, can be used as human food (fruits and vegetables), medicine and ornamental plants, fodder, natural wind barrier, soil stabilizer, re-vegetation resource to control water and wind erosion in eroded soils (Nobel, 1994, Rodriguez-Felix, 2002, Russell and Felker. 1987; Pimienta-Barrios and Munoz-Urias, 1995). The sweet fruit of this plant is known as "Tuna" useful as human food and entire plant can be cultured as raw-industrial material to produce several sub-products such as jam, wine, candies and jellies *etc.* (Hegwood, 1990, Flores-Valdez, 1995, Saenz-Hernandez, 1995). Besides, *O. ficus-indica* has been grown from pre-Columbian times as a host plant for cochineal insects (*Dactylopius coccus*) for the production of valuable, vivid red and purple dyes (Donkin, 1977; Nobel, 1994). It is also source of natural-dye, carminic acid which is used for coloring fabrics, food, and cosmetics (Flores-Flores and Tekelenburg, 1995). In addition the tree is also known for its medicinal properties as it is a rich source of flavonoids having affirmative health-benefits (Knishinsky, 1971; Iwashina *et al.*, 1984; Wang, 1988; Goycoolea and Cárdenas, 2003; Tesoriere *et al.*, 2004). It also contains antioxidants, primarily pectin, carotenes, betalains, ascorbic acid, quercetina and quercetin derivatives which is useful in the protection of mammalian cells and organs and slow the aging process, illness and disease (Kuti, 2004). Stem of this plant has been reported to treat diabetes and useful in the cure of hyperlipidemy (excess of lipids in the blood) and obesity (Saenz, 2000). Alcoholic extract of this plant is useful in treating a number of diseases and conditions, including anti-inflammatory effects (Park *et al.,* 1998; Sahelian, 2001), hypoglycemic effects (Frati *et al.,* 1990), inhibition of stomach ulceration (Galati *et al.,* 2003) and neuroprotective effects (Dok-Go *et al.,* 2003). In Chinese medicine, fruits of this species are considered weak poisons and are used for treatment of inflammation, pain and as detoxification agents for snake bites (Wang, 1988). A decoction of the bark, which contains about 10 per cent of tannin, is used as an astringent lotion in leucorrhoea. The leaves of the plant are heated and applied as a poultice to abscesses to promote suppuration and discharge of pus (Galati *et al.,* 2003). *O. ficus-indica* is well recognized for their wound healing properties and anticancer effects (Zou *et al.,* 2005). Conventionally, *Opuntia ficus-indica* is propagated by seeds but physiological limitations of the seeds, such as; low germination rate, genetic segregation, less guarantee of genetic stability, a long juvenile stage, less availability and low viability, and slow seedling growth rate are major constraints in the mass production of this species in the nature (Escobar *et al.,* 1986; Soltero 1996). In addition, it is asexually propagated by "areoles" (main meristematic tissue in stems) and cladodes which is time taking process. Furthermore, propagation by cuttings is not always feasible and often leads to a low multiplication rate (Akram *et al.,* 2013). Vegetative propagation, which is widely used, can be performed through the rooting of single or multiple cladodes, small portions of mature cladodes comprising two or more areoles (Estrada-Luna *et al.,* 2008). All these methodologies require large spaces for propagation and has threat of a low propagation rate. Therefore, the micro propagation is a feasible alternative option for the rapid multiplication and maintenance of germplasm, because it provides high propagation rates, reduced requirements for space, the production of healthy and pathogen-free plants (Johnson and Emino, 1979; Smith *et al.,* 1991). The first study on *Opuntia* (prickly pear cactus) micropropagation reported by Sachar and Iyer (1959) and various successful strategies have been described for different species of Opuntia, including *O. dillenii* Haw, *O. amyclaea* Tenore, *O. ficus indica* Linn Mill, *O. streptacantha* Lemaire, *O. robusta* Wendl, *O. cochinera* Griffiths, *O. leucotricha* De Candolle, *O. albicarpa* Scheinvar and *O. ellisiana* Griff. (Mauseth and Halperin, 1975, Mauseth, 1977, 1979, Estrada-Luna, 1988, Mohamed-Yasseen *et al.,* 1995, Estrada- Luna and Davies, 2001), however, an efficient *in vitro* protocol is not yet available for commercialization because most cactus plant responses to tissue culture are highly dependent on the genotype. Some important modifications and adjustments might be needed when a new species or cultivar is considered for tissue culture, especially to optimize the overall environmental culture conditions, media, plant regulators (type, concentration, and combination), during the shoot proliferation stage (Aliyu and Mustapha, 2007). Rooting and plantlet acclimatization conditions might also be investigated since they may limit the success of micropropagation (Hartmann *et al.,* 1997). Acclimatization is an important step in micropropagation study to produce a well-developed of *ex vitro* plantlets that can successfully survive in the field environment (Pospóšilová *et al.,* 1999). It determines the quality of the end product and, in commercial production, the economic viability of the enterprise (Conner and Thomas, 1982). Thus, *O. ficus-indica* is an economically important plant in Ethiopia and undoubtedly, several difficulties are in its cultivation. In this paper authors propose a novel method of mass multiplication and acclimatization of *O. ficus-indica.* The *in vitro-* raised plants are subjected to treatment of various temperature and soil condition which yielded good height, width and per cent of survival in Ethiopia.

15.2 Methodology

15.2.1 Plant Material for Tissue Culture

One-year-old *in vitro*-raised plantlets of Beles were collected from Tigray Biotechnology Centre, Mekelle (Ethiopia) in the second week of December 2018 which in turn were procured from Michael Technology Charitable Organization, USA. Few more plantlets were collected from Mekelle University which procured by Dry Grow Foundation Company of Italy in 2017. All the plant materials were maintained in the shade house of Beles Institute, Adigrat University, Adigrat (Ethiopia) prior to the experiment.

15.2.2 Initiation of Growth Experiment

To know the effects of microclimatic conditions and substrate on acclimatization the *in vitro*-raised plantlets were washed carefully in running tap water to remove the adhering agar and transferred to earthen pots as procedures reported by Shukla and Khare (2012). The experiment was carried out in laboratory of Beles Institute, Adigrat University, Adigrat with six clones of cultures, *i.e.*, X1 (4.26 and 1.56), X2 (8.91 and 4.93), X3 (5.73 and 2.89), X4 (9.57 and 4.52), X5 (10.66 and 5.56) and X6 (11.93 and 4.73) (average length and width in cm). On 15-12-2018, 60 *in vitro*-raised rooted shoots with different length (10 shoots of each clone) were transferred in to earthen pots containing two types of substrate which composed of sand only as substrate I and materials listed in Table 15.1 as substrate II, respectively.

15.2.3 Soil Preparation

The soil is composed of different materials which have great importance in the growth of plantlets. The soil was homogenized using spade, then filled to black bottom perforated polyethylene bag with its diameter of 50cm. The single filled bag weighed 7.4 kg. The percentage of the soil composition are shown below in the Table 15.1. The plants were treated with 16:8hr of light and dark conditions and watered twice a week.

Table 15.1. Material Type and their Proportion Used for Soil Preparation

Sl.No.	*Materials*	*Percentage (per cent)*	*Weight (kg)*
1	Sand	19.2	25.93
2	Compost	22.2	29.99
3	Ash	7.5	10.12
4	Clay	18.5	24.99
5	Crushed stones	2.22	2.99
6	Banana peel	2.22	2.99
7	Egg shell	2.22	2.99
	Total	**74.06**	**100**

15.2.4 End of Experiment

The growth experiment was completed on 15-03-2019 (after three months). Plantlets were initially covered with polythene sheets in order to maintain a high humidity (~ 90 per cent) and kept in the culture room at temperature range of 20-32°C and photoperiod of 16h/day and watered twice weekly. After one week, the polythene covers were removed and plantlets were moved to glass house after transferring to thirteen pots containing sand and soil respectively. The survival rate of plantlets was recorded. Here, some of plantlets get light from Florence lamp and the others get from cactus growth lamp.

15.2.5 Statistical Analysis

The collected data's were analyzed statistically to determine the level of significance through Analysis of Variance (ANOVA) among the growth of plantlets through the effects of temperature and soil by using SPSS version 16 software.

15.3 Results and Discussion

In the present study out of 6 used *in vitro* -raised clones, maximum average length of the plantlet (13 cm.) and width (8.5 cm.) has been observed in X6 and X5 clones at 30.29 and 29.15°C temperatures, respectively on substrate II composition (Figure 15.1, Table 15.2). After three months of acclimatization highest survival rate of the plantlets was 95 per cent. Most of the plantlets grew well and showed similar phenotypic homogeneity such as, color, yield *etc.* as in the case of plants grown in natural environment.

Statistically, the effects of temperature and duration were significant factor in the increment of all clones, expect X4 with $P<0.05$ was 0.08 and X6 with $P<0.05$ was 0.296 respectively (Table 15.3). Here, the increased temperature and prepared soil from different materials has significant in clones' height. The optimum temperature for the growth of cactus plant found in between 25-35°C (FAO, 2017). Similarly, temperatures and soil for the width growth of the plantlets had also effect. From

Table 15.2. Effect of Microclimatic Conditions on the Growth of *In vitro*-Raised Plantlets of Beles (After 40 days)

Clones	*Plantlet Growth*							
	Substrate I (Sand only)				*Substrate II (Soil prepared from the seven materials as Table 15.1)*			
	Average Length of Shoots (cm.)	*Average Width of Shoots*	*Temperature*	*Per cent Culture Response*	*Average Length of Shoots (cm.)*	*Average Width of Shoots*	*Temperature*	*Per cent Culture Response*
X1	3.75±0.43	2.75±0.41	22.62±0.13	35.2±0.12	6.75±0.17	4.75±0.16	24.78±0.08	40.3±0.02
X2	1.67±0.18	2.13±0.00	21.45±0.15	40.4±0.11	3.67±0.13	5.13±0.12	25.92±0.12	45.2±0.34
X3	3.66±0.19	1.50±0.12	20.23±0.14	60.3±0.01	4.66±0.17	2.00±0.42	25.14±0.14	65.4±0.21
X4	6.50±0.39	4.00±0.63	22.34±0.12	65.1±0.23	8.00±1.54	6.00±1.09	28.55±0.13	70.6±0.43
X5	9.35±0.48	3.50±0.44	23.20±0.08	70.4±0.13	12.33±1.43	8.50±1.04	29.15±0.11	80.5±0.54
X6	12.00±1.26	6.43±0.54	26.16±0.05	75.6±0.21	13.00±1.41	6.83±0.70	30.29±0.10	95.4±0.04

Xa - the "X" refers to the name of plantlet; "a"- subscript refers to treatment number.

the result and observation the temperature fluctuation and plant grew in soiled pot shown good progress than plant grew in sand pot only. The Analysis of variance (ANOVA) results also shown as X1(P P<0.05=0.00), X2(P<0.05=0.00), X4(0.012) and X5(P<0.05=0.00) (Table 15.3).

Figure 15.1. Acclimatization of *In vitro*-Raised Plantlets of *Opuntia ficus-indica* (L.) Mill. in Culture Room.

(a) Multiple shoots; (b) Rooted plantlet; (c) Plantlets transferred from the medium to the polyethylene bags; (d) Plantlets after three weeks (e and f) after 40 days.

Besides, the increment of width in clones of X3 (P<0.05=0.06) and X6 (0.218) had not significant difference (Table 15.3), this may be due to placement and obtained the same light energy and soil materials. Thus, this showed better acclimatization percentage than earlier report on prickly pear cactus species (Escobar *et al.,* 1986; Estrada-Luna, 1988; Mohamed-Yasseen *et al.,* 1995) and most micropropagated cacti (Johnson and Emino, 1979; Ault and Blackmon 1987). Several research papers on the micropropagation of Cactaceae can be found, but *in vivo* acclimatization of the plantlets is not always reported, and in some cases this is shown to be critical (Smith *et al.,* 1991; Rodriguez-Garay and Rubluo, 1992; Rubluo *et al.,* 1993). Due to its recalcitrant property, cactus is not a good choice for tissue culture. Only a few scattered studies have been reported on *in vitro* regeneration and acclimatization of this plant. Rooting and *in vivo* establishment have proven difficult for several species of micropropagated cacti (Smith *et al.,* 1991; Rodriguez-Garay and Rubluo, 1992; Rubluo *et al.,* 1993; Llamoca- Zarate *et al.,* 1999). The method of acclimatization presented in this article is different from earlier reported work in terms of plant growth regulators or biostimualtors and aeration were not used. Air humidity was also not maintained at a certain level mechanically or otherwise.

Table 15.3. Analysis of Variance (ANOVA) Summarized Results

Effects of Temperature and Soil		
Clones	*Height ($P<0.05$)*	*Width ($P<0.05$)*
X1	0.000	0.000
X2	0.000	0.000
X3	0.000	0.060
X4	0.080	0.012
X5	0.011	0.000
X6	0.296	0.218

15.4 Conclusions

The acclimatization methods presented in this paper are economical and feasible which can be used by local people of the Tigray region of Ethiopia because it eliminates the necessity of special fertilizers and plant growth regulators. They do not even require reliable irrigation systems, because the constant availability of water is ensured by its presence in the tubs. By this method we can produce a huge amount of acclimatized plantlets of Beles in a limited place and shorter time. In Tigray about 83 per cent of the population lives in rural areas and their main source of livelihood is based on agriculture. Regular income and food security is highly dependent on the annual rainfall resulting in Tigray region with the highest harvest failures and lowest per capita income/expenditure (Giuseppe De Bac, 2009). Present study will be helpful in order to strengthen the link between conservation and utilization. Cactus plant has high potential as commercial crop that it can be processed so easily and it has potential in international market and be source of foreign currency for the country. Our developed efficient acclimatization techniques of this crop will ensure the supply of the *in vitro*-raised plantlets throughout the year to the consumers.

Acknowledgments

The authors are thankful to the Horticulture Department and Beles Institute of Adigrat University, Adigrat, Ethiopia (Africa) for providing the necessary facilities to carry out this work. Thanks are also due to Tigray Biotechnology Centre, Mekelle to provide the *in vitro*-raised cultures.

REFERENCES

Akram, G., Tahereh, H. and Mojtaba, K.N. (2013) Micropropagation of tuna (*Opuntia ficus- indica*) and effect of medium composition on proliferation and rooting. International Journal of Biosciences 3(11): 129-139.

Aliyu, B.S. and Mustapha, Y. (2007) Effect of different media on the *in vitro* growth of cactus (*Opuntia ficus-indica*) explants. African Journal of Biotechnology 6(11): 1330-1331.

Ault, J.R. and Blackmon, W.J. (1987) *In vitro* propagation of *Ferocactus acanthodes* (Cactaceae). Hort Science 22: 126-127.

Ben, H., Nefzaoui, A., Abdouli, H. and Orskov, E.R. (1996) Effect of increasing level of spineless Beles (*Opuntia ficus-indica* var. *inermis*) on intake and digestion by sheep given straw based diets. Journal of Animal Sciences 62: 293-299.

Bravo-Hollis, H. (1997) Introducción. In: Valles, S.C. (Ed.), Suculentas Mexicanas, Cactáceas, CVS Publications, Mexico DF, p 12

Conner, A.J. and Thomas, M.B. (1982) Re-establishing plantlets from tissue culture: A review. *International Plant Propagators' Society.* Combined Proceedings of Annual Meetings 31: 342-357.

Dok-Go, H., Lee, K.H., Kim, H.J., Lee, E.H., Song, J.L.Y., Lee, Y.H. and Jin, C. (2003) Neuroprotective effects of antioxidative flavonoid, quercetin, (1)-dihydroquercetin and quercetin 3-methyl ether, isolated from *Opuntia ficus-indica* var. *saboten*. Brain Research 965: 130-136.

Donkin, R. (1977) Spanish red: An ethnogeographical study of cochineal and the *Opuntia* cactus. Transactions of the American Philosophical Society 67: 1-77.

Escobar, H.A., Villalobos, V.M. and Villegas, A. (1986) *Opuntia* micropropagation by axillary proliferation. Plant Cell, Tissue and Organ Culture 7: 269-277.

Estrada-Luna, A.A. (1988) Produccion de brotes e injertacion *in vitro* de seis species de nopal (*Opuntia* spp.) originarias del Altiplano Potosino- Zacatecano. Tesis de MC. Colegio de Postgraduados, Montecillo, Edo, De Mexico, Mexico, pp. 160.

Estrada-Luna, A.A. and Davies, F.T. (2001) Mycorrhizal fungi enhance growth and nutrient uptake of pricklypear cactus (*Opuntia albicarpa* Scheinvar "Reyna") plantlets after ex vitro transplantation. Journal of Horticultural Science and Biochemistry 76(6): 739-745.

Estrada-Luna, A.A., Martínez-Hernández, J.J., Torres-Torres, M.E. and Chablé-Moreno, F. (2008) *In vitro* micropropagation of the ornamental prickly pear cactus *Opuntia lanigera* Salm-Dyck and effects of sprayed GA_3 after transplantation to *ex vitro* conditions. Sci. Hortic. 117(4): 378-385.

FAO (2017) FAO Crop Ecology, Cultivation and Uses of Cactus Pear 2017; Chile. Available on http: //www.fao.org/3/a-i7012e.pdf.

Flores-Flores, V. and Tekelenburg, R. (1995) Dacti (*Dactilipius coccus* Costa) dye production. In: Barbera, G., Inglese, P. and Pimienta-Barrios, E. (Eds.), Agro-Ecology, Cultivation and Uses of Cactus Pear. FAO, Rome, pp. 167-185.

Flores-Valdez, C.A. (1995) Nopalitos production, processing and marketing. In: Barbera, G., Inglese, P. and Pimienta-Barrios, E. (Ed.), Agroecology, Cultivation and Uses of Cactus Pear, FAO, Rome, pp. 92-99.

Frati, A.C., Jiménez, E. and Ariza, C.R. (1990) Hypoglycemic effect of *Opuntia ficus-indica* in non-insulin-dependent diabetes mellitus patients. Phytotherapy Research 4: 195-197.

Galati, E.M., Mondello, M.R., Monforte, M.T., Galluzzo, M., Miceli, N. and Tripodo, M.M. (2003) Effect of *Opuntia ficus-indica* (L.) Mill. cladodes in the wound-healing process. Journal of the Professional Association for Cactus Development 5: 1-16.

Giuseppe De Bac (2009) Improved utilization of cactus pear for food, feed, soil and water conservation and other products in Africa Proceedings of International Workshop held in Mekelle, Ethiopia, 19-21 October (Strengthening cactus pear production in Tigray and North Wollo).

Goycoolea, F.M. and Cárdenas, A. (2003) Pectins from *Opuntia* spp.: A short review. Journal of the Professional Association for Cactus Development 5: 17-29.

Hartmann, H.T., Kester, D.E., Davies, F.T. and Geneve, R.L. (1997) Plant Propagation-Principles and Practices, 6th ed. Prentice-Hall, Englewood Cliffs, NJ, USA.

Habtu, L. (2005) Beles in southern Tigray: Current status, potential uses, utilization and threat. M.Sc. Thesis, Addis Ababa University.

Hegwood, D.A. (1990) Human health discoveries with *Opuntia* sp. (prickly pear). Hort Science 25(12): 1515-1516.

Iwashina, T., Ootani, S. and Hayashi, K. (1984) Neochilenin, a new glycoside of 3-O-methylquercetin and other flavonols in the tepals of *Neochilenia, Neoporteria* and *Parodia* species (Cactaceae). Journal of Plant Research 97(1): 23-30.

Johnson, J.L. and Emino, E.R. (1979) Tissue culture propagation in the Cactaceae. Cactus and Succulent Journal 51: 275-277.

Knishinsky, R. (1971) Prickly pear cactus medicine. Healing Arts Press, Rocheste.

Kuti, J.O. (2004) Antioxidant compounds from four *Opuntia* cactus pear fruit varieties. Food Chemistry 85(4): 527-533.

Llamoca-Zarate, R.M., Studart-Guimarães, C., Landsmann, J. and Campos, F.A.P. (1999) Establishment of callus and cell suspension cultures of *Opuntia ficus-indica*. Plant Cell, Tissue and Organ Culture 58: 155-157.

Mauseth, J.D. nd Halperin, W. (1975) Hormonal control of organogenesis in *Opuntia polyacantha* (Cactaceae). American Journal of Botany 62(8): 869-877.

Mauseth, J.D. (1977) Cactus tissue culture: a potential method of propagation. Cactus and Succulent Journal 49(2): 80-81.

Mauseth, J.D. (1979) A new method for the propagation of cacti: Sterile culture of axillary buds. Cactus and Succulent Journal 51: 186-187.

Mohamed-Yasseen, Y., Barringer, S.A., Splittstoesser, W.E. and Schnell, R. (1995) Rapid propagation of tuna (*Opuntia ficus-indica*) and plant establishment in soil. Plant Cell, Tissue and Organ Culture 42: 117-119.

Neumann, L. (1997) Opening speech. In: Proceedings of the International Workshop on "Opuntia in Ethiopia: State of Knowledge in Opuntia Research. February 23-27, 1997. Mekelle University, Ethiopia and Wiesbaden Polytechnic, Germany, pp. 5-9.

Nobel, P.S. (1994) Remarkable agaves and cacti. Oxford University Press, New York, New York, USA.

Park, E.H., Kahng, J.H. and Paek, E.A. (1998) Studies on the pharmacological actions of cactus: Identification of its anti-inflammatory effect. Archives of Pharmacal Research 21: 30-34.

Pimienta-Barrios, E. and Munoz-Urias, A. (1995) Domestication of *Opuntia* and cultivated varieties. In: Barbera, G., Inglese, P. and Pimienta- Barrios, E. (Eds.), Agro-ecology, Cultivation and Uses of Cactus Pear. FAO International Technical Cooperation Network on Cactus Pear, pp. 58-63.

Pospóšilová, J., Tichá, I., Kadleèek, P., Haisel, D. and Plzáková, Š. (1999) Acclimatization of micropropagated plants to *ex vitro* conditions. Biologia Plantarum 42(4): 481-497.

Rodriguez-Felix. A. (2002) Postharvest physiology and technology of cactus pear fruits and cactus leaves. Proceedings of the 4th International Conference on Cactus Pear and Cochineal. Acta Horticulturae 581: 191-199.

Rodriguez-Garay, B. and Rubluo, A. (1992) *In vitro* morphogenetic responses of the endangered cactus *Aztekium ritteri* (*Boedeker*). Cactus and Succulent Journal 64: 116-119.

Rubluo, A., Chavez, V., Martinez, A.P. and Martinez-Vazquez, O. (1993) Strategies for the recovery of endangered orchids and cacti through *in vitro* culture. Biological Conservation 63: 163-169.

Russell, C.E. and Felker, P. (1987) The prickly-pears (*Opuntia* spp., Cactaceae): A source of human and animal food in semi-arid regions. Economic Botany 41(3): 433-445.

Sachar, R.C. and Iyer, R.D. (1959) Effect of auxin, kinetin and gibberellin on the placental tissue of *Opuntia dellenii* Haw. cultured *in vitro*. Phytomorphology 9: 1-3.

Saenz, C. (2000) Processing technologies: An alternative for cactus pear (*Opuntia* spp.) fruits and cladodes. Journal of Arid Environments 46: 209-225.

Saenz-Hernandez, C. (1995) Food manufacture and by-products. In: Barbera, G., Inglese, P. and Pimienta-Barrios, E. (Eds.), *Agroecology,* Cultivation and Uses of Cactus Pear. FAO, Rome, pp. 137-143.

Sahelian, R. (2001) An anti-inflammatory principle from prickly pear cactus. *Fitoterapia* 72(3): 288-290.

Shetty, A.A., Rana, M.K. and Preetham, S.P. (2012) Cactus: A medicinal food. Journal of Food Science and Technology 49(5): 530-536.

Shukla, S.P. and Khare, P.B. (2012) *In-vitro* mass multiplication of a threatened tree fern, *Cyathea spinulosa* Wall. ex Hook. International Journal of Genetic Engineering and Biotechnology 3(1): 15-23.

Smith, R.H., Burdick, J.P., Anthony, J. and Reilley, A.A. 1991. *In vitro* propagation of *Coryphantha macromeris.* HortScience 26(3): 315.

Soltero, Q.R. (1996) Micropropagación de dos especies de la Línea B Strombocacti (Cactaceae). MS Thesis in Biotechnology Process, CUCEI, Universidad de Guadalajara, pp. 51.

Tesoriere, L., Butera, D., Am, P., Allerga, M. and Liverea, M.A. (2004) Supplementation with cactus pear (*Opuntia ficus-indica*) fruit decreases oxidative stress in healthy humans: a comparative study with vitamin C. The American Journal of Clinical Nutrition 80(2): 391- 395.

Wang, P.Z. (1988) Chinese medicine surgery. Ancient Chinese Medicine Press, pp. 164-183.

Zou, D., Brewer, M., Garcßa, F., Feugang, J.M., Wang, J., Zang, R., Liu, H. and Zou, C. (2005) Cactus pear: a natural product in cancer chemoprevention. Nutrition Journal 4: 25-36.

Chapter 16

The Key of our Good Health: Kiwi Fruit (*Actinidia deliciosa* L.)

Kiwis are a powerhouse of vitamins that can help keep your heart and digestive system healthy. Eating kiwis can also increase the amount of protective plant compounds you consume. Kiwis are small fruits that pack a lot of flavor and plenty of vitamins, minerals, and powerful plant compounds. They're sweet, easy to eat, and highly nutritious. Plus, these small fruits offer a few impressive health benefits. This article covers 4 health benefits of kiwi and gives you tips on how to incorporate them into your diet. Kiwis, also known as kiwifruit, are a type of fruit native to the mountains and hillsides of Southwest China. They're now grown in many other areas of the world, including New Zealand, which is a top producer of this popular fruit. Kiwis are also grown in many other areas of the world, including the United States. Kiwis were first brought to New Zealand in 1904 by a schoolteacher named Isabel Fraser, who brought kiwi seeds back home with her after traveling through China. Once known as Chinese gooseberry, kiwis were given the name "kiwi" by New Zealand fruit exporters after the flightless kiwi birds, which are endemic to New Zealand.

Keywords: *Good health, Kiwi, Actinidia deliciosa, Vitamin C, E, Carotenoids, Antioxidants.*

16.1 Introduction

Botanically, kiwis are classified as berry fruits. *Actinidia deliciosa* and *A. chinensis* are the two most commonly consumed kiwi species. Kiwifruit are a nutrient-dense fruit and extensive research over the last decade on the health benefits of kiwifruit has linked their regular consumption to improvements not only in nutritional status, but also benefits to digestive, immune and metabolic health. The health benefits of consuming fruit are well documented (Boeing

2012). Kiwifruit are exceptionally high in vitamin C and contain an array of other nutrients, notably nutritionally relevant levels of dietary fibre, potassium. The most popular type of *A. deliciosa* kiwi is the Hayward kiwi, a green kiwi named after a man named Hayward Wright, who popularized kiwis in New Zealand in the 1920s. Hayward kiwis are oval in shape and about the size of a large hen's egg. They have fuzzy brown skin, green flesh, and tiny black seeds. Kiwis also come in yellow-fleshed or golden varieties (*A. chinensis*). The difference between green-fleshed and yellow-fleshed kiwis is that green kiwis contain a green pigment called chlorophyll while yellow kiwis do not. It should be noted that kiwis are not the same species as kiwiberry. Even though they belong to the same plant family, *Actinidia*, kiwiberries are small and have edible, grape-like skin. Kiwis are larger and have thicker, fuzzy skin. The skin of most varieties of kiwi should be peeled off before you eat them. However, some people like to eat the skin of golden or yellow kiwi varieties because the skin is softer than the skin of green kiwis. Kiwi possess vitamin E and folate, as well as various bioactive components, including a wide range of antioxidants, phytonutrients and enzymes, that act to provide functional and metabolic benefits. The contribution of kiwifruit to digestive health is attracting particular attention owing to a growing body of evidence from human intervention studies. There are several plausible mechanisms of action that are likely to act together including the fibre content and type, the presence of actinidin (a natural proteolytic enzyme unique to kiwifruit which breaks down protein and facilitates gastric digestion (Kaur *et al.,* 2010a; 2010b), and other phytochemicals which may stimulate motility (Ciardiello *et al.,* 2008). The kiwifruit of commercial cultivation are large-fruited selections of predominantly *Actinidia deliciosa* cv Hayward (green kiwifruit) and an increasing range of gold varieties of various *Actinidia* species. The Hayward cultivar is an oval shaped berry with a dull brown hairy skin, however, one of its most attractive features is the strikingly beautiful appearance of the bright translucent green flesh interspersed with several rows of small black seeds. The flesh of the green Hayward cultivar is described as a tangy, sweet and sour combination providing a unique flavour combination, whereas the gold cultivar is described as having a sweet and tropical taste (Nishiyama 2007; Singletary 2012).

16.2 Historical Account

Kiwifruit are native to the temperate forests of the mountains and hills of southwest China. Missionaries in the nineteenth century made many contributions to the advancement of botany and the distribution of horticultural plants (Ferguson and Bollard 1990). The first botanical specimens of *A. chinensis* were sent to Europe by the Jesuit priest Père Pierre Noël Le Chéron d'Incarville around the 1750s and later by Robert Fortune, a plant collector. Robert Fortune was sent to China by the Horticultural Society of London (1843–1845) to "collect seeds and plants of an ornamental or useful kind", and one of Fortune's specimens of *A. chinensis* was held at the Royal Botanic Gardens at Kew, London. The first fruits of *A. chinensis* to be seen in Europe were sent, preserved in spirit, to Kew in 1886. Today New Zealand is a major producer of kiwifruit, and all early commercial varieties of kiwifruit plants in New Zealand and around the world can be traced back to a Church of Scotland mission station in Yichang, China, in 1878. Early in the twentieth century, the seeds and plants were regarded as ornamental curiosities with no mention of the edible fruit. The introduction of kiwifruit to New Zealand can be traced to a school teacher, Isabel Fraser, who in 1904 returned from a visit to China with seeds (Ward and Courtney 2013). Around 1922, Hayward Wright, a nurseryman living near Auckland, New Zealand, offered plants in his catalogue, listing the plant as "a wonderful fruiting climber" and promoting it as a highly valuable new fruit because it ripens in the winter over a long period, thus making the fruit a valuable addition to the short supply of winter fruits. The Hayward cultivar has been sold widely from the late 1930s and the dominance of this cultivar worldwide is now complete. The first commercial orchards and large-scale plantings occurred around this time. Orcharding kiwifruit required brave and courageous decisions as the work was hard, there were no proven patterns of management by growers and agronomic problems were faced as they occurred. World War II and then agricultural and marketing incentives from the 1950s to the present day resulted in the rapid geographical expansion of orchards in New Zealand, Australia. Whilst the kiwifruit season requires winter growing, the fruit can be stored very well once harvested and is available throughout the year (Vaughan and Geissler 2009).

16.3 Nutritional Constituents

Comprehensive and independent data on the nutritional composition of kiwifruit can be found in the USDA National Nutrient Database for Standard Reference (US DA 2016) and the New Zealand Food Composition Database (NZFCD) (Boland 2013). Chemical analyses are conducted on fruit ripened to the "ready-to-eat" state to ensure that the data are reflective of what would normally be consumed. Typically, kiwifruit (*A. deliciosa* and *A. chinensis*–"green" and "gold" cultivars, respectively) are eaten with the skin removed, and hence the analysed. A recent update to this information in the NZFCD now includes nutritional composition of the skin, as there is anecdotal evidence of growing number of consumers who choose to eat the skin, particularly of the gold varieties since it is smoother, thinner, and hairless. Consumption of whole Sun Gold kiwifruit (including the skin) increases the fibre, vitamin E and folate contents by 50, 32 and 34 per cent, respectively (Sivakumaran and Sivakumaran 2017). The total ascorbic acid content is the most distinctive nutritional attribute of kiwifruit (Boland 2013). The levels in the Hayward green cultivar are typically between 80 and 120 mg per 100 g fresh weight (Beever and Hopkirk 1990). This natural variation of amounts of vitamin C in fruit, including kiwifruit, is due to numerous factors including growing region and conditions, use of fertilizers, maturity at harvest, time of harvest, storage and ripening conditions (Lee and Kader 2000). In terms of nutritional value, using scoring models that rank and compare the amount of important nutrients present in foods, kiwifruit score well against other fruit. This provides a useful means for communicating those nutritional benefits to consumers (Drewnowski 2005; Darmon *et al.,* 2005; Drewnowski 2010), and should be noted that the high nutrient density score is largely driven by their high vitamin C content (Boland 2013). The SunGold kiwifruit contains 161.3 mg vitamin C per 100 g–almost three times the amount found in oranges and strawberries on an edible flesh weight basis. From the technical and sensory perspectives, the high ascorbic acid and low tannin content in kiwifruit are thought to explain why the cut fruit does not develop the typical browning reaction that occurs in most other fruits (Beever and Hopkirk 1990). Vitamin C is an essential dietary nutrient for humans, as we lack the terminal enzyme l-gulono- γ-lactone oxidase in the ascorbate biosynthetic pathway (Nishikimi *et al.,* 1994). There is an absolute requirement for Vitamin C for a range of biological functions. Vitamin C is a cofactor of metallo-enzymes necessary for the biosynthesis of collagen, l-carnitine, catecholamine, neurotransmitters, and peptide hormones such as oxytocin (Englard and Seifter 1986; Arrigoni and De Tullio 2002). Vitamin C in involved a in the regulation of transcription factors (Massip *et al.,* 2009). The strong antioxidant properties of Vitamin C are well documented, it scavenges free radicals and other reactive oxygen and nitrogen species, with a capacity to regenerate other small molecule antioxidants from their respective radicals (Carr *et al.,* 2013; Mandl *et al.,* 2009). Thus, it protects biomolecules such as lipids and DNA against oxidative damage (Mandl *et al.,* 2009; GNS 2015). There is evidence from *in vitro*, as well as animal and human intervention studies that supports the role of vitamin C in the functioning of the immune system.

16.3.1 Sugars

As kiwifruit develop and ripen, the concentrations of chemical components in the tissue change. The most marked change in the physiology of the fruit during ripening leads to a rapid decrease in starch concentration and a consequent increase in fructose and glucose. Kiwifruit tissue is very hard while the fruit is developing on the vine, but flesh firmness decreases during the later stages of development (Beever and Hopkirk 1990).

Fortunately, kiwifruit that are physiologically mature but have barely started to ripen can be harvested and will continue to ripen successfully off the vine. Cool storage immediately after harvest reduces the rate of ripening. It is these particular characteristics of kiwifruit that allow producing countries such as New Zealand to store unripe fruit and ship to it distant markets over an extended period. Suitable indicators of maturity for kiwifruit are used to ensure that fruit reaches an appropriate stage of development before harvest. A "maturity value" is important, and three changes in kiwifruit are taken into account–decreasing flesh firmness, conversion of starch to sugar and soluble solids concentration (to measure sugar concentration) are all used to provide an accurate assessment of final eating quality. The predominant sugars present in *Actinidia* are glucose and fructose with a small amount of sucrose present when the fruit is ripe and ready-to-eat. The amount of total sugars and ratios of these sugars vary not only as a function of maturity but also with the variety of kiwifruit (Pérez *et al.,* 1997; Nishiyama *et al.,* 2008). The ratio of fructose: glucose is important in terms of digestive health and preferably should be around 1:1 to reduce symptoms of gastrointestinal discomfort, such as bloating, caused by rapid fermentation by gut bacteria. Interestingly, as they ripen, many fruits undergo a marked

16.3.2 Antioxidants

In addition to the various nutrients in kiwifruit described above, for which there are dietary intake recommendations and well described physiological functions, kiwifruit contain a complex network of minor compounds that may also be associated with beneficial physiological functions. Various *Actinidia* species have been extensively analysed for their antioxidant chemical profiles (Wojdy³o *et al.,* 2017). As well as vitamins C and E, the other antioxidants include the carotenoids lutein, zeaxanthin and β-carotene, chlorophylls, quinic acid, caffeic acid glucosyl derivatives, β-sitosterol, chlorogenic acid, phenolics, including flavones and flavonones, to name but a few (Montefiori *et al.,* 2005). The antioxidant capacity of kiwifruit constituents has been measured by means of various in vitro chemical assays that monitor the quenching, scavenging or retarding of free radical generation (Singletary 2012). For example, the total antioxidant capacity of kiwifruit was reported to be higher than apple, grapefruit and pear, but less than raspberry, strawberry, orange and plum (Wang *et al.,* 1996).

16.3.3 Vitamin C

The total ascorbic acid content is the most distinctive nutritional attribute of kiwifruit (Boland 2013). The levels in the Hayward green cultivar are typically between 80 and 120 mg per 100 g fresh weight (Beever and Hopkirk 1990). This natural variation of amounts of vitamin C in fruit, including kiwifruit, is due to numerous factors including growing region and conditions, use of fertilisers, maturity at harvest, time of harvest, storage and ripening conditions (Lee and Kader 2000). In terms of nutritional value, using scoring models that rank and compare the amount of important nutrients present in foods, kiwifruit score well against other fruit. This provides a useful means for communicating those nutritional benefits to consumers (Drewnowski 2005), and should be noted that the high nutrient density score is largely driven by their high vitamin C content (Boland 2013). The SunGold kiwifruit contains 161.3 mg vitamin C per 100 g–almost three times the amount found in oranges and strawberries on an edible flesh weight basis. From the technical A recent Cochrane systematic review (Hemila and Chalker 2013) upholds the role of vitamin C in improving immune function and reducing the duration of common cold symptoms in the ordinary population. A Gold kiwifruit intervention study showed enhanced plasma vitamin C concentration and reduced severity and duration of upper respiratory infection symptoms in 32 elderly people supplemented with four kiwifruit per day for 4 weeks (Hunter *et al.,* 2011).

16.3.4 Vitamin E

Kiwifruit contain relatively high levels of vitamin E (Boland 2013), compared to other commonly consumed fruit. SunGold and green kiwifruit contain 1.40 and 1.46 mg per 100 g (USDA 2016), respectively, of the main form, α-tocopherol present in the flesh (Fiorentino *et al.,* 2009). These levels are sufficient to permit the use of nutrient function claims for Vitamins E in the EU. Fiorentino *et al.* (2009) showed that α-tocopherol is found in the flesh of kiwifruit, possibly associated with cell membranes and therefore potentially bioavailable. Fiorentino *et al.* (2009), also identified a new form of vitamin E in kiwifruit, Δ-tocomonoenol, noting that its radical scavenging and antioxidant capacity contributed to the total antioxidant activity of kiwifruit. Studies showing that the consumption of both green and gold kiwifruit correlates with increased plasma vitamin E concentrations, suggest the vitamin E in kiwifruit is bioavailable, (Chang and Liu 2009).

16.3.5 Folate

Kiwifruit are often referred to as being a good source of dietary folate. The folate content of 31 μg per 100 g in gold kiwifruit meets the criteria of EU Regulation to make a "source" claim as it exceeds the 15 per cent of the Reference Intake of 200 μg/day. In other countries, where the recommended daily intake is often higher (*e.g.,* 400–500 μg/day in Nordic counties, 400–600 μg/day in the USA Australia and NZ), such nutrient content claims cannot be made.

16.3.6 Potassium

Green and gold kiwifruit are good sources of potassium, containing typically around 301–315 mg per 100 g. These amounts are sufficient to meet the criteria of EU Regulation (EC) no. 1924/2006 on nutrition and health claims made on foods to make a natural "source" claim, as it exceeds the 15 per cent of the Reference Intake of 2000 mg/day as authorized by health claims in the EU for potassium. The potassium content of kiwifruit as compared to other fruit is high.

16.3.7 Dietary fibre

The dietary fibre of kiwifruit comes almost entirely from the plant cell walls, and particularly the polysaccharides that form the major structural components of these walls. Kiwifruit contain about 2–3 per cent of fresh weight non-starch polysaccharides (Ferguson and Ferguson 2003) that make up the fruit cell walls, providing a valuable contribution of both soluble and insoluble fibre to the diet. Analysis of dietary fibre of green and gold kiwifruit has shown they comprise about one-third soluble and two-thirds insoluble fibres, although kiwi gold fruit contain considerably less total fibre than green (Mishra and Monro 2012). The soluble fibre fraction contains almost exclusively pectic polysaccharides, whereas the insoluble fibre is mostly cellulose and hemicelluloses.

Changes occur in the composition and structure of kiwifruit cell walls during development and ripening. These structural changes in cell wall polysaccharides are reviewed in detail by Sims and Monro (2013). Cell wall polysaccharides are generally resistant to digestion and absorption in the human small intestine and are considered to be delivered to the colon in a chemically unaltered state. However, even minor chemical or structural changes can impact on the physicochemical properties and fermentability that determine their impact on health.

16.3.8 Actinidin and Minor Proteins

Kiwifruit contain several unique proteins and the cysteine protease actinidin, the most abundant protein in kiwifruit, of interest for their bioactive potential. The characterization and biochemical properties of actinidin have been extensively studied (Arcus 1959; Lewis and Luh 1988), and more recently its potential role in human health (Grozdanovic *et al.,* 2014; Bayer *et al.,* 2017). Actinidin is active over a wide range of pH, including those of the GI tract (Sun *et al.,* 2016) thus having the potential to influence protein digestion, and intestinal permeability (Cavic *et al.,* 2014). In contrast to potential benefits, actinidin is also the major kiwifruit allergen (Palacin *et al.,* 2008). Green and gold kiwifruit have been known to cause allergic reactions ranging from mild symptoms localized to the oral mucosa in the majority of individuals to anaphylactic reactions, particularly in children (Lucas *et al.,* 2004). Very little information is available in the literature on the prevalence of kiwifruit allergy, and human intervention studies with kiwifruit have shown that kiwifruit are well tolerated without any adverse side effects (Beck *et al.,* 2010; Gammon *et al.,* 2012).

16.4 Medicinal Uses of Kiwi

16.4.1 Digestive Health

Early Chinese pharmacopoeia from the Tang Dynasty onwards (AD 618–907) list a whole variety of medicinal uses for "mihoutao" fruit, the Chinese name generally used for *Actinidia* species, including aiding digestion, reduction of irritability and curing of dyspepsia and vomiting. Functional gastrointestinal disorders (FGIDs) are common and distressing (Talley *et al.,* 2015). FGIDs include functional dyspepsia (FD) and irritable bowel syndrome (IBS), affecting an estimated 3–28 per cent of the global population (Foxx-Orenstein 2016), particularly the elderly and women, and may severely affect the individual's quality of life and wellbeing (Talley *et al.,* 2015). Upper gastrointestinal disorders include gastric reflux, stomachache, delayed gastric emptying, nausea and vomiting, and lower gastrointestinal disorders include constipation, indigestion, bloating and diarrhoea. Current interventions for FGIDs include lifestyle and dietary modifications as well as pharmacological interventions targeting pain, motility, laxation and the gut microbiota (Foxx-Orenstein 2016). The worldwide growth in the incidence of FGIDs has created an immediate need to identify safe and effective food-based interventions. For example, constipation may be present in up to 29 per cent of the population, depending on the definitions used (Bharucha *et al.,* 2013). Food ingredients such as *Psyllium* and wheat bran are the most studied for maintaining a healthy gut and to manage abdominal discomfort. Additionally, it is generally regarded that adequate intakes of fibre rich fruits and vegetables daily with sufficient water will prevent constipation. Whole green kiwifruit have been used and promoted for many years to maintain abdominal comfort

Kiwifruit are certified as low FODMAP fruits by the Monash University low-FODMAP diet digital application (https://www.monashfodmap.com/i-have-ibs/get-the-app/), based on their relatively low proportions of fructose and fructans per single fruit serve. A recent pilot study demonstrated that the consumption of two green kiwifruit is not associated with clinically significant evidence of colonic fermentation as shown by hydrogen and methane on breath testing (Chen *et al.*, 2017), lending support for the low FODMAP status for kiwifruit.

16.4.2 Metabolic Health

Metabolic abnormalities such as dyslipidaemia [increased blood total cholesterol (TC), low density lipoprotein cholesterol (LDL-C), triglycerides (TG), lower high density lipoprotein cholesterol (HDL-C)], hypertension, vascular inflammation, abnormal glucose metabolism and haemostatic disorders all play important roles in the pathophysiology of the major causes of morbidity and mortality such as diabetes, cardiovascular disease (CVD), stroke and dementia (Gorelick *et al.*, 2011). A number of studies have investigated the effects of green and gold kiwifruit on some of these metabolic markers, including the effects of kiwifruit on glucose and insulin balance, and on body weight maintenance and energy homeostasis. Green and gold kiwifruit have clinically measured glycaemic indices (GIs) of 39 and 48, respectively, which puts them in the GI "low" category (GI<55). The glycaemic response to a fruit depends not only on GI, but also the amount of carbohydrate consumed in the fruit. As kiwifruit contains only about 12 per cent available carbohydrate and a low GI; the impact kiwifruit produces on plasma glucose levels is low enough for the fruit to be suitable in managing diets for people with reduced tolerance to glucose. The fibre content of kiwifruit may cause a delay in carbohydrate digestion and absorption by way of swelling action that reduces the rate of glucose diffusion (Monro 2013). This reduction in glycaemic response by 200 g kiwifruit (approximately two fruits) has been demonstrated in a human intervention study conducted by Mishra *et al.* (2016). The authors concluded that the low in vivo glycaemic impact could partly be attributed to the carbohydrate in kiwifruit being fruit sugars (fructose) and partly to the non-digested fibre components reducing the rate of intestinal processes such as digestion, sugar diffusion and mixing of intestinal contents. This partial substitution of starch-based staple foods, such as a high carbohydrate breakfast cereal, with kiwifruit could be an effective and healthy way to improve glucose homeostasis (Mishra *et al.*, 2016). Further exploration of this effect was investigated by Mishra *et al.* (2017) to better understand the role of non-sugar components in kiwifruit in modulating glycaemic response.

Kiwifruit consistently reduced the amplitude of the glycaemic response of participants following a series of wheat-based cereal meals adjusted to match the available carbohydrates from kiwifruit leading the investigators to conclude that components other than the available carbohydrate in kiwifruit, such as cell wall remnants or phenolic compounds, may be involved in the improved glycaemic response to co-ingested foods (Mishra *et al.,* 2017). The energy value of foods is also an important dietary aspect in managing risk factors for metabolic syndrome. Using an *in vivo–in vitro* model that determines the available energy (AE) content based on ATP yield at the cellular level (Coles *et al.,* 2010), Henare *et al.* (2012) found the AE of green and gold kiwifruit was significantly less than that predicted by the traditional

At water system, suggesting kiwifruit are useful in dietary weight management strategies. Further studies will explore the use of kiwifruit as an effective dietary strategy to reduce postprandial hyperglycaemia while at the same time increasing the amount of essential nutrients consumed.

Regular consumption of green and gold kiwifruit can also affect beneficially certain physiological biomarkers, particularly in individuals with metabolic abnormalities related to major causes of morbidity and mortality, such as diabetes, cardiovascular disease (CVD), stroke and dementia (Stonehouse *et al.,* 2013). For example, Chang and Liu (2009) investigated the effects of two kiwifruits on the lipid profile, antioxidants and markers of lipid peroxidation in hyperlipidaemic adult men and women in Taiwan. After 8 weeks of the intervention, the HDL-C concentration was significantly increased, whilst the LDL-C/HDL-C ratio and TC/HDL-C ratio were significantly decreased. Vitamin C and vitamin E, the antioxidant nutrients, together with plasma antioxidant status, also increased significantly in fasting blood samples.

16.4.3 Managing Blood Pressure

In 2012, Karlsen *et al.* (2012) demonstrated that intake of three kiwifruit per day for 3 weeks promoted pronounced anti-hypertensive effects, as well as antithrombotic effects in male, middle-aged and elderly smokers. The authors commented that this dietary approach may be helpful in postponing pharmacological treatment in individuals with high-normal blood pressure or hypertension. From a further randomized, controlled study over a period of 8 weeks, Svendsen *et al.* (2015) concluded that among men and women aged between 35 and 69 years with moderately elevated BP, the intake of 3 kiwifruit added to the usual diet was associated with lower systolic and diastolic 24-h BP compared with one apple a day. The authors observed these results were in contrast those of Gammon *et al.* (2014), noting the differences in study population criteria (normotensive versus moderately elevated BP) may have been a contributory factor. Although Svendsen *et al.* (2015) found no differences in measures of endothelial function in their study, they suggested that an increase in plasma antioxidant status (lutein), and in increased dietary potassium, resulting from the kiwifruit intervention, could be an explanation for the improvements in BP observed. In vitro studies on antioxidant and fibrinolytic activities. *In vitro* studies on antioxidant and fibrinolytic activities have also indicated the potential cardiovascular protective properties of kiwifruit extracts (Jung *et al.,* 2005). Evidence that consumption of kiwifruit can modulate platelet reactivity towards collagen and ADP in human volunteers was provided in a study by Duttaroy and Jørgensen (2004). The authors concluded that kiwifruit may have the potential to increase the effectiveness of thrombosis prophylaxis.

REFERENCES

Arcus AC (1959) Proteolytic enzyme of *Actinidia chinensis*. Biochim Biophys Acta 33(1): 242–244

Arrigoni O, De Tullio MC (2002) Ascorbic acid: much more than just an antioxidant. Biochim Biophys Acta 1569(1–3): 1–9

Bayer SB, Gearry RB, Drummond LN (2017) Putative mechanisms of kiwifruit on maintenance of normal gastrointestinal function. Crit Rev Food Sci Nutr. https: //doi.org/10.1080/10408 398.2017.1327841

Beck K, Conlon CA, Kruger R, Coad J, Stonehouse W (2010) Gold kiwifruit consumed with an iron-fortified breakfast cereal meal improves iron status in women with low iron stores: a 16-week randomised controlled trial. Br J Nutr 105(1): 101–109. https: //doi.org/10.1017/S0007114510003144

Beever DJ, Hopkirk G (1990) Fruit development and fruit physiology. In: Warrington IJ, Weston GC (eds) Kiwifruit: science and management. The New Zealand Society for Horticultural Science and Ray Richards Publisher, Auckland, pp 97–126

Bharucha AE, Pemberton JH, Locke GR, 3rd (2013) American gastroenterological association technical review on constipation. Gastroenterology 144 (1): 218–238. https: //doi.org/10.1053/j. gastro.2012.10.028

Boeing H, Bechthold A, Bub A, Ellinger S, Haller D, Kroke A, Leschik-Bonnet E, Müller MJ, Oberritter H, Schulze M, Stehle P, Watzl B (2012) Critical review: vegetables and fruit in the prevention of chronic diseases. Eur J Nutr 51: 637– 663

Boland MJ (2013) Kiwifruit proteins and enzymes: actinidin and other significant proteins. In: Boland M, Moughan PJ (eds) Advances in food and nutrition research: nutritional benefits of kiwifruit, vol 68. Academic Press, pp 59–80 Carr AC, Bozonet SM, Pullar JM, Simcock JW, Vissers MC (2013) Human skeletal muscle ascorbate is highly responsive to changes in vitamin C intake and plasma concentrations. Am J Clin Nutr 94(4): 800–807. htt: //doi.org/10.3945/ajcn.112.05320

Carr A, Frei B (1999) Does vitamin C act as a pro-oxidant under physiological conditions? FASEB J Off Publ Fed Am Soc Exp Biol 13 (9): 1007–1024

Cavic M, Grozdanovic MM, Bajic A, Jankovic R, Andjus PR, Gavrovic Jankulovic M (2014) The effect of kiwifruit (*Actinidia deliciosa*) cysteine protease actinidin on the occludin tight junction network in T84 intestinal epithelial cells. Food Chem Toxicol Int J Publ Br Ind Biol Res Assoc 72c: 61–68. https: //doi.org/10.1016/j.fct.2014.07.012

Ciardiello MA, Meleleo D, Saviano G, Crescenzo R, Carratore V, Camardella L, Gallucci E, Micelli S, Tancredi T, Picone D, Tamburrini M (2008) Kissper, a kiwi fruit peptide with channel-like activity: structural and functional features. J Pept Sci 14(6): 742–754. https: //doi.org/10.1002/psc.992

Chang W-H, Liu J-F (2009) Effects of kiwifruit consumption on serum lipid profiles and antioxidative status in hyperlipidemic subjects. Int J Food Sci Nutr 1–8. https: //doi.org/10.1080/09637 480802063517

Chen A, Offereins MSL, Mulder CJ, Frampton CM, Gearry RB (2017) A pilot study if the effect of green kiwifruit on intestinal fermentation in humans as measured by hydrogen and methane breath testing.

Coles LT, Moughan PJ, Awati A, Darragh AJ, Zou ML (2010) Predicted apparent digestion of energy-yielding nutrients differs between the upper and lower digestive tracts in rats and humans. J Nutr 140(3): 469–476. https: //doi.org/10.3945/jn.109.116293

Darmon N, Darmon M, Maillot M, Drewnowski A (2005) A nutrient density standard for vegetables and fruits: nutrients per calorie and nutrients per unit cost. J Am Diet Assoc 105(12): 1881–1887. https: //doi.org/10.1016/j.jada.2005.09.005 Drewnowski A (2005) Concept of a nutritious food: toward a nutrient density score. Am J Clin Nutr 82(4): 721–732

Drewnowski A (2010) The nutrient rich foods index helps to identify healthy, affordable foods. Am J Clin Nutr 91(4): 1095S–1101S. https: //doi.org/10.3945/ajcn.2010.28450D

Duttaroy AK, Jørgensen A (2004) Effects of kiwi fruit consumption on platelet aggregation and plasma lipids in healthy human volunteers. Platelets 15(5): 287–292

Englard S, Seifter S (1986) The biochemical functions of ascorbic acid. Annu Rev Nutr 6: 365–406. https: //doi.org/10.1146/annurev.nu.06.070186.002053 Ferguson AR, Bollard E (1990) Domestication of the kiwifruit. In: Warrington IJ, Weston GC (eds) Kiwifruit: science and management. Ray Richards in Association with the New Zealand Society of Horticultural Science, Auckland, pp 165–246

Ferguson AR, Ferguson LR (2003) Are kiwifruit really good for you? Acta Hort 610: 131–138

Ferguson AR, Huang H (2007) Genetic resources of kiwifruit: domestication and breeding. Horticult Rev 33: 1–121

Fiorentino A, Mastellone C, D'Abrosca B, Pacifico S, Scognamiglio M, Cefarelli G, Caputo R, Monaco P (2009) Δ-Tocomonoenol: a new vitamin E from kiwi (*Actinidia chinensis*) fruits. Food Chem 115(1): 187–192

FODMAP (https: //www.monashfodmap.com/i-have-ibs/get-the-app/)

Foxx-Orenstein AE (2016) New and emerging therapies for the treatment of irritable bowel syndrome: an update for gastroenterologists. Ther Adv Gastroenterol 9(3): 354–375. https: //doi. org/10.1177/1756283x16633050

Gammon CS, Kruger R, Minihane AM, Conlon CA, von Hurst PR, Stonehouse W (2012) Kiwifruit consumption favourably affects plasma lipids in a randomised controlled trial in hypercholesterolaemic men. Br J Nutr FirstView: 1–11. https: //doi. org/10.1017/S0007114512004400 doi

Gammon CS, Kruger R, Brown SJ, Conlon CA, von Hurst PR, Stonehouse W (2014) Daily kiwifruit consumption did not improve blood pressure and markers of cardiovascular function in men with hypercholesterolemia. Nutr Res 34(3): 235–240. https : //doi.org/10.1016/j.nutres.2014.01.005

German Nutrition Society (DGE) (2015) New reference values for vitamin C intake. Ann Nutr Metab 67(1): 13–20. https: //doi. org/10.1159/000434757

Gorelick PB, Scuteri A, Black SE, Decarli C, Greenberg SM, Iadecola C, Launer LJ, Laurent S, Lopez OL, Nyenhuis D, Petersen RC, Schneider JA, Tzourio C, Arnett DK, Bennett DA, Chui HC, Higashida RT, Lindquist R, Nilsson PM, Roman GC,

Grozdanovic MM, Ostojic S, Aleksic I, Andjelkovic U, Petersen A, Gavrovic- Jankulovic M (2014) Active actinidin retains function upon gastro-intestinal digestion and is more thermostable than the E-64-inhibited counterpart. J Sci Food Agric. https: //doi.org/10.1002/jsfa.6656

Hemila H, Chalker E (2013) Vitamin C for preventing and treating the common cold. Cochrane Database Syst Rev (1): Cd000980. https: //doi. org/10.1002/14651858.CD000980. pub4

Henare SJ, Rutherfurd SM, Drummond LN, Borges V, Boland MJ, Moughan PJ (2012) Digestible nutrients and available (ATP) energy contents of two varieties of kiwifruit (*Actinidia deliciosa* and *Actinidia chinensis*). Food Chem 130(1): 67–72

Hunter DC, Skinner MA, Wolber FM, Booth CL, Loh JMS, Wohlers M, Stevenson LM, Kruger MC (2011) Consumption of gold kiwifruit reduces severity and duration of selected upper respiratory tract infection symptoms and increases plasma vitamin C concentration in healthy older adults. Br J Nutr 108(7): 1235–1245. https: //doi.org/10.1017/S0007114511006659

Jung K-A, Song T-C, Han D, Kim I-H, Kim Y-E, Lee C-H (2005)

Cardiovascular protective properties of kiwifruit extracts in vitro. Biol Pharmaceut Bull

Karlsen A, Svendsen M, Seljeflot I, Laake P, Duttaroy AK, Drevon CA, Arnesen H, Tonstad S, Blomhoff R (2012) Kiwifruit decreases blood pressure and whole-blood platelet aggregation in male smokers. J Hum Hypertens. https: //doi. org/10.1038/jhh.2011.116

Kaur L, Rutherfurd SM, Moughan PJ, Drummond L, Boland MJ (2010a) Actinidin enhances protein digestion in the small intestine as assessed using an in vitro digestion model. J Agric Food Chem 58(8): 5074–5080.

https: //doi.org/10.1021/jf903835g

Kaur L, Rutherfurd SM, Moughan PJ, Drummond L, Boland MJ (2010b) Actinidin enhances gastric protein digestion as assessed using an in vitro gastric digestion model. J Agric Food Chem 58(8): 5068–5073. https: //doi.org/10.1021/jf903332a

Latocha P, Krupa T, Wo³osiak R, Worobiej E, Wilczak J (2010) Antioxidant activity and chemical difference in fruit of different Actinidia sp. Int J Food Sci Nutr 61(4): 381–394. https: //doi.org/10.3109/09637480903517788

Lee SK, Kader AA (2000) Preharvest and postharvest factors influencing vitamin C content of horticultural crops. Postharvest Biol Technol 20: 207–220

Lewis DA, Luh BS (1988) Development and distribution of Actinidin in kiwifruit (*Actinidia chinensis*) and its partial characterization. J Food Biochem 12(2): 109–116. https: //doi. org/10.1111/j.1745-4514.1988.tb00363.x

Leontowicz H, Leontowicz M, Latocha P, Jesion I, Park YS, Katrich E, Barasch D, Nemirovski A, Gorinstein S (2016) Bioactivity and nutritional properties of hardy kiwi fruit *Actinidia arguta* in comparison with *Actinidia deliciosa* ' Hayward' and *Actinidia eriantha* 'Bidan'. Food Chem 196: 281–291. https: // doi. org/10.1016/j.foodchem.2015.08.127

Lucas JSA, Grimshaw KEC, Collins K, Warner JO, Hourihane JOB (2004) Kiwi fruit is a significant allergen and is associated with differing patterns of reactivity in children and adults. Clin Exp Allergy 34(7): 1115–1121

Mandl J, Szarka A, Banhegyi G (2009) Vitamin C: update on physiology and pharmacology. Br J Pharmacol 157(7): 1097– 1110. https: //doi.org/10.1111/ j.1476-5381.2009.00282.x

Massip L, Garand C, Paquet ER, Cogger VC, O'Reilly JN, Tworek L, Hatherell A, Taylor CG, Thorin E, Zahradka P, Le Couteur DG, Lebel M (2009) Vitamin C restores healthy aging in a mouse model for Werner syndrome. FASEB J. https: //doi. org/10.1096/fj.09-137133

Mishra S, Monro JA (2012) Kiwifruit remnants from digestion in vitro have functional attributes of potential importance to health. Food Chem 135(4): 2188–2194. https: //doi. org/10.1016/j.foodchem.2012.06.102

Mishra S, Willis J, Ansell J, Monro JA (2016) Equicarbohydrate partial exchange of kiwifruit for wheaten cereal reduces postprandial glycaemia without decreasing satiety. J Nutr Sci 5: e37. https: //doi.org/10.1017/jns.2016.30

Mishra S, Edwards H, Hedderley D, Podd J, Monro J (2017) Kiwifruit non-sugar components reduce glycaemic response to co-ingested cereal in humans. Nutrients. https: //doi. org/10.3390/nu9111195

Monro JA (2013) Fiber: composition, structures, and functional properties. In: Boland M, Moughan PJ (eds) Advances in food and nutrition research: nutritional benefits of kiwifruit, vol 68. Academic Press, pp 257–272

Montefiori M, McGhie TK, Costa G, Ferguson AR (2005) Pigments in the fruit of red-fleshed kiwifruit (*Actinidia chinensis* and *Actinidia deliciosa*). J Agric Food Chem 53(24): 9526–9530

Nishiyama II, Fukuda T, Shimohashi A, Oota T (2008) Sugar and organic acid composition in the fruit juice of different actinidia varieties. Food Sci Technol Res 14 (1): 67–73

Nishiyama II (2007) Fruits of the actinidia genus. Adv Food Nutr Res 52: 293–324

Nishikimi M, Fukuyama R, Minoshima S, Shimizu N, Yagi K (1994) Cloning and chromosomal mapping of the human nonfunctional gene for l-gulono- gamma-lactone oxidase, the enzyme for l-ascorbic acid biosynthesis missing in man. J Biol Chem 269(18): 13685–13688

North CS, Hong BA, Alpers DH (2007) Relationship of functional gastrointestinal disorders and psychiatric disorders: implications for treatment. World J Gastroenterol 13(14): 2020–2027

Palacin A, Rodriguez J, Blanco C, Lopez-Torrejon G, SanchezMonge R, Varela J, JimÈnez MA, Cumplido J, Carrillo T, Crespo JF, Salcedo G (2008) Immunoglobulin E recognition patterns to purified Kiwifruit (*Actinidinia deliciosa*) allergens in patients sensitized to Kiwi with different clinical symptoms. Clin Exp Allergy 38(7): 1220–1228

Park YS, Namiesnik J, Vearasilp K, Leontowicz H, Leontowicz M, Barasch D, Nemirovski A, Trakhtenberg S, Gorinstein S (2014) Bioactive compounds and the antioxidant capacity in new kiwi fruit cultivars. Food Chem 165: 354–361. https: //doi. org/10.1016/j.foodchem.2014.05.114

Pérez-Burillo S, Oliveras MJ, Quesada J, Rufián-Henares JA, Pastoriza S (2018) Relationship between composition and bioactivity of persimmon and kiwifruit. Food Res Int 105: 461–472. https: //doi.org/10.1016/j.foodres.2017.11.022

Pérez AG, Olías R, Espada J, Olías JM, Sanz C (1997) Rapid determination of sugars, nonvolatile acids, and ascorbic acid in strawberry and other fruits. J Agric Food Chem 45(9): 3545– 3549. https: //doi.org/10.1021/jf9701704

Sellke FW, Seshadri S (2011) Vascular contributions to cognitive impairment and dementia: a statement for healthcare professionals from the American Heart Association/American Stroke Association. Stroke J Cerebral Circ 42(9): 2672–2713. https: //doi. org/10.1161/STR.0b013e3182299496

Sims IM, Monro JA (2013) Fiber: Composition, structures, and functional properties. In: Boland M, Moughan PJ (eds) Advances in food and nutrition research: nutritional benefits of kiwifruit, vol 68. Academic Press, pp 81–99

Singletary K (2012) Kiwifruit: overview of the potential health benefits. Nutr Today 47(3): 133–147

Sivakumaran S, Sivakumaran S (2017) Partial nutritional analysis of Zespri® SunGold kiwifruit skin (HN1745). In: Confidential Report for Zespri International Ltd. Plant and Food Research, Palmerston North

Stonehouse W, Gammon CS, Beck KL, Conlon CA, Von Hurst PR, Kruger R (2013) Kiwifruit: our daily prescription for health. Can J Physiol Pharmacol 91(6): 42–47. https: //doi.org/10.1139/cjpp-2012-0303

Sun Q, Zhang B, Yan QJ, Jiang ZQ (2016) Comparative analysis on the distribution of protease activities among fruits and vegetable resources. Food Chem 213: 708–713. https: //doi. org/10.1016/j.food chem.2016.07.029

Svendsen M, Tonstad S, Heggen E, Pedersen TR, Seljeflot I, Bohn SK, Bastani NE, Blomhoff R, Holme IM, Klemsdal TO (2015) The effect of kiwifruit consumption on blood pressure in subjects with moderately elevated blood pressure: a randomized, controlled study. Blood Press 24(1): 48–54. https: //doi. org/10 .3109/08037051.2014.976979

Talley NJ, Holtmann G, Walker MM (2015) Therapeutic strategies for functional dyspepsia and irritable bowel syndrome based on pathophysiology. J Gastroenterol 50(6): 601–613. https: //doi. org/10.1007/s00535-015-1076-x

Vaughan JG, Geissler CA (2009) The New Oxford book of food plants. Oxford University Press, Great Clarendon Street Oxford X2 6DP

US Department of Agriculture (2016) Green Kiwifruit. USDA National Nutrient Database for Standard Reference, Release 28 (slightly revised). Version: May 2016. US Department of Agriculture (USDA), Agricultural Research Service (ARS), Nutrient Data Laboratory, Beltsville (MD). http: //www.ars.usda.gov/ ba/bhnrc/ndl.

Ward C, Courtney D (2013) Kiwifruit: taking its place in the global fruit bowl. In: Boland M, Moughan PJ (eds) Advances in food and nutrition research: nutritional benefits of kiwifruit, vol 68. Academic Press, pp 1–15

Wang H, Cao G, Prior RL (1996) Total antioxidant capacity of fruits. J Agric Food Chem 44(3): 701–705

Wojdyo A, Nowicka P, Oszmiański J, Golis T (2017) Phytochemical compounds and biological effects of Actinidia fruits. J Funct Foods 30: 194–202. https: //doi. org/10.1016/j. jff.2017.01.018

Chapter 17

Composition and Health Benefits of different Parts of Hisalu (*Rubus ellipticus* Sm.)

Rubus ellipticus Sm. possesses many health benefits due to presence of anthocyanin, phenolics, antioxidants, ascorbic acid, flavonoids as well as minerals (P, K, Ca, Mg). Because of these, it is traditionally utilized for the cure of fever, cough, sore throat and CHD (coronary heart diseases). Beside fruits, folks also utilize the paste/powder of leaves, shoots and roots for the effective treatments of wounds, bone fracture, stomach-ache, bacterial infection and tumors. Beside this, the plant exhibit various biological properties such as anti-inflammatory, analgesic, antipyretic, antitumor, anticonvulsant, etc. and has immense ethnomedicinal value because of its numerous compounds. Although fruits of R. ellipticus are shown to be highly nutritious, delicious, and rich in vitamins and sugars, their antioxidant and anti proliferative potentials remain under explored. Therefore, the present review has summed up to date knowledge on chemical composition, with particular emphasis on therapeutic functionality and traditional uses. of R. ellipticus.

Keywords: *Anthocyanin, Phenolics, Ethnomedicinal, Pharmacological, Nutritional value, Rubus ellipticus.*

17.1 Introduction

Rubus ellipticus Sm. belongs to the family Rosaceae and comprises of about 1500 species. It is distributed all over India and is listed in top ten wild edible medicinal plants (Figure 17.1). *R. ellipticus*, commonly called yellow Himalayan raspberry is also known by various other names *viz.* Indian raspberry tree,

Cheeseberry *etc.* Indian names: huftoo, kale hinure (Himachal Pradesh); kala hinsalu, kali anchhi (Hindi); kandiari, kharmuch, surganch, oche (Kashmir); kalawar, kalahisalu (Kumaon); kaloaselu (Nepal); gunacha, kandiara (Punjab) (Amjad and Arshad 2014).

Figure 17.1. (A) Fruiting Plant; (B) Fruits; (C) Various parts Flower, Fruits, and Leaves of *Rubus ellipticus.*

Due to useful ethnomedicinal and pharmacological properties; *Rubus* species has been used in folk medicine. The phytochemical, antioxidant and medicinal attributes and health promoting constituents of cultivated *Rubus* berries are usually well recognized. The fruit (Figure 17.1A) is round yellow cluster of druplets type and contains numerous bioactive moieties *viz.* protein, lipids, carbohydrates, fibre and several. Phytochemicals. Because of its well known pharmacological properties, this plant has been categorized under top ten wild edible medicinal plants. This fruit offers excellent opportunities of cultivation as a hedge offence plant.It is thus being used traditionally but it is losing its identity in the modern era due to the negligence of researchers and processors. This review majorly focus on its composition, nutritive value, medicinal properties (Wójcik 1989; Parmar and Kaushal 1982).

17.2 Chemical Composition and Pharmacological Attributes

The major components contributing towards the composition of *Rubus ellipticus* fruit are moisture and carbohydrates. Beside these, the fruit also contain numerous minerals. The qualitative analysis of amino acid screening of *R. ellipticus* fruit showed the presence of L-Hydroxyproline, DL Iso-leucine, DL Valine, DL-2-Aminobutyric acid, L-Cysteine, DL-Nor-leucine, DL Alanine, L-Glutamic acid, L-Arginine, DL-Aspartic acid, L-Cystein hydroxychloride, L-Leucine, L-Lysine monochloride, DL-Methionine, DL-Threonine, L-tyrosine. Phytochemical screening of fruit yields flavonoids, glycosides, steroids, phenols, tannins, antioxidants, anthocyanin, ascorbic acid, and resin (Wang and Lin 2000; Sharma and Kumar 2011; Ahmad *et al.,* 2015; Saini *et al.,* 2014; Badhani *et al.,* 2015). The fruit is edible medicinally have astringent, febrifuge, kidney, miscellany, stomachic properties. The juice of the fruit is used in the treatment of fever, colic, coughs and sore throat. The inner bark is used in Tibetan medicine, it is said to have a sweet and sour flavor plus a heating potency. A renal tonic and ant diuretic, it is used in the treatment of weakening of the senses, vaginal/seminal discharge, polyuria and micturation during sleep. In recent years, multiple drug/chemical resistance in both human and plant pathogenic microorganisms have been developed due to indiscriminate use of commercial antimicrobial drugs/chemical commonly used in the treatment of infectious diseases.

Modern scientific studies have found that alcoholic extract of roots has antioxidant properties and antimicrobial ones. Ten new triterpenoid saponins have been found and research is continuing in this plant. Additionally, some of the constituents show promising characteristics in terms of functionality.

17.3 Genetic Diversity of *Rubus ellipticus*

Genetic relationship among 21 *R. ellipticus* accessions, collected from different sites in Himachal Pradesh and Uttrakhand, was evaluated by using 20 Expressed sequence tag-Simple sequence repeats and 35 Inter simple sequence repeats. The ESR-Stirs were designed by extracting seven EST sequences from *R. ellipticus* and 3184 sequences from other species of *Rubus,* such as; *Rubus ulmifolius* and *R. ideaus* obtained from national center for biotechnology information. The EST containing simple sequence repeat motifs were extracted using SSR Identification Tool and 20 primers were designed with primer 3 software. These EST-SSR primers generated 40 polymorphic bands, whereas ISSRs generated 173 polymorphic bands. Polymorphic information content (PIC) for ISSRs was higher (0.886) than that of EST-SSRs (0.667). Effective multiplex ratio and marker index values for EST-SSRs were 1.09 and 0.62 and, for ISSRs, these values were 8.24 and 5.83, respectively, which were used to develop dendrograms. With EST-SSR two cluster 'A' contained 19 collections, whereas cluster 'B' contained two collections with maximum similarity of 67 per cent. For ISSRs, 21 collections were divided into two clusters, *i.e.*, cluster 'C', which contained six collections and cluster 'D' that contained 15 collections with maximum similarity of 56 per cent. Thus EST-SSRs and ISSRs used in the present study revealed a high level of polymorphism in the 21 collections of *R. ellipticus*, revealing their efficiency for diversity analysis studies (Samriti *et al.,* 2017).

17.4 Traditional Uses of *Rubus ellipticus*

*Rubus ellipticus*is rich in phytochemicals and is traditionally being used as a source of medicine (Kafkas *et al.,* 2008; Sharma *et al.,* 2014). Every part of the plant is rich in one or more essential component which can be used either in manufacturing of drugs or the part itself can be used in various forms such as laxative, decoction and many more. The complete plant is astringent in nature and is used to lower down the body temperature during fever, especially during typhoid. The fruit possess laxative properties and is thus used in traditional medicinal system of Tibet and in folk medicinal system. The young roots and shoots of the plant are effective during stomach ache, abdominal pain, colic pain, *etc.* (Dutta *et al.,* 1997). In Unani and Ayurveda, antifertility property of plant *R. ellipticus* has been reported. The plant is also used for wound healing, gastralgia, dysentery, ulcer, diabetes mellitus, antifertility, analgesic, antimicrobial and epilepsy (George *et al.,* 2013). The fruit has laxative properties, and is used as traditional medicine in Tibet for a number of purposes (Hazarika and Pongener 2018). The whole plant has astringent properties and has been used to reduce fevers, especially typhoid. The inner bark of the Yellow Himalayan Raspberry is used as kidney tonic and an anti-diuretic. The juice extracted from the root has also been used for fevers, gastric problems (including infant colic when the young shoots are used too), diarrhea and dysentery and the root paste, applied to wounds promotes healing. In summer it is used to promote sweating as a diaphoretic, and as a diuretic, and as the fruit is fibrous it aids the digestive processes. Due to useful medicinal properties of *Rubus* species, it has been used in folk medicine (Amjad and Arshad 2014). Roots and young shoots of *R. ellipticus* are used for colic pain. The inner bark of the *R. ellipticus* plant is valued as a medicinal herb in traditional Tibetan medicine, including its use as a renal tonic andanti-diuretic. Its fruits are edible and can also be used to produce a purplish blue dye. Traditionally it is used for gastralgia, wound healing, dysentery, antifertility, antimicrobial, analgesic, epiepsy, diabetes mellitus and ulcer (Saklani *et al.,* 2012; George *et al.,* 2015). For curing various ailments they use the medicinal plants from the wild since ages. *R. ellipticus* is one among the shortlisted plants used as antipyretics. The root bark is also used in diarrhea, dysentery, as abortificient, emmenagogue and in fracture bones. *R. ellipticus* root paste is used as poultice for the treatment of bone fracture, applied on forehead during severe headache; fruit is edible. Ripe fruits as laxative and are used in the case of constipation, paste of young fruits are taken in case of gastritis, diarrhea and dysentery. The root juice drunk against urinary tract infection and its fruits are edible and were listed in The inner root bark of the plant is valued as a medicinal herb in traditional Tibetan medicine, including its use as a renaltonic and anti-diuretic. *R. ellipticus* root paste is used as poultice for the treatment of bone fracture, applied on forehead during severe headache; fruit is edible. The root juice drunk against urinary tract infection and its fruits are edible and were listed in the top ten wild edible medicinal plants in Tanahun district of Western Nepal (Bhakuni *et al.,* 1987; Ringmichon *et al.,* 2013). The young shoot is chewed raw to relieve sudden stomach pain. Root decoction given to the children to get rid of stomach warm. The inner root bark of the plant is valued as a medicinal herb in traditional Tibetan medicine, including its use as a renal tonic and antidiuretic.

17.5 Culinary Uses

The fruits of *R. ellipticus* are not only edible but are also delicious and have the ability to be utilized in food industry as a very good substrate (Ahmad *et al.,* 2015).The antioxidant potential of this fruit *i.e., R. ellipticus* makes it useful for the development of various supplemented probiotic products (Ahmad *et al.,* 2015). Owing to its composition, berries can be utilized for development of the numerous food products (juice, squash, jam, jelly, preserve and candy) and pharmaceutical products with the intervention of researchers and processors for their wide popularization. The juice of *R. ellipticus,* which has an attractive colour and rich flavour, can be preserved as such and can also be used for squash making. Fruit juice is also used to bring down the temperature of a fever and for colic, and is good for sore throats and colds too (Patel *et al.,* 2004).

17.6 Conclusion

Hisalu is commonly known as golden Himalayan raspberry. Missing the taste of this yellow colored, nutrients packed fruits which is found in High Reaches of Himalyan States (Kala 2005; Pandey and Bhatt 2016). This tasty fruit can only be found here in large quantities in Himachal and Uttarakhand; Kumaon in places like Patal Bhuvaneshwar, Nainital, Bhimtal and Mukteshwar *etc.* this small fruit is a commonly occurring fruit during the summer months. Considered to be freely available packets of energy it tastes sweeter than nectar and makes up for one of the most loved fruits grown in this magnificent hilly region. Although not commonly found in the plains these Fruits are commonly occurring delicacies. This can also be referred to as a naturally occurring raw fruit berry and is also one of the best sources of multivitamins.

The berries of *Rubus* are rich source of Nutraceuticals (Singh *et al.,* 2023). The fruits of yellow *R. ellipticus* are an important source of natural antioxidants and their consumption may play vital role in reducing the oxidative stress and preventing the degenerative diseases including cancer, diabetes *etc.* Owing to the good blend of acids (malic, citric, tartaric) and sugars, it have ability for its utilization in food product development *i.e.* jams, squash, wine, vinegar, juice, *etc.* However, it is still unable to get its identity due to its negligence by the processors and researchers. Different part of the plant have been claimed to be useful in ailments like diabetes, diarrhoea, gastralgia, dysentery, epilepsy and as wound healing agent,anti-fertility agent, antimicrobial, analgesic and as renal tonic (Rana *et al.,* 1990). Beside all these, nowadays it is also used for manufacturing of purplish blue coloured dyes. But, the scientific intervention in the processing of this crop is a lacking issue and this crop is still unexplored, and needs to be exploited with the passage of time for the development of wide range of food products.

REFERENCES

Ahmad M., Masood S., Sultana S., HaddaTB., Bader A., Zafar M. (2015). Antioxidant and nutraceutical value of wild medicinal Rubus berries. Pak.J. Pharmaceut. Sci., 28(1): 241- 247.

Amjad MS., Arshad M. (2014) Ethnobotanical inventory and medicinal uses of some important woody plant species of Kotli, Azad Kashmir, Pakistan. Asian PacificJ. Trop. Biomed., 4(12): 952-958.

Badhani A., Rawat S., Bhatt ID., Rawal RS. (2015) Variation in Chemical Constituents and Antioxidant Activity in Y ellow H imalayan (*R ubusellipticus* Smith) and Hill Raspberry (*Rubus niveus* T hunb.). J. food biochem.,39(6): 663-672.

Bhakuni RS., Shukla YN., Thakur RS. (1987) Chemical examination of the roots of *Rubus ellipticus.* Ind. Drugs, 24: 272.

Dutta S., Ghatak KL. Ganguly KN. (1997). Isolationand Structure Elucidation of New Pentacyclic Triterpene Acid from the Leaves of *Rubus ellipticu.* Nat.Prod. Sci.,32: 108 110.

George BP., Parimelazhagan T., Kumar YT., Sajeesh T. (2015). Antitumor and Wound Healing Properties of Rubusellipticus Smith.*J. Acupunct. Merid. Stud.,* 8(3): 134-141.

George BP., Parimelazhagan T., Saravanan S. (2013).Anti-inflammatory, Analgesic and Antipyretic activities of *Rubus ellipticus* Smith. Leaf Methanol Extract. Int. J. Pharm. Pharmaceut. Sci.,5(2): 220-224.

Hazarika TK., Pongener M. (2018). Potential wild edible fruits of Nagaland, North-east India and its significance in the livelihood and nutritional security of rural, indigenous people. Genetic Resour. Crop Evolut., 65(1): 199-215.

Kala CP. (2005). Ethnomedicinal botany of the Apatani in the Eastern Himalayan region of India. J. Ethnobiol. Ethnomed., 1(1): 11. doi: 10.1186/1746-4269-1-11

Kafkas E., Ozgen M., Ozogul Y., Turemis N. (2008). Phytochemical and fatty acid profile of selected red raspberry cultivars: A comparative study. J. Food Qual.,31: 67-78.

Pandey Y., Bhatt SS. (2016). Overview of Himalayan yellow raspberry (*Rubus ellipticus* Smith.): A nutraceutical plant. J. Appl. Nat. Sci., 8(1): 494-499.

Parmar C., Kaushal MK. (1982). *Rubus ellipticus*. In: Wild fruits. Kalyani Publishers, New Delhi, India,84–87.

Patel AV., Vera RJ., Dacke CG. (2004). Therapeutic constituents and actions of Rubus species. Curr. Med. Chem., 11(11): 1501-1512.

Rana AC., Santana DD., Saluja AK. (1990). Pharmacological screening of the Alcohlic Extract of the leaves of *Rubus ellepticus* Ind. J. Pharmaceut.Sci., 52: 174-177.

Ringmichon CL., Gopalkrishnan B., Dixit AP.(2013). Ethnopharma - cognostical studies on root bark of *Rubus ellipticus* Smith. from Manipur. J. Pharmacog. Phytochem. 2(2): 223-228

Saini R., Dangwal K., Singh H., Garg V. (2012). Anti-oxidant and Antiproliferative activities of phenolicsisolated from fruits of Himalayan Yellow Raspberry(*Rubus ellipticus*). J. Food Sci. Technol., 5: 22-23.

Saini R., Dangwal K., Singh H., Garg V. (2014). Antioxidant and antiproliferative activities of phenolics isolated from fruits of Himalayan yellow raspberry (*Rubus ellipticus*). J. food Sci.Technol., 51(11): 3369-3375.

Saklani S., Chandra S., Badoni PP., Dogra S. (2012). Antimicrobial activity, nutritional profile and phytochemical screening of wild edible fruit of *Rubus ellipticus*. Int. J. Med.Aromat. Plants, 2(2): 269-274.

Samriti, Kaur, R., Chaudhary, S., Vaidya, E., Poonam, Thakur, D., Kumar, K. (2017). Assessment Of Genetic Diversity in Rubus ellipticus (Smith) Using Molecular Markers. Proceedings of the Indian National Science Academy. 92. 10.16943/ptinsa/2017/49120.

Sharma M., Kaur J., Kumar V., Sharma K. (2019). Nutraceutical potential of *Rubus ellipticus*: a critical review on phytochemical potential, health benefits, and utilization *Rubus ellipticus:* a review Think Ind. J., 22(37): 878-8.

Sharma M., Neerajani G., Kumar A., Basak D. (2014).Evaluation of total Phenolic content and Antioxidant properties of different leaf extract of *Rubus ellipticus* Ethiop. Int. J. of Multidiscipl. Res., 2(2): 14-17.

Sharma US., Kumar A. (2011). In vitro antioxidant activity of *Rubus ellipticus* fruits. J.Adv.Pharmaceut. Technol. Res.,2(1): 47-50 doi: 10.4103/2231-4040.79805

Singh B., Singh L., Kewlani P., Vinod Chandra Joshi V.C., Indra D. Bhatt I.D. (2023), *Rubus* spp. (*Rubus armeniacus, Rubus ellipticus, Rubus fruticosus, Rubus nepalensis, Rubus niveus, Rubus occidentalis*), In: Himalayan Fruits and Berries, (eds): Tarun Belwal, Indra Bhatt, Hari Devkota, Academic Press, 2023,pp. 381-394,

Wang,SY., Lin HS. (2000). Antioxidant activity in fruits and leaves of blackberry, raspberry and straw-berry varies with cultivar and developmental stage. J. Agri. Food Chem.,48(2): 140-146.

Wójcik E. (1989) Phytochemical investigation of *Rubus ellipticus* blackberry. Acta Poloniae Pharmceut., 46: 386-390.

Chapter 18

Edible and Medicinal Ferns Adding Diversity to World Food Basket

Edible ferns are the familiar plant food used by people all over the world as they are easily available in large biomass. This review mainly focuses on edible and medicinal ferns and their fern allies, that are consumed in different regions of India by various tribal communities but whose economic importance has not been fully scrutinized. So far, non-comprehensive ethnobotanical account on edible ferns in India has been reported with no information of ferns and their allies species are being consumed as food by tribal people in India. This review provides an outline of edible ferns and fern allies of traditional food uses in the tribal communities of India that can be exploited for consumption in other regions of India. Summarizing the reports, we gather that the most of the fern's species are consumed in North East as compare to other regions of India. Thus, these would provide a vital source for healthcare and economic revenue and it could add diversity in the world food basket for human population in the changing climate. The chapter describes the edible value of fresh and processed ferns. The ethnobotanical uses of different ferns found in Nigeria are described.

***Keywords**: Edible, Medicinal, Ferns, Himalayas, Nigeria, Fiddlehead fern.*

18.1 Introduction

Humans have been eating fern members of Pteridophyte group, for thousands of years, but not all fern species are edible. Edible ferns are mentioned in Chinese literature as early as 3000 years ago, and it is believed to have been a common item in Medieval Europe. Today, ferns are eaten mainly in North America (the USA and Canada) and in New Zealand (Nekrasov *et al.*, 1985) and Japan. There are three main species of edible fern: the ostrich fern, the lady fern, and the bracken fern.

Edible ferns can be delicious but need to be boiled or steamed to become safe to eat. Ostrich ferns are believed to be the tastiest and safest to eat. The Indian subcontinent has rich pteridophytic flora which constitutes about 191 genera comprising 1250 species (Dixit 1984), nearly 10 per cent of recorded 13,600 species all over the World (Moran 2008). In India, diversity of Pteridophytes is rich in Himalayas, Western Ghats and Eastern Ghats (Sathiyaraj *et al.,* 2015).The medicinal importance of the fern and fern allies studies were made by frequent field surveys during July2008-April 2010 covering various remote localities of the district in rainy seasons. All the specimens collected from the area for medicinal use are deposited in the herbarium of Botanical Survey of India, Dehra Dun. The Fern and fern allies has an immense share in the inventory of medicinal plants and contributes an important and effective part in Ayurvedic, Homeopathic and Allopathic systems of medicine (Singh *et al.,* 2021).The edible ferns, called fiddleheads, are differentiated in the three main species. It will also explain how to identify edible ferns and harvest them from the wild. Strictly speaking, a "fiddlehead fern" is not a type of fern but the earliest stage in fern growth. As ferns mature, new fronds unroll and grow upward: the fiddleheads are fronds that have yet to unfurl and are still tightly wound and close to the ground. Fiddleheads are characteristic of young ferns. Even though all ferns have a fiddlehead stage, edible ferns are called fiddlehead ferns because they are harvested as young plants. The fronds of mature ferns are bitter and preferably should not be eaten (Caius 1935).

Figure 18.1. *Nephrolepis* Fern in Fern House of the Botanical Garden of MS University of Baroda.

Several fern species are edible, but fiddlehead fern generally refers to the ostrich fern (*Matteucia struthiopteris*). The ostrich fern is safest for consumption and has the lowest toxicity; in contrast, other edible species are toxic or carcinogenic when raw. The ostrich fern grows in damp, temperate areas and is widespread throughout the northern hemisphere, mainly thriving in cooler climates. As with other edible ferns, ostrich ferns are harvested in the spring before the fronds unfurl. The fiddlehead stage lasts only for a few days, and ferns are not cultivated but foraged only in the wild.

Southern Nigeria represents one of the richest zones of vegetational wealth of Nigeria. A total of 36 plant species belonging to 22 families were identified. Folk utilization of Pteridophytes by the southern Nigerians was studied. The results of this survey display an interesting diversity and distribution of ferns and fern-allied plants used as medicines, food, cosmetics, and manure in southern Nigeria. These pteridophytes could provide a vital source for healthcare and economic revenue after appropriate pharmacological analysis by patents. Susheela and Devi reported occurrence of AM spores in roots of *Adiantum peruviavianum* (Sucheela and Devi 1997).

18.2 Identification of Edible Ferns

The ostrich fern grows mainly in the northern parts of the United States and Canada; however, it is also found in Eurasia and East Asia. In early spring, they appear along streams and riverbanks or in hardwood forests, preferring shady areas. The mature ostrich fern is three to six feet tall, and its fronds are arranged in a rosette to form a vase shape. The fronds grow out of rhizomes in the ground (which look like mounds or clumps) and are plume-like, resembling the tail feathers of an ostrich.

The fiddleheads of the ostrich fern are about an inch in diameter and have smooth, green stems. In contrast, the coiled tops are covered in brown, paper-thin scales that fall off as the fronds unfurl. The inside of each stem also has a deep U-shaped track similar to the groove in celery. The brown coverings on the tops of ostrich fern fiddleheads can make them difficult to find by directly looking for the coiled structures. Fiddleheads are more easily found by looking for the deep green stems, which stand out against the dark soil. They should be about one to four inches in height. In addition to the ostrich fern, the other two main species of edible fern are the lady fern (*Athyrium filix-femina*) and the bracken fern (*Pteridium aquilinum*). Like the ostrich fern, the lady fern grows mainly in the temperate northern hemisphere; on the other hand, the bracken fern is more widely distributed and can be found in both tropical and temperate climates.

The lady fern is found in moist forests and meadows, as well as in swamps. Like the ostrich fern, it prefers to grow in the shade; however, it is only two to three feet tall at maturity, and its stems are darker at the base. While lady fern stems exhibit a groove like the ostrich fern's, it is proportionally smaller. The bracken fern grows in various habitats, including forests (dry or wet), meadows, clearings, and roadsides. Mature plants are two to four feet tall, and the stems are fuzzy and lack U-shaped grooves. Unlike the ostrich fern, the fiddleheads of the bracken fern are not covered in a brown husk. Each bracken fern has only a single frond.

18.2.1 Harvesting of Fiddleheads

To harvest fiddleheads, pinch and snap the stem about a half-inch to an inch from the tightly coiled head. Never pick a clump clean as this may kill the fern: leaving a few fiddleheads intact also assures next year's crop. However, because poisonous ferns can be mistaken for edible species, foraging should be left to the professionals, and it is better to buy edible ferns at the market. Choose fiddleheads with tightly coiled discs and firm, bright green stems. Avoid those that have started to unravel or appear discolored.

18.2.2 Are Edible Ferns Dangerous?

The lady fern and the bracken fern are both edible but can be dangerous when consumed raw. The lady fern contains thiaminase, an enzyme that depletes vitamin B1 in the body. Thiaminase is not toxic when consumed in small quantities by people whose diet is rich in vitamin B1, but it can induce vitamin B1 deficiency in large doses. Symptoms of vitamin B deficiency (or "beriberi") include weakness and muscle loss, mental confusion or speech difficulties, and tingling in the hands and feet (also known as "pins and needles"). On the other hand, bracken fern contains Ptaquiloside (PTA) which is a known carcinogen. PTA causes DNA damage, leading to cancers of the digestive tract; furthermore, consumption of bracken fern has been shown to induce intestinal cancer in rats and bladder cancer in cattle. In addition to PTA, bracken fern also contains thiaminase. Both thiaminase and PTA are broken down by heat, which is why fiddleheads should never be eaten raw or undercooked. To neutralize these harmful elements, you can either boil fern fiddleheads for 15 min or steam them for 10-12 min. After this step, fiddleheads can be used in recipes like any other vegetable.

18.3 Ferns in World Market

The fern (Pteridophyte) saprophyte is usually the dominant life stage both in terms of size and lifespan. The saprophyte typically has conspicuous leaves called fronds, which are compound and display a great diversity in size and shape. The global economic value of ferns is much lower than that of seed plants; however, there are a number of examples of the ethnobotanical usage of ferns (Keller and Prance 2015). The traditional and modern usage of ferns for food has been observed in different regions (Maroyi 2014; Keller and Prance 2015). Fronds (mostly young), rhizomes, stems and tubers can be eaten as fresh greens, cooked as vegetables or processed for starch (Liu *et al.,* 2012).

The most popular edible ferns available on the market include the bracken fern (*Pteridium aquilinum*) and the ostrich fern (*Matteuccia struthiopteris*). According to an earlier estimate, sales of the ostrich fern fiddleheads might be in the range of USD 7–10 million in North America (DeLong and Prange 2008). Edible ferns available on the market in Japan are the bracken fern, the ostrich fern and *Osmunda japonica*, which are priced at JPY 1.2 million (ca. USD 11,000), JPY 5.4 million (ca. USD 49,500), and JPY 22.1 million (ca. USD 202,700) per ton, respectively (Koide and Kadoya 2019). Furthermore, in China, where about 52 species of ferns are consumed as food, fronds on the market are mostly represented by *P. aquilinum* var. *latiusculum* and *O. japonica*, with the bracken fern occupying more than 80 per cent of the fern food market (Liu *et al.,* 2012). The bracken fern is also the most harvested fern across Siberia and the Far East of Russia. In the three regions of the Russian Far East (Khabarovsky Krai, Primorsky Krai, and Sakhalin Oblast), its total annual harvest reached 2500 ton in 1976–1980 due to its high demand in Japan. Subsequently, the export of the salted fern shoots decreased; however, it may still be an essential part of the local economy. According to customs data, the Far Eastern region of Russia exported 94.1 tons of ferns in 2016–2019. The young fern fronds can be preserved in different ways: by freezing and canning (in North America), salting (mostly in Russia) and drying (in China and East Asia) (Gerasimenko and Korol, 2003; DeLong and Prange 2008; Liu *et al.,* 2012).

There are at least 13 edible fern species in the Russian Far East (Khrapko 2007). Unlike the most harvested bracken fern, which is widely sold and seemingly may be found on a shelf in any supermarket in the region, the ostrich fern is less common for the local market. However, the ostrich fern is promoted as being safer than other edible ferns. The fern *Osmunda japonica*, which is popular in East Asia, grows in Russia only on Sakhalin Island. Instead, *Osmundastrum asiaticum* is used for food consumption and is also consumed in China (referred to as *Osmundastrum cinnamomeum* - Liu *et al.,* 2012).

18.4 Biochemicals Present in Ferns

Fern green tissue is of particular interest due to its fatty acid composition. Unlike flowering plants, fern fronds contain long-chain polyunsaturated fatty acids (LC-PUFAs)–arachidonic acid (ARA), eicosapentaenoic acid (EPA), sciadonic acid (SCA), juniperonic acid (JA) and other acids (Kawashima 2019). LC-PUFAs are considered to be valuable nutrients due to their physiological roles and limited biosynthesis in humans. The most common and important among them are ARA, EPA, and docosahexaenoic acid (DHA) (Kawashima 2019; Harayama and Shimizu 2020). LC-PUFAs are precursors of different signaling molecules and are important for the regulation of membrane properties. These fatty acids and their metabolites are involved in normal and pathological processes, ranging from inflammation and its resolution to the functioning of the nervous system (Harayama and Shimizu 2020). LC-PUFAs have to be obtained from food, and they also can be synthesized endogenously from the corresponding essential fatty acids (18:2n-6 and 18:3n-3). However, both dietary sources and the endogenous conversion appear to be important for normal metabolism (Harayama and Shimizu 2020).

The edible ferns grown in the Russian Far East were analyzed for dietary sources of LC-PUFAs. The hypothesis behind this study was that commercially available products of these ferns, which are processed using methods traditional for the East Asian region, preserve the valuable fatty acids. The obtained results aim at the establishment of ferns as a plant source of LC-PUFAs, and are thus of interest in the formulation of new vegetarian diets containing these fatty acids.

18.4.1. Contents in Raw Young Fronds

The fern species belong to different families: Onocleaceae (*Matteuccia struthiopteris*), Dennstaedtiaceae (*Pteridium aquilinum*) and Osmundaceae (*Osmundastrum asiaticum*). For each species, fronds were from two Russian Far Eastern regions–Primorsky Krai and Amur Oblast–with the collection sites being about 850 km apart. The content of total lipids ranged from 0.8–0.9 per cent of wet weight in *O. asiaticum* to 1.0–1.2 per cent in *P. aquilinum*. The lipid content in the fiddleheads of *P. aquilinum* was significantly higher ($p < 0.05$) than in the other two species. The lipid content was higher in the ferns harvested in Amur Oblast; however, the difference between the two regions was significant only for *P. aquilinum*.

The major fatty acids of the fiddleheads were palmitic 16:0 (25.0–30.0 per cent of total fatty acids), linoleic 18:2n-6 (17.0–27.0 per cent), α-linolenic 18:3n-3 (10.8–24.0 per cent), ARA 20:4n-6 (6.4–13.5 per cent), and oleic 18:1n-9 (5.0–10.0 per cent) acids (Etenko *et al.,* 2011). Significant portions were found for 24:0 (1.3–3.2 per cent), EPA 20:5n-3 (0.8–3.2 per cent), 16:3n-3 (1.3–2.8 per cent), γ-linolenic 18:3n-6 (0.9–2.7 per cent) and dihomo-γ-linolenic 20:3n-6 (0.9–2.6 per cent) acids, 18:0 (1.0–2.0 per cent), 22:0 (0.6–2.5 per cent) and 18:1n-7 (0.6–1.6 per cent). All other fatty acids were usually below 1 per cent. Although the fern species did not differ significantly in terms of the sum of polyunsaturated fatty acids (PUFAs), they were different in terms of the content of omega-6 (n-6) and omega-3 (n-3) fatty acids. *P. aquilinum* had the highest ratio of n-6/n-3 fatty acids (2.6–3.2) due to the high levels of linoleic acid and, to a lesser extent, ARA, and lower percentages of α-linolenic acid and EPA.

Regional differences in the fatty acid percentages were not similar for the fern species. The sum of PUFAs was significantly higher in *M. struthiopteris* from Primorsky Krai and *O. asiaticum* from Amur Oblast, whereas there was no difference between the two regions for *P. aquilinum*. In the literature, data on the fatty acid composition of young fronds of edible ferns cover *M. struthiopteris* of North American origin (Bushway *et al.,* 1982; DeLong *et al.,* 2013), *P. aquilinum* from Russia (Shalisko *et al.,* 2016), as well as these and other species of the European origin (Dvorakova *et al.,* 2021). In general, the contents of fatty acids in the young fronds of *P. aquilinum* and *M. struthiopteris* have a similar tendency: high percentages of palmitic and linoleic acids. followed by α-linolenic and/or ARA, which have been found previously for these (DeLong *et al.,* 2011). A very high content of γ-linolenic acid (18:3n-6) was found for fern fiddleheads in a previous study (Dvorakova *et al.*, 2021), which was comparable or even exceeded the content of palmitic acid. This was not observed in other studies (DeLong *et al.*, 2011; Shalisko *et al.*, 2016).

18.4.2 Chemicals present in Processed Fronds

Processed fronds (dried or salted) were not significantly different from freshly harvested fronds of corresponding species in terms of lipid content on the dry weight basis. A significant difference was found only for the salted fronds of the bracken fern from Amur Oblast; however, the value was closer to the freshly harvested fronds (7.9 per cent vs. 8.4 per cent dry weight) as compared to the two other products (7.7 per cent and 7.8 per cent), which had slightly higher deviations of their means.

When commercially available fern fronds processed in different ways were compared in terms of their fatty acid contents, the biggest difference was noted for salted fronds of *P. aquilinum*. The salted products from two different producers were characterized by higher percentages of saturated fatty acids (mostly due to palmitic acid), monounsaturated fatty acids (due to oleic acid), and a lower one of polyunsaturated fatty acids (both n-6 and n-3) as compared to the freshly harvested fronds. The difference in the weight content of fatty acids was less pronounced for saturated and monounsaturated fatty acids, but was reproduced for PUFAs. All the major PUFAs, including ARA and EPA, were significantly lower in the salted fronds.

The dried bracken fern product from China contained less total lipids, total fatty acids and individual fatty acids on a dry weight basis than the raw fronds harvested in Amur Oblast; however, the differences were often less pronounced than in the case of the salted products. Moreover, the raw fronds and the dried product were quite similar in their percentages of fatty acids. Significant differences were found only for saturated (more palmitic acid in the dried fronds), monounsaturated (less oleic acid in the dried fronds), and some minor PUFAs (18:3n-6, 18:4n-3, 20:2n-6, SCA). Similar results were observed for the dried ostrich fern product from Sakhalin Island and the raw fronds of *M. struthiopteris* harvested in Amur Oblast– the samples contained similar levels of total lipids (the content was insignificantly higher in the dried fronds), as well as total and individual fatty acids. The dried fronds from Sakhalin Island contained more α-linolenic acid and less oleic acid (both in percentage and weight content). Other differences were related to some minor fatty acids.

To investigate the effect of drying on fatty acid content in fern fronds, fatty acids were analyzed in fiddleheads of *O. asiaticum* dried under two conditions: at 110 °C for 26 h or at room temperature under exposure to sunlight for 6 days. The dried fiddleheads were found to contain less total fatty acids, as well some major fatty acids (16:0, 18:2n-6, ARA) than the raw fronds; however, the differences were insignificant. The percentage of fatty acids was almost identical for the raw and dried fiddleheads. They can be dried quickly at a high temperature or slowly at a low temperature (providing that the fronds are preliminary boiled to inactivate enzymes).

Fern fronds processed in different ways have been analyzed previously in terms of their nutrient composition: canned and frozen ostrich fern fiddleheads (Bushway *et al.,* 1985), salted (with and without preliminary blanching) ostrich fern fiddleheads, ostrich fern fiddleheads stored in cold water (DeLong *et al.,* 2013) and frozen and salted bracken fern fronds. A small decrease in fat (lipid) content in the canned and frozen fiddleheads was observed, as compared to raw ones, accompanied by increased water content (Bushway *et al.,* 1985). The storage of raw ostrich fern fiddleheads in cold water led to the accumulation of some fatty acids, including saturated, mono- and polyunsaturated acids (but not EPA), an effect that was attributed to cold temperature acclimation and caused by fatty acid upregulation (DeLong *et al.,* 2013). Preliminary thermal treatment (boiling in water for 1 min) of ostrich fern fiddleheads had a positive effect on the quality and shelf-life of the salted product and was suggested as a preferable method for enzyme inactivation.

Quick enzyme inactivation after boiling in water or heating at a high temperature during the drying process provides better preservation of PUFAs than the protracted process of salting, even at a high salt concentration. The influence of salt on lipid oxidation in meat and seafood products is under discussion and possible mechanisms of salt pro-oxidant activity have been suggested (Mariutti and Bragagnolo 2017).

18.4.3 Chemicals Present in Cooked Ferns

There are different dishes with ferns in the Russian Far East. Usually, fern fronds are boiled, then fried or stewed and used in a hot dish with meat, in salads or appetizers. The purpose of this part of the study was to confirm the preservation of the valuable PUFAs (ARA, EPA, γ-linolenic and dihomo-γ-linolenic acids) in cooked ferns. Since meat contains all the fatty acids, we analyzed dishes with ferns that were free of any meat: a salad with the bracken fern and an appetizer with *Osmunda* fern.

The percentages of fatty acids in the cooked ferns were quite different from all other samples (raw and processed) due to the high levels of oleic and linoleic acids. The major portion of the fatty acids obviously originated from vegetable oils used for the frying or dressing of the ferns. Indeed, the content of total lipids in the parts of the bracken fern from the salad was almost twice as high as that measured in its raw fronds in wet and dry weights. The weight content of oleic and linoleic acids in the fern from the salad also significantly exceeded their content in the uncooked fronds (more than six times higher than in the fronds of the salted bracken fern from Amur Oblast). The weight contents of the valuable PUFAs (ARA, EPA, γ-linolenic, and dihomo-γ-linolenic acids), as well as some major fatty acids (palmitic and α-linolenic), were close to those of the salted fern from Amur Oblast.

In contrast, the fern in the appetizer was not significantly different from the crosiers of *O. asiaticum* in terms of the total lipid content. Although the percentages for oleic and linoleic acids were also twice as high than those observed in the raw fiddleheads, their weight contents (along with palmitic acid) were not different between the cooked and raw material. The majority of other fatty acids, including the polyunsaturated ones of interest, were significantly lower in weight content in the cooked fern parts (up to 7.5 times for ARA). A likely reason for this observation may be the parts of the fern used in the dish. The appetizer was found to contain only pieces of frond petioles, which can be depleted in lipids as compared to leafy fiddleheads (Jarvis and Duncan 1974; Ansorena *et al.,* 2010).

Altogether, these results demonstrate the preservation of the valuable PUFAs in processed and cooked ferns; thus, fatty acids become part of a diet when ferns are consumed. These results are in agreement with the literature data. Changes in the fatty acid composition of food during cooking have been the subject of a number of studies (Gladyshev *et al.,* 2007; Leung *et al.,* 2018). There is accumulation of the oxidized products of PUFAs under cooking conditions (Leung *et al.,* 2018).

18.4.4 Comparison of Ferns with other Sources of LC-PUFAs

Low Carbon -PUFAs are considered important for human physiology and have bioactive potential (Tsai *et al.,* 2018). Fish is the major source of long-chain omega-3 fatty acids (EPA, DHA), whereas long-chain omega-6 fatty acids (ARA and its precursor, 20:3n-6) are consumed mainly with meat and other foods of animal origin (Marventano 2015; Kawashima 2019).

The ARA intake for normal healthy adults was estimated to be 100–250 mg/day in advanced countries, whereas vegetarians may consume much less: 3–44 mg/day (Kawashima 2019). The edible ferns in our study were found to contain ARA: 80.1 mg/100 g wet weight (5.5 mg/g dry weight) for *P. aquilinum*, 52.5 mg/100 g wet weight (4.1 mg/g dry weight) for *M. struthiopteris* and 40.2 mg/100 g wet weight (2.15 mg/g dry weight) for *O. asiaticum*. In the cooked ferns, the ARA content was found to be 72.0 ± 4.3 mg/100 g wet weight for the bracken fern from the salad and 7.1 ± 1.9 mg/100 g wet weight for the *Osmunda* fern from the appetizer. Thus, the edible ferns contain ARA in quantities comparable with meat (4.8–97 mg/100 g wet weight) (Haak *et al.,* 2007; Dugan *et al.,* 2015; Ye *et al.,* 2020), many fish species (0.29–0.85 mg/g dry weight, 30–212 mg/100 g wet weight) (Czerner 2015) and edible macroalgae (0.13–5.4 mg/g dry weight) (Biancarosa *et al.,* 2018).

The average daily intake of EPA is 33 mg in the USA, though there are significant differences depending on age, gender and pregnancy status (Thompson *et al.,* 2019). On the wet weight basis, its content in the ferns was: 7.1 mg/100 g for *P. aquilinum*, 14.1 mg/100 g for *M. struthiopteris* and 14.5 mg/100 g for *O. asiaticum*. In terms of EPA content, the ferns are obviously inferior to fish (3.1–7.2 mg/g dry weight and up to 1.7 g/100 g wet weight) (Gladyshev *et al.,* 2007), some macroalgae (0.39–8.3 mg/g dry weight) (Biancarosa *et al.,* 2018) and some meat products (29, 28, 27, and 46 mg/100 g wet weight for beef, veal, lamb and mutton, respectively (Droulez *et al.,* 2002), 21–33 mg/100 g wet weight for lamb (Ye *et al.,* 2020). However, the ferns are comparable in terms of EPA content with other kinds of meat, such as pork (2.2–5.3 mg/100 g wet weight (Dugan *et al.,* 2015), and broiler chicken (1.4–2.4 mg/100 g wet weight (De Souza *et al.,* 2021).

Unlike animal sources, terrestrial ferns do not contain DHA (Dvorakova *et al.*, 2021). However, the edible ferns contain precursors of ARA and EPA which can contribute to their dietary value. Potentially bioactive (Tsai *et al.*, 2018) Δ5-unsaturated polymethylene-interrupted fatty acids, such as 5,11,14-20:3 and 5,11,14,17-20:4, are minor components in the young fronds of the edible ferns with the maximum content of only 0.2 and 0.1 mg/g dry weight (or 0.8 per cent and 0.5 per cent of total fatty acids) for SCA and JA, respectively, in *O. asiaticum*. Seeds of some conifer and flowering species contain much more these fatty acids (Wolff and Christie 2002) though not all of them are usually consumed as food.

In general, this study has proven the dietary value of edible ferns from the Russian Far East as sources of LC-PUFAs. However, the processing of these ferns' young fronds should be improved in order to avoid the loss of these acids, as was found with the fiddleheads after dry salting.

18.5 Edible Ferns from Himalayas

After the small village of Vashist across the Beas River from touristic Manali to hike up to a scenic waterfall. Apple orchards were in full bloom around us. As we were gaining elevation, we entered the Deodar (*Cedrus deodara)* forest (Bernard 2016). Up there by the waterfall viewing the valley below, a small crowd of pilgrims were paying respect to Lord Shiva. In all its simplicity a shelter more than a temple, made of stones and covered with a slate roof, was telling us about the intimate link between the source of the waters and the origin of a religion. On our way downhill, we met two young hill ladies who were gathering something from the open meadows between the tall cedars. We approached them, asking what they were doing and if we could take some photos. They opened their basket, producing handfuls of fern shoots to sell at the local market where they were considered a delicacy.

Our next stop on our quest for edible ferns would take us to the city of Palampur in the Kangra Valley. In the Institute of Himalayan Bioresources Technology we met Dr. Alka Kumari of the Biodiversity department. Dr Alka is the fern expert in India. After welcoming us, her first words to us were: "No fern is useless," and she went on enumerating a list of beneficial properties of ferns. Our visit started with the Institute's fern herbarium where 300 species of native ferns are stored and actively maintained and reproduced through their spores, a sophisticated technique that Dr. Alka is one of the few to master. It is uncontroversial that plants have profoundly influenced the culture,and different stages of civilization of human race in many parts of the world. We then headed to the fern house where *Adiantum* and the majestic Royal Fern (*Osmunda*) that carries spores on its spikes were growing.

But what about edible ferns? What about the fern shoots that those ladies we met in the mountains were gathering? Only two species of Himalayan ferns are edible and safe: *Diplizium esculentum* and *D. maximus.* Both are found at elevations between 1828 m and 3050 m in the Himalayas, often in clearings among Deodar cedars. They closely resemble ferns of the *Pteridium* family that are also edible, but not recommended as they contain carcinogenic components. It takes real skills, comparable to identifying edible mushrooms, to differentiate edible ferns from toxic ferns at the early vegetative stage of shoots, also called fiddleheads. The common name in the Western Himalayas for edible ferns is "lingri". They are a seasonal source of income for the hill people and a seasonal delicacy in the months of April and May on Northern Indian tables. They are usually sautéed and eaten as a snack. They can also be added to curry sauces.

The other part of the world where edible ferns are gathered is in North America in the humid mountains of New England and of the Pacific Northwest. Those were known and gathered by the Native Americans in the same way of the hill people of the Himalayas. One question lingered after having been seeded in our mind by Dr. Alka: Could those ferns be grown and made available to consumers, so they don't become depleted from their natural environments? Could edible ferns be reproduced from spores to produce "shoot greens" in protected environments like greenhouses in temperate climates? Efforts are on to bring fern shoots to our tables.

The fern collected from area and the literature consulted Khullar ferns of Western Himalayas as documented is 107,out of which 34 are found to be of medicinal use (Singh *et al.,* 2021). Since the area is a tourist spot the ferns are under the threat. The data collected shows that ferns need to be conserve as they have many important medicinal properties as they can be used in the fields of clinical research for making medicines, also in treatment of common ailments like stomach ache, diarrhea, dysentery skin problems *etc.* The plant parts used by the inhabitant in different form such as juice, extract, decoction and paste *etc.* The *Adiantum venustum* is the single species has wide spectrum which is used in different disease such as cough, fever, tumour, antimicrobial activity and for wounds, similarly *Dryopteris cochleata* is used in diarrohea, anthelminthic activity and snake bite. While *Asplenium adiantum-nigrumis* used in eyes, jaundice, cough. Due to the increase in deforestation for the development of tourist spots present number of taxa in ferns and fern allies species have been eradicated from Mussoorie. First historical effort was made by Caius in 1935 to describe the medicinal uses of ferns in India, therefore, get recognized as the first man to take this very initiative kind of investigations. Important contribution on studies of ferns in this region have been carried out by several investigators (Clarke 1880; Beddome 1883; Hope 1899). Ferns are lesser known for their economic utilization and not much documented several workers have emphasized this aspect of the group (Bir 1983; Tryon and Tryon 2012).

18.6 How to use Fiddleheads ?

Humans have been eating ferns for thousands of years. Edible ferns are mentioned in Chinese literature as early as 3000 years ago.

Several fern species are edible, but fiddlehead fern generally refers to the ostrich fern (*Matteucia struthiopteris*). The ostrich fern is safest for consumption and has the lowest toxicity; in contrast, other edible species are toxic or carcinogenic when raw. The ostrich fern grows in damp, temperate areas and is widespread throughout the northern hemisphere, mainly thriving in cooler climates. As with other edible ferns, ostrich ferns are harvested in the spring before the fronds unfurl. The fiddlehead stage lasts only for a few days, and ferns are not cultivated but foraged only in the wild.

To maximize shelf life, loosely wrap unwashed fiddleheads in a paper towel and refrigerate them in a sealed plastic bag. Fiddleheads can remain edible for up to 10 days when refrigerated; however, their flavor fades over time, so it is still best to consume them shortly after purchase. Before cooking fiddleheads, should be washed in cold, running water to remove any dirt or residual soil; afterward, place them in a bowl of water and rub off the brown husk. To cook fiddleheads, one should either boil them for 15 min or steam them for 10-12 min. until tender. Discard the water used for boiling or steaming the fiddleheads.

It is essential to thoroughly cook fiddleheads before using them in a recipe to neutralize any harmful elements. In outbreaks of foodborne illness associated with fiddleheads, the ferns eaten were usually either raw or only lightly cooked (sautéed, parboiled, or microwaved). Once rendered safe to eat through boiling or steaming, fiddleheads can be prepared like any other vegetable. Many describe its taste as a cross between *Asparagus* and green bean with an additional hint of nuttiness. Edible ferns can give you a taste of spring, but only certain species. The ostrich fern is the safest species to eat, but the lady fern and the bracken fern can also be eaten if prepared properly. Regardless of species, you should only consume ferns in the "fiddlehead" stage, when the plant is still young.

18.7 Use of Ferns in Human Health

Thirty-six plants belonging to 23 families were documented in a field survey. This study is based on structured and unstructured interviews with 45 herbalists and 20 healers or medicine men and local inhabitants of southern Nigeria. Taxonomic identification was compiled with aid of keys (Nwosu 1997).

Adiantum aethiopicum L./Adiantaceae/True Maidenhair *Fern/Ntutu umu/ Decoction* of whole plant used as tonic against bruises, gangrenous wounds; expectorant; diuretic disorders/MON 010.

Adiantum capillus-veneris L./Adiantaceae/Net Hair *Fern/Ofu isi, Akite, Apa/ Infusion* of leaves drunk against infertility and other women's diseases such as vaginal discharge caused by fungal growth, *e.g., Candida* spp.; decoction of roots used for washing hair to remove dandruff; powdered rhizome is mixed with camphor and dissolved in palm-kernel oil and applied externally to hair infested by lice known locally as *Jiga* (*Pedicullus* spp.)/MON 002.

Adiantum caudatum L./Adiantaceae/Trailing Maidenhair *FernlNtutu nwanyi, Akpado/*Decoction used for treating jaundice in animals; against scabies; potherb ta ken internally against abdominal pain, constipation/MON 003.

Adiantum incisum Forsk./Adiantaceae/no common *name/Afufu, Akpado/ lnfusion* of young fronds used as treatment against malaria and bronchial diseases/ MON 004.

Adiantum venustum G.Don./Adiantaceae/Venus Hair *Fern/Aban; Nya/Extract* of rhizome and fronds drunk against diabetes and liver problems; diuretic; infusion taken internally for circulatory disorders/MON 005.

Aleuriopteris albomarginata Clarke/Sinoptereridaceae/Glade *FernlOgbogbo, lgba, Doka/De*coction of whole plant taken internally against peptic ulcer; fronds are dried under the sun and powdered, mixed with pulverized ginger rhizomes (*Zingiber officinale* L.) and used for treatment of infertility in women; infusion taken orally against dysentery/MON 006.

Ampeloptreris prolifera Its ethnobotanical use was recorded in India. Local tribals of Tehri region of Uttar Pradesh and Pachmarhi in Central India use this fern as a vegetable.

Arthromeris wallichiana (Spr.) Ching/Polypodiaceae/no common *name/ Abua ocha, Ab~lila/*Paste prepared from leaves applied externally as treatment for sprains; also paste is mixed with egg albumen to obtain an homogenous mixture and applied to fractured bone (bone resetting process); for calcinosis/MON 007.

Asplenium achillerfolium (Lamb.) C. Ch./Aspleniaceae/Common Mother *Spleenwort/Icheki, Kpaye/Decoction* of leaves and roots drunk as health tea; purge; paste of leaves used externally as facial beauty mask; powder from roots mixed with palm-kernel oil (*Elaeis guineensis* Jacq.) taken internally to relief stomachache and intestinal worms/MON 008.

Asplenium adiantum-nigrum L./Aspleniaceae/Black *Spleenwort/Fara, lkpa/ Extract* of fronds drunk with castor oil (*Ricinus communis* L.) as purge to remove intestinal worms especially tapeworm (*Taenia* spp.); decoction of leaves taken against urinary disorders; tonic used as antiseptic after ear-piercing (female babies and young girls)/MON 009.

Asplenium bulbiferum L./Aspleniaceae/Mother *Spleenwort/Ubeze, Okrokro/ Paste* prepared from leaves applied externally for hemorrhoids; decoction of fronds taken for liver problems/MON 010.

Botry~chium lanuginosum Wall. ex Hook et Grev./Botrychiaceae/Grape *Fern/ lya, Oziza ato/*Young leaves eaten as a green vegetable; boiled rhizomes used in treating pneumonia, catarrh, expectorant; paste from roots used as facial mask (cosmetics) by young girls preparing for traditional marriage/MON 011.

Cheilanthes albomarginata Clarke/Sinopteridaceae/Bristly Cloak Fern, Slender Lip *Fern~lie Nza/Leaf* extract used externally for maturation of abscess; decoction of rhizomes administered internally against diarrhea/MON 012.

Coniogramme caudata (Wall. ex Ettingsch) Ching/Hemionitidaceae/Tree *Fern/Marga, Apasha/Extract* of fronds taken internally to relieve urinary disorders, backache, and gall stones/MON 013.

Diplazium esculentum (Retz.) Sw./Athyriaceae/Vegetable *Fern/Akwukwo Nni/Young* leaves eaten as a vegetable with yam (*Dioscorea* spp.); infusion of fronds used for toothache (caries); fronds used by pregnant women as protection against difficult childbirth/MON 014.

Dryopteris filix-mas L./Dryopteridaceae/Common Wild *Fern/Ihi, Erinji/ Young* circinately coiled fronds used as vermifuge; The robust rhizomes of the perennial fern *D. authamantica* were used as an effective vermifuge particularly in the treatment of tapeworm. *D paleaceae* was used as an antithelmentic in some regions of Columbia (Padala 1988). infusion of leaves used as aphrodisiac/MON 015.

Dryopteris wallichiana (Spreng) Hyl./Dryopteridaceae/Mountain Wood *Fern/ Mbabe, Ire, Ajanmuo/Leaves* are used as fodder for goats and sheep; fronds used as manure for homestead gardens; tonic from roots drunk against rheumatism/ MON 016.

Equisetum diffusum D. Don./Equisetaceae/Branched *Horsetail/Aziza, Eru/ Decoction* of roots used for treatment of psychosis; dried leaves powder mixed with water used as insect repellent (*e.g.*, mosquitoes)/MON 017. Herb *E. arvense* was a home remedy for kidney disorders in many parts of Europe. The tender basal attachment was used as vegetable in raw form by Kiowa Indians of Mexico. *E. bogeotense* was used as an effective diuretic in Colombia and Peru, besides as an astringent, antihemmoragic in both raw and dried form. It was used since decades for diabetic in Venezuelan Peninsula. The fragile as well as older stem tips were dried and a saturated decoction was prepared from *E. hiemale*. The elixir was used for polishing wooden musical instruments and furniture by many tribals as well as professional wood workers.

***Glaphyropteridopsis erubescens* (Wall ex Hook.) Ching**/Thelypteridaceae/ No common/*Nzu/Decoction* of young fronds used against indigestion for adults; paste prepared from leaves applied externally for treatment of rheumatism; powder from dried roots mixed with water used as antidote for scorpion bite/MON 018.

Gleichenia linearis L./Gleichenaceae/Savannah *Fern/Ami Ogwu, Usele/Extract* of whole plant administered internally to children suffering from convulsion after which a cold bath is given to reduce high body temperature. This therapy is sometimes followed up by incantations by the healer to chase away evil spirits that are usually associated with this illness/MON 019.

Lycopodium clavatum L./Lycopodiaceae/Running *Pine/Agwo/Decoction* of young fronds applied externally to deteriorating wounds or ulcer; pulverized sun-dried leaves mixed with mashed unripe dried plantain (*Musa acuminata*) to which milk is added and given in small doses to infants suffering from diarrhea, dysentery/MON020.

Lycopodium selago L./Lycopodiaceae/Fir Se*lago/Agwo, Ude, Otuen/Spores* and root powder mixed with palm kernel oil (*Elaeis guineensis*) is applied externally to treat eczema, cuts; extract of leaves drunk against snake bite as antidote; whole plant is hung in front of doors to chase away evil spirits/MON 021.

Lygodium flexuosum L. Sw./Lygodiaceae/Climbing *Fern/Agbiligbi, Usehe/ Infusion* of leaves for treating female infertility; paste is applied to fix fractured bones; against hemorrhoids/MON 022.

Lygodium palmatum L./Lygodiaceae/American Palm *Fern/Nyie awu, Eyanyan/Aqueous* extract of young fronds applied externally to boils, abscess, swellings; used as astringent, emollient, and cosmetics by women/MON 023.

Marsilea quadrifolia L./Marsileaceae/Water *Shamrock/Akwa nmili, Nze/ Whole* plant extract used as aphrodisiac (for fertility by married couple); this plant is therefore planted within homestead gardens for protection; also used as ornamental/MON 024.

Nephrolepsis cordifolia (L.) Presl./Nephrolepidaceae/Erect *Swordfern/Nma ozo/Tubers* are cleaned and boiled with salt and water and eaten as food; infusion of fronds is administered to elderly folks for treatment of amnesia/MON 025.

Ophioglossum vulgatum L./Ophioglossaceae/Common Tongue *Fern/lle nza/ Decoction* of rhizome taken internally to treat pulmonary/bronchial and heart diseases; dried pulverized rhizome externally applied to ulcerated sores and wounds; for burns/MON 026.

Ophioglossum grande L./Ophioglossaceae/Ribbon *Fern/Achu, Tsage/Young* fronds used as fodder for livestock (goats); retted leaves used as garden manure; extract of rhizomes used as antidote for snake bite/MON 027.

Osmunda regalis L./Osmundaceae/Royal *Fern/Eboshi/Whole* plant extract taken internally for psychosis; it is believed that tonic can chase away evil spirits; also used for treatment of moon-madness; infusion of roots used against malaria and jaundice/MON 028.

Polypodium microrhizoma Clarke ex Baker/Polypodiaceae/Golden *Locks/ Achandu, Akoti/*Decoction of rhizome and leaves used for the relief of gastrointestinal disorders; backache; jaundice; paste prepared from dried leaves (leaves are dried over the open fire) are applied externally to fissures on hand and wound healing. The paste is mixed with palm-kernel oil and is applied externally to domestic animals such as sheep or cattle/MON 029. The tender tips of fronds and the rhizome of *P. angustifolium* were used to cure high temperature and dropsy among the tribes of Peru. Similarly, the concentrated decoction of rhizomes of *P. aureum* was used as remedy for cough and cold both for infants as well as for adults. Root rhizome and tender parts of *P. fimbriatum* as an effective diuretic and an expectorant. This was also used as analgesic by some tribals of El Salvador.

Pteridium aquilinum (L.) Kuhn v. Decken/Pteridaceae/Eastern Bracken, Turkey Foot Fern/*Ukwunnume/Decoction* of rhizome drunk as herbal health tea; purge; young fronds eaten as vegetables; retted fronds used as manure in homestead gardens MON 030.

Pteris quadriaurita Retz./Pteridaceae/*Ntie/Crushed* rhizome used as astringent, emollient; extract of leaves prepared with local gin (locally distilled gin prepared from palm wine (*Elaeis guineensis*) is used in households as insect repellent especially for red ants or termites during the rainy season/MON 031.

Pteris vittata L./Pteridaceae/Rusty Brake, Chinese *Brake/Beke/Whole* plant is poisonous; it is usually planted on village borders; local inhabitants believe it keeps enemies away/MON 032.

Selaginella cinerascens L./Selaginellaceae/Pygmy *Selaginella/Akakpo, Idea~Dried* rhizomes and spores are pulverized, mixed with boiled water and used by elderly women as douche; emollient; infusion of leaves drunk to relieve gastrointestinal disorders such as heart burns and atrophy caused by excessive acid from meals prepared from pulses, *e.g.*, black-eyed beans (*Vigna unguicullata* (L.) Wasp.) or drinking of unskimmed fresh milk/MON 033.

Selaginella serpens Spring/Selaginellaceae/Snake *Selaginella/Agwo elili/ Whole* plant is grown near houses. Its presence is believed to chase away evil spirits; retted fronds used as manure; decoction drunk against liver problems; hepatoma; paste prepared from leaves mixed with ashes from burnt inflorescence of oil palm (*Elaeis guinenses*) is applied externally to tattooed skin as an antiseptic (cosmetics)/MON 034.

Selaginella pallescens (C. Prest) Spring/Selaginellaceae/Moss *Fern/Njule/ Paste* of crushed leaves applied externally on joints to treat rheumatism; infusion of rhizomes is drunk against insomnia; young fronds are steamed over char coal fire and mixed with Shear Butter Fat *lVitellaria paradoxa,* syn. *Butyrospermum paradoxum*) and applied for treatment of hemorrhoids as ointment. It is a applied on fistulas/MON 035.

Tectaria macrodonta (Fee) C. Chr./Tectariaceae/No common *name/Ose ala/ Dried* leaves are powdered and mixed with castor oil (*Ricinus communis*) and given to household animals (goats and sheep) to stop running stomach; young fronds are chewed by livestock (cow) after delivery of their young ones to accelerate the expulsion of the afterbirth (placenta); paste prepared from ground roots externally applied to stop foot itching caused by fungal growth; infusion of roots drunk against venereal diseases, *e.g.*, gonorrhea/MON 036.

Woodwardia unigemmata (Mak.) Nakai/Blechnaceae/Chain *Fern/Ntue/ Infusion* of spores and rhizomes administered internally to treat skin diseases, *e.g.*, ringworm (*Tinea vesicolor*); decoction of leaves used as tea as well as bath for curing infertility in women; for abdominal pain, constipation, and sore throat/MON 036.

REFERENCES

Ansorena, D.; Guembe, A.; Mendizabal, T.; Astiasaran, I. (2010) Effect of fish and oil nature on frying process and nutritional product quality. J Food Sci. 75(2): H62-67.

Beddome RH. A Handbook to the Ferns of British India, Ceylon and Malaya Peninsula, Supplement. Thacker Spink and Co., Calcutta, 1883.

Bernard R. (2016) Edible Ferns: No fern is useless. In Plant profile, Mother earth gardner. Pub on August 3, 2016

Biancarosa I., Belghit I., Bruckner C.G., Liland N.S., Waagbo R., Amlund H.; Heesch S.; Lock, E.J. (2018) Chemical characterization of 21 species of marine macroalgae common in Norwegian waters: Benefits of and limitations to their potential use in food and feed. J. Sci. Food Agric. 98: 2035–2042.

Bir S S. (1983) Pteridophytic flora of Garhwal Himalaya (Mussoorie, Dehradun, Chakrata, and adjoining hills): an annotated catalogue of the recorded species, their ecology, distribution in the region, and common synonyms. P.K. Rastogi for Jugal Kishore, Dehradun,

Bushway A.A.; Serreze D.V.; McGann D.F.; True R.H.; Work T.M.; BushwayR.J. (1985) Effect of processing method and storage time on the Nutrient Composition of Fiddlehead Greens. Food Science.50(5): 1491-1492

Bushway A. A., Wilson A.M., McGann D.F., Bushway R.J. (1982) The nutrient composition of fresh fiddlehead greens. J. Food Sci. 47: 666–667.

Caius J.F. (1935) Medicinal and poisonous ferns of India. Bombay Nat. Hist. Soc. (1935).

Clarke CB. (1880) A review of the ferns of Northen India. Trans. Linn. Soc. London II Bot,1880: 1: 425-611.

Czerner, M.; Agustinelli, S.P.; Guccione, S.; Yeannes, M.I. (2015) Effect of different preservation processes on chemical composition and fatty acid profile of anchovy (*Engraulis anchoita*). Int. J. Food Sci. Nutr. 66: 887–894.

DeLong J. M., HodgesD. M, PrangeR. K., Forney C. F., FanL., BishopM. C., ElliotM. L., JordanM. A., and Doucette C. (2013) The influence of cold water storage on fatty acids, antioxidant content and activity, and microbial load in ostrich fern (*Matteuccia struthiopteris*) fiddleheads. Canadian Journal of Plant Science. 93(4): 683 697. https: //doi.org/10.4141/cjps2012-165

DeLong, J.M.; Prange, R.K. (2008) Fiddlehead fronds: Nutrient rich delicacy. Chron. Hortic. 48: 12–15.

DeLong, J.M.; Hodges, D.M.; Prange, R.K.; Forney, C.F.; Toivenon, P.M.A.; Bishop, M.C.; Elliot, M.L.; Jordan, M.A. (2011) The unique fatty acid and antioxidant composition of ostrich fern (*Matteuccia struthiopteris*) fiddleheads. Can. J. Plant Sci. 91: 919–930.

Dixit R.D. (1984) A census of the Indian Pteridophytes. Report by BSI Hawra.

Droulez, V.; William, P.G.; Levy, G.; Stobaus, T.; Sinclair, A. (2006) Composition of Australian red meat 2002. 2. Fatty acid profile. Food Australia, 58: 335–341.

Dugan, M.E.; Vahmani, P.; Turner, T.D.; Mapiye, C.; Juarez, M.; Prieto, N.; Beaulieu, A.D.; Zijlstra, R.T.; Patience, J.F.; Aalhus, J.L. (2015) Pork as a source of omega-3 (n-3) fatty acids. J. Clin. Med. 4: 1999–2011.

Dvorakova, M.; Pumprova, K.; Antoninova, Z.; Rezek, J.; Haisel, D.; Ekrt, L.; Vanek, T.; Langhansova, L. (1996) Nutritional and Econ. Bot. 50: 57–70.

Etenko, O.A.; Ivankina, N.F.; Anikina, E.A.(2011) Fatty acid composition of bracken fern fronds from the southern and northern zones of Amur Oblast. In Botanical Investigation in Priamurie and Borderline Territory, Proceedings of the Region Conference, Blagoveshchensk, fatty acid and antioxidant composition of ostrich fern (*Matteuccia struthiopteris*) fiddleheads. Can. J. Plant Sci. 91, 919–930.

Gerasimenko, N.; Korol, A.; Pikhanova, S.; Gochachko, S. (2003) Marketing evaluation of non-wood products of Far Eastern south forests. Pract.Mark. 3: 17–25. (In Russian)

Gladyshev, M.I.; Sushchik, N.N.; Gubanenko, G.A.; Demirchieva, S.M.; Kalachova, G.S. (2007) Effect of boiling and frying on the content of essential polyunsaturated fatty acids in muscle tissue of four fish species. Food Chem. 101: 1694–1700.

Haak L., Sioen I. Raes K.; Van Camp J., De Smet S. (2007) Effect of pan-frying in different culinary fats on the fatty acid profile of pork. Food Chem. 102: 857–864.

Harayama, T.; Shimizu, T. (2020) Roles of polyunsaturated fatty acids, from mediators to membranes. J. Lipid Res. 61: 1150–1160.

Hope C W. (2013) The ferns of North Western India including Afganistan, the Trans Indus protected states and influence of cold water storage on fatty acids, antioxidant content and activity, and microbial load in ostrich fern (*Matteuccia struthiopteris*) fiddleheads. Can. J. Plant Sci. 93: 683–697.

Jarvis, M.C.; Duncan, H.J.(1974) Distribution of glycolipids and phospholipids in *Pteridium aquilinum*. Phytochemistry 13: 979–981.

Kawashima, H. (2019) Intake of arachidonic acid-containing lipids in adult humans: Dietary surveys and clinical trials. Lipids Health Dis. 18: 101.

Keller, H.A., Prance, G.T. (2015) The ethnobotany of ferns and lycophytes. Fern Gaz. *20:* 1–13.

Khrapko, O.V. (2007) Far Eastern ferns: Opportunities for application. Bull. Bot. Gard. Inst. 1: 81–87. (In Russian)

Koide, D.; Kadoya, T. (2019) Resource amount and cultural legacy affect spatially unbalanced human use of Japan's non-timber forest products. Ecol. Indic. 97: 204-210

Leung, K.S.; Galano, J.M.; Durand, T.; Lee, J.C.Y. (2018) Profiling of omega-polyunsaturated fatty acids and their oxidized products in salmon after different cooking methods. Antioxidants 7: 96.

Liu, Y.; Wujisguleng, W.; Long, C. (2012) Food uses of ferns in China: A review. *Acta Soc.* Bot. Polon. 81: 263–270.

Mariutti, L.R.B.; Bragagnolo, N. (2017) Influence of salt on lipid oxidation in meat and seafood products: A review. Food Res. Int. 94: 90–100.

Maroyi, A. (2014) Not just minor wild edible forest products: Consumption of pteridophytes in sub- Saharan Africa. J. Ethnobiol. Ethnomed. 10: 78.

Marventano, S.; Kolacz, P.; Castellano, S.; Galvano, F.; Buscemi, S.; Mistretta, A.; Grosso, G. (2015) A review of recent evidence in human studies of n-3 and n-6 PUFA intake on cardiovascular disease, cancer, and depressive disorders: Does the ratio really matter? Int. J. Food Sci. Nutr. 66: 611–622.

Moran, R.C. (2008) Diversity, biogeography and floristics. In: Ranker, T. A. and Haufler C. H. (eds.), Biology and Evolution of Ferns and Lycophytes. Cambridge University Press, New York 367–394

Nwosu. M.O.(1997)"Ethnobotanical Studies on Some Pteridophytes of Southern Nigeria." Economic Botany 56(3) 2002, pp. 255–59. *JSTOR*, http: //www.jstor.org/stable/4256579.

Nekrasov, E.V.; Tallon, S.J.; Vyssotski, M.V.; Catchpole, O.J. (1985) Extraction of lipids from New Zealand fern fronds using near-critical on the nutrient composition of fiddlehead greens. J. Food Sci. 50: 1491–1492.

Sathiyaraj G., Muthukumar T., Ravindran K.C. (2015) Ethnomedicinal importance of fern and fern allies traditionally used by tribal people of Palani Hills (Kodaikanal), Western Ghats, South India. J. Med. Herbs Ethnomed. 1(1), 4-9.

Shalisko, I.V.; Dmitrichenko, M.I.; Pelenko, V.V.; Shakhov, S.V. (2016) Changes in consumer properties of bracken using different storage methods. *Proceed.* Voronezh State Univ. Eng. Tech. 3: 151–158. (In Russian)

Singh C, Chauhan N, Rani A, Kishore K, Singh K, Singh R. (2021) Ethnobotanical and pharmacological studies of pteridophytes (Ferns) from Mussoorie region, Dehradun district, Uttarakhand. International Journal of Botany Studies. 6(6): 694-702

Sucheela MR and Devi S.(1997)Intercellular VAM spores in *Adiantum peruvianum*. In : XXI National Conf. Of Electron Microscopy Society of India, RRL. Thiruvanthapuram, Dec. 17-191997.

Thompson, M.; Hein, N.; Hanson, C.; Smith, L.M.; Anderson-Berry, A.; Richter, C.K.; Stessy Bisselou, K.; Kusi Appiah, A.; Kris-Etherton, P.; Skulas-Ray, A.C.; *et al.* (2019) Omega-3 fatty acid intake by age, gender, and pregnancy status in the United States: National health and nutrition examination survey 2003–2014. Nutrients 11: 177.

Tryon RM, Tryon AF. (2012) Ferns and allied plants: with special reference to tropical America. Springer Science and Business Media, 2012

Tsai P.J.; Huang W.C.; Lin S.W.; Chen S.N.; Shen H.J.; Chang H.; Chuang L.T. (2018) Juniperonic acid incorporation into the phospholipids of murine macrophage cells modulates pro-inflammatory mediator production. Inflammation 41: 1200–1214.

Tsevegsuren, N.; Aitzetmuller, K.(1997) Unusual Δ5*cis*-fatty acids in seed oils of *Cimicifuga* species. J. High Res. Chrom. 20: 237–241.

Wolff, R.L.; Christie, W.W. (2002) Structures, practical sources (gymnosperm seeds), gas- liquid chromatographic data (equivalent chain lengths), and mass spectrometric characteristics of all-*cis* Δ5-olefinic acids. Eur. J. Lipid Sci. Tech. 104: 234–244.

Ye, Y.; Eyres, G.T.; Reis, M.G.; Schreurs, N.M.; Silcock, P.; Agnew, M.P.; Johnson, P.L.; Maclean, P.; Realini, C.E. (2020) Fatty acid composition and volatile profile of *M. longissimus thoracis* from commercial lambs reared in different forage systems. Foods 9: 1885.

Chapter 19

Biochemical Assessment of Tropical Fruit Diets and their Applications

In spite of unprecedented human achievements in science and technology, including the Green Revolution 20th century, we still face the challenges of widespread poverty, chronic hunger, diseases and malnutrition. About 17 million children under 5 die with malnutrition, contributing to at least one third of these deaths. Fruits provide nutritional security and are needed for maintaining good health. Fruits are sources of different sugars, minerals, vitamins and other useful ingredients. About 1,000 ha is under fruit crops in arid regions of Gujarat. Important fruits being, ber, pomegranate, guava, sweet orange, date palm, fig, Jamun, and aonla. The productivity of fruits has doubled from 5.52 in 1961 to 10.23 tones per ha (in 1993) and the total production has increased by 32 per cent. In USA the losses from fruits and vegetables amount to 24 per cent. In India the postharvest losses may be as high as 40 per cent or even more. Taking the loss at 50 per cent the net availability of fruits and vegetable should be about 66 million tones. The change in our diet can make only as a result of conscious efforts, supported by knowledge age of dietary values of fruits and their scientific cultivation. Horticultural crops covering 8 per cent the total crop area contribute 24.5 per cent of the GDP. We have reached at the number 2 in production of fruits. We produce the highest quantity of mango, while productivity of grapes in India at 56 tones per hectare is a world record. This chapter discusses certain biochemical aspects of tropical fruits.

Keywords: *Biochemicals, Tropical fruits, Applications, Mango, Bioflavonoids, Flavour.*

19.1 Introduction

India is one of the leading producers of fruits and vegetables in the world but unfortunately, average Indian does not get the basic daily requirement of fruits and vegetables (Roy 2001). Although India has largest population of vegetarians yet fruits are consumed on very few occasions like festivals, parties or celebrations. In past few decades situation has changed. Now these are consumed by common man but more often not in the minimum amounts required daily. For our healthy and happy life balanced diet is recommended. Fruit is an important component of balance diet. These provide us necessary minerals, carbohydrates, and much needed vitamins. As per the specifications of the National Institute of Nutrition, at least 300 g of fruits and vegetables are to be consumed by an individual per day for a balance diet (Arya 2004). Fruit is a available source of antioxidants, bioflavonoids, and vitamin C, which may help to protect against degenerative diseases including cancer and heart diseases. As fruits are low in calories it is an excellent food for anyone trying to lose weight. I cannot think of any prosperity with hungry stomach. The availability of food must be ensured for increasing population to reduce the level of food insecurity that already exists in parts of the world food crops are viable alternative. Efficient and effective measures should be adapted to promote more fruit tree plantation and manage the postharvest losses, during storage, and marketing. Cultivation of fruits in a systematic manner should be practiced, as it is the need of the hour.

Of all the fruits Babur encountered in Hindustan, the only one that appealed to him was mango, although with the caveat that mangoes when good are very good, but as many are eaten, few are first rate. Babur planted gardens and plants were imported from Central Asia and Persia (Sen 2015). Fruits and vegetables indigenous to India include some varieties of pumpkins, melons and gourds. The Indian gooseberry, ber, jamoon, bael and jackfruit and one of India's culinary gifts to the world is mango (Achya 2002). Two indigenous nuts are the chironji (*Buchanania lanzan*) the seeds of which are used to flavour sweet dishes and the chilgoza (*Pinus gerardiana*) the nut of an evergreen tree. Coconut palm was domesticated long back and banana is grown for early times. Grapes, peaches, plums, apricot, pomegranate came to India from China. *Ain-I-Akbari* written by Abu'l- Fazl (1873) mentioned that a kitchen garden provided a continuous supply of fresh vegetables and especially fruit for the King Akbar. ' His Majesty looks upon fruits as one of the greatest gifts of the Creator, and pays much attention to them. Akbar brought horticulturists from Central Asia and Iran to supervise his orchards. They cultivated many varieties of melons, peaches, apricots, walnuts, pistachios, pomegranate, almonds, plums, apples, pears, cherries, chestnuts and grapes (Sen 2015).

Grapes occupy a predominant position in terms of world fruit production, accounting for about 16 per cent of the global fruit production. China is the leading producer of grapes in the world with a contribution of 15.09 per cent of the global grape production during the year 2012-13. Italy (10.38 per cent) is the second largest producer followed by U.S.A (10.03 per cent) and Spain (9.69 per cent). The world production during the period stood at 77181.12 MT. Indian stands in the ninth position with 3.22 5 of the global grape production during the same year. The major grape producing states in India are Maharashtra, Karnataka, and Tamil Nadu (Nithin 2016). In India, mango (*Mangifera indica* L.) is grown on the area of 1.52 million hectare with an annual production of 10.23 million tonnes with productivity of 6.7 tonnes ha (NHB 2003). It occupies first position among farmers who grow fruit crops and 'Dashehari' is one of the choicest cultivar of North India (Panwar 2007).

Charaka Samhita is a composite work that may have been written by many authors over centuries. It consists of 120 chapter divided into eight parts covering diet, the cause of disease, anatomy, embryology, diagnosis, therapy and pharmacy. More than 60 fruits and nuts are cited, including apples, almonds, bananas, pistachios, four varieties of jujube, two varieties each of grape and pomegranate, the nagaranga (a kind of orange) and the starfruit or carambola, which is native to eastern India and Southern Asia (Sen 2015). As early as1791, Roxburgh has suggested two plants *Phoenix pusilla* and *Caryota urens* as alternative for food (Roxburgh 1791). Other fruits suggested were coconut, jack fruit, wood apple, *Madhuca*, tamarind and mango tree. Mango fruits and seed kernel are consumed after boiling and dried flowers of *Madhuca* are edible and sweet (Robinson 2008).

Jamun is one of the delicious and highly priced fruits for its nutritive and nutraceutical values. Its juice is attributed for its anti-diabetic properties and prominently finds mention in the Indian pharmacopeia. The fruit is highly perishable but has good demand in domestic market. The Indian subcontinent presents good amount of diversity that offers scope for genetic enhancement and improvement of traits. It is largely fruit propagated through seeds, though vegetative propagation is being pursued in recent years. There is great consumer preference for genotypes with higher pulp percentage and seedlessness. Seedless accession of jamun was identified by the Central Institute for Subtropical, Lucknow, during 2008-2009 in a countrywide survey, primarily based on collection, conservation and documentation of indigenous germplasm resources for underutilized fruits. The seedless accession, CISH J–42 (IC0587714; INGR11024 also registered with NBPGR), was identified in wild road side avenue at Navgarh Road (Near Vindhya Hills), Chakia, Chandauli in Uttar Pradesh (Singh *et al.,* 2014).

Malnutrition is one of the greatest problems. Surveys conducted by WHO has brought out the fact that children of India and other Asian-African countries are suffering from malnutrition. Our intake of dairy products per head is one of the poorest in the world, all these, have far reaching consequences, particularly with the increasing population. It is here that a banana, an apple or even a cut piece of mango can make a difference. The change in our diet can make only as a result of conscious efforts, supported by knowledge age of dietary values of fruits and their scientific cultivation (Arya 1996; González-García *et al.,* 2018; Chen *et al.,* 2019). Horticultural crops covering 8 per cent the total crop area contribute 24.5 per cent of the GDP. We have reached at the number 2 in production of fruits after Brazil. We are leading producer of the mango, banana, sapota and acid lime (Dutt 2001). The productivity of grapes in India at 56 tones per hectare is a world record. Export of mango from 25000 tones to 45010 tones (25 per cent share of World Trade). According to 2021 production of fruits and vegetables during as per the national Horticulture board was over 341 million tones. Fruit production is 103 million metric tones (32 Mt banana and 20 million tones of mango). Of the 3.2 km^2 of hot arid zone in India 61 per cent is in western districts of Rajasthan. Of this about 4,000 ha is under orchards, 73 per cent area is in Ganganagar district alone, mostly under citrus. About 1,000 ha is under fruit crops in arid regions of Gujarat. Important fruits being, ber, pomegranate, guava, sweet orange, date palm, fig, Jamun, and aonla. The productivity of fruits has doubled from 5.52 in 1961 to 10.23 (in 1993) tones per ha and the total production has increased by 32 per cent. In USA the losses from fruits and vegetables amount to 24 per cent. In India the postharvest losses may be as high as 40 per cent or even more. Taking the loss at 50 per cent the net availability of fruits and vegetable should be about 66 million tones. Constraints in food production and microbial infections and methods to control postharvest diseases are important to analyze (Lal and Arya 1982; Rai *et al.,* 1982). Transgenic technology to develop disease resistant trees may provide permanent solution to diseases. Availability of fruits and cultivation of newly promoted promising source in different parts of the world may provide additional nutritional sources and ultimately lead to healthy human life.

19.2 Nutrients Present in Fruits

Fruit is a valuable source of antioxidants such as bioflavonoids and vitamin C, help to protect against degenerative diseases including cancer and heart diseases. As fruit is low in calories, it is an excellent food for anyone trying to lose weight World Health Organisation recommended try to eat at five portions of different fruits vegetables each day fruits provide Vitamin A and C and rich source of potassium which helps to regulate blood pressure. Fruits such as orange, apple, bananas, apricot, and fig help in lowering cholesterol. Ready to eat daily levels supply of vitamin A for an adult is set at above 700 micrograms the requirement of ascorbic acid vitamin C, it is 40 mg a day which should be provided by small orange, or one large peach. Papain present in papaya helps digestion and in relieving pain.

Nobel Laureate Gurudev Rabindranath Tagore has said:

Let the earth, and the water, the air and
the fruits of my country be sweet my GOD.
Let the homes, and the marts, the orchards
and the fields of my country be full my GOD.

19.2.1 Carbohydrates in Fruits

Fruit is high in fructose which can only be metabolized in the liver, where it is turned into fat. We all know the term "beer belly", otherwise known as alcoholic fatty liver disease, where fat deposits itself around the abdomen (the most dangerous type of fat). But, you can also develop NON-alcoholic fatty liver disease – from too much fructose. That fructose may come from table sugar (half fructose half glucose) in a high sugar or high carb diet, or it may come from the fructose found in fruit. They have vitamin C, but so do lower-carb vegetables, which we eat in abundance without the fructose to accompany them.

19.2.2 Fruit Juice and Smoothies

A smoothie or juice based on fruit can be higher than a can of fizzy drink. Yes, there will be some nutrients in it, but many persons think this somehow presence of the high sugar in the smoothie. Remember, a glass of orange juice isn't the same as the goodness from 6 oranges, it's the same as the sugar from 6 oranges. If you would like to enjoy a juice or a smoothie, based them on leafy vegetables and healthy fat such as coconut milk or cream.

19.2.3 Carbs in Fruits

Many regard bananas as the perfect fuel for the gym, and to grab some extra potassium. Let's take a closer look.

- ☆ **A small banana** (and most are huge these days) has 24g carbs (21.4g net), 1.2g of protein, 9mg Vitamin C, and 379mg of potassium.
- ☆ **A cup of chopped broccoli** will give you 6g carbs (3.6g net), 2.6g of protein, 81mg Vitamin C, and 287mg of potassium.
- ☆ **A medium tomato** is worth 4.8g carbs (3.3g net), 1.1g of protein, 16.8mg Vitamin C, and 291mg potassium.

As you can see, it's actually better to eat broccoli than a banana. It has fewer carbs, more protein, and more vitamin C.

Lower-Sugar Fruits

- ☆ Strawberries – 6g net carbs per 100g
- ☆ Raspberries – 5g net carbs per 100g
- ☆ Blackberry – 5g net carbs per 100g
- ☆ Watermelon – 8g net carbs per 100g

Higher-Sugar Fruits

- ☆ Apples – 12g net carbs per 100g
- ☆ Blueberries – 12g net carbs per 100g
- ☆ Kiwi – 12g net carbs per 100g
- ☆ Pear – 12g net carbs per 100g
- ☆ Pineapple – 12g net carbs per 1oog

The fruits with the most natural sugars are: Dates, Raisins, Dried cranberries, Bananas. Avoid these as much as possible if you are on a low-carb or keto diet also, not a surprise, lemons and limes have the least amount of carbs. The sweeter a food is, the more sugar and carbs it will contain. The fruit with the most carbs is mango (50 g per fruit).

Table 19.1. Nutrients and Calories Present in different Fruit

Fruit	*Measurement*	*Weight*	*Calories*	*Carbs*	*Protein*	*Fat*
Apple	1 medium	182 g	95	25	0.5	0.3
Apricot	1 apricot	35 g	17	3.9	0.5	0.1
Avocado	1 avocado	201 g	322	17	4	29
Bananas	1 medium	118 g	105	27	1.3	0.4
Blackberries	1 cup	144 g	62	14	2	0.7
Blueberries	1 cup	148 g	85	21	1.1	0.5
Cantaloupe	1 melon	552 g	186	45	4.6	1
Cherries	1 cup	155 g	77	19	1.6	0.5
Cranberries (fresh)	1 cup	100 g	46	12.2	0.4	0.1
Dates	1 date	24 g	66	18	0.4	0
Grapefruit	1/2 fruit	123 g	52	13	0.9	0.2
Grapes	1 cup	92 g	62	16	0.6	0.3
Honeydew	1 cup	170 g	61	15	0.9	0.2
Kiwi	1 fruit	69 g	42	10	0.8	0.4
Lemon	1 fruit	58 g	17	5	0.6	0.2
Lime	1 fruit	67 g	20	7	0.5	0.1
Mandarin oranges	1 fruit	75 g	40	10	0.6	0.2
Mango	1 fruit	336 g	201	50	2.8	1.3
Nectarine	1 fruit	142 g	63	15	1.5	0.5
Oranges	1 fruit	96 g	45	11	0.9	0.1
Peaches	1 fruit	150 g	59	14	1.4	0.4
Pear	1 fruit	178 g	102	27	0.6	0.2
Pineapple	1 cup	165 g	82	22	0.9	0.2
Plums	1 fruit	66 g	30	7.5	0.5	0.2
Pomegranate	1/2 C seeds	87 g	72	16	1.5	1
Raspberries	1 cup	123 g	65	15	1.5	0.8
Strawberries	1 cup	152 g	49	12	1	0.5
Watermelon	1 cup	152 g	46	11	0.9	0.2

19.3 Vitamins Present in Fruits

Vitamins in fruit are complex chemical substances which are mainly sourced from food with the exception of K and D vitamins. Our bodies are unable to produce vitamins itself so it relies on obtaining them from food. One exception is vit. D, also known as the sunshine vitamin as the body obtains it from sunlight. There are foods that contain vit. D, mainly being fish, meat and dairy products. There is no evidence of any fruit containing vit. D. Vitamins are wonderfully beneficial to our overall well-being, health and bodies, in actual fact, eating fruit with lots of essential vitamins can even be life changing! A good diet, rich in vitamins, can strengthen our immune system, protect our health and bodies by helping us to fight off illnesses and major diseases, increase our life expectancy, keep us looking younger and feeling more energetic, relieve us from anxiety and stress, help with depression by improving our mood! The health benefits from vitamins are truly amazing. If you would like to read further interesting facts and information about this topic, please refer to the 'benefits of Fruits' link in the main index to find out exactly how fruits can improve our health and lifestyles. Vitamins in fruit keep the mind and body strong and healthy. Indian gooseberry and guava are very rich sources of vit. C. Pineapple is a good source of vit. C found to vary with season like most other fruits.

19.3.1 Fat-Soluble and Water-Soluble Vitamins in Fruit

Vitamins can be separated into two groups - either fat-soluble or water-soluble which is usually an indication of the types of food where these vitamins are found - either watery or fatty foods. There are thirteen vitamins for humans and nine of them are water-soluble - eight B vitamins and vitamin C. Four vitamins are fat-soluble and they are vit. A, D, E and vit. K. Many of the vitamins in fruit are water-soluble as fruit is generally high in B and C vitamins. This means that these types of vitamins can dissolve in water so they are lost very quickly and easily through cooking processes and time lapsed between the Fruit being picked and being eaten. Most fruits are eaten raw and remain unaffected by any cooking methods. None of the fruits contain any preformed vit. A, but only carotene. Mangoes are rich source of carotene. Papaya, apricots and prunes are rich sources of carotene.

Table 19.2. Vitamin Present in Fruits and their Health Benefits

Vitamins in Fruit	*Fruit with Vitamins*	*Health Benefits from vitamins in Fruit*
Vitamin A	Plums, Prunes, Mango, Apricots, Cantaloupe Melon, Grapefruit, Passion Fruit, Guava, Papaya, Watermelon	A vitamins in Fruit function as an antioxidant, strengthen the immune system, helps vision and protects skin health *i.e.* ulcers, spots, acne.
Vitamin B1 (thiamine)	Grape, Pineapple, Pomegranates, Avocados, Watermelon	B1 vitamins in Fruit help the body to turn glucose into energy which keeps our brains and minds stimulated and energized; increasing our ability to concentrate and focus for longer periods of time. Good for poor mind function and lack of concentration.

Vitamins in Fruit	*Fruit with Vitamins*	*Health Benefits from vitamins in Fruit*
Vitamin B2 (riboflavin)	Banana, Pineapple, Blueberries, Strawberries, Raspberries, Plum, Prune, Apricots, Pomegranate, Avocado	B2 vitamins in Fruit improves hair, nail and eye health. Good for poor eye, hair and nail condition. A lack of B2 vitamins can result in skin conditions such as eczema, dermatitis and sore tongue.
Vitamin B3 (niacin)	Lemons, Grapefruits, Bananas, Apples, Pears, Figs, Grapes, Pineapples, Kiwi Fruit, Mangos, Papaya, Blueberry, Cherries, Strawberry, Raspberry, Plum, Prune, Apricots, Pomegranate, Cantaloupe Melon, Avocado	B3 vitamins in Fruit helps the body to produce melatonin to encourage better sleep, serotonin (happy neurotransmitter). B3 vitamins help to balance blood sugar levels, promotes good mood and energy. Good for depression, low mood and tiredness. Lack of B3 vitamins can cause acne and eczema, headaches and inability to sleep properly.
Vitamin B5(pantothenic acid)	Lemon, Grapefruit, Banana, Figs, Pineapple, Kiwi Fruit, Papaya, Blueberry, Cherry, Cranberries, Strawberry, Raspberry, Plum, Prune, Apricots, Pomegranate, Avocado, Watermelon	B5 vitamins in Fruit decreased levels of stress and increases the power of the memory. This anti-stress vitamin is great for those who suffer from anxiety and memory loss, low energy and poor concentration. A lack of B5 vitamins can cause aches, cramps, tender muscles.
Vitamin B6 (pyridoxine)	Lemon, Banana, Figs, Grapes, Pineapple, Kiwi Fruit, Strawberry, Plum, Prune, Pomegranate, Cantaloupe Melon, Avocado, Watermelons	B6 vitamins in Fruit helps the body to produce serotonin (happy neurotransmitter), another anti-stress vitamin that is good for depression, anxiety, stress relief and low energy.
Vitamin B12 (cobalamin)	There is no evidence of any fruits containing vitamin B12	B12 vitamins in Fruit are required for DNA synthesis and to help the blood circulation to function by helping oxygen to be carried around the blood. Therefore, B12 vitamins protects the skin and hair, reduces feelings of tiredness, fatigue, irritability, tension, low mood, stress and anxiety. This vitamin helps to put color in your cheeks.
Vitamin C (ascorbic acid)	Lemon, Grapefruit, Banana, Apple, Pear, Fig, Grapes, Pineapples, Kiwi Fruit, Mango, Papaya, Blueberry, Cherry, Cranberry, Strawberry, Raspberry, Plum, Prune, Apricots, Pomegranate, Cantaloupe Melon, Avocado, Guava (The	C vitamins in Fruit function as an antioxidant, strengthen the immune system. This vitamin is renown for its amazing affects on the health and body as it assists in fighting colds, flu, viruses, illnesses and infections. C vitamins help to form collagen in the skin, joints and bones. This vital vitamin is essential and it helps the body turn food into energy.

Vitamins in Fruit	*Fruit with Vitamins*	*Health Benefits from vitamins in Fruit*
	vit. C content of a kiwi is so high that it provides almost double the RDA of vitamin C)	
Vitamin D (chole-calciferol)	There is no evidence of any fruits containing vitamin D	D vitamins promote healthy bones and teeth as it assists the body with its use of calcium. D vitamins promotes good mood, protects the heart and helps the body to fight against cancer.
Vitamin E (tocopherol)	Lemon, Grapefruit, Apple, Pear, Pineapple, Kiwi Fruit, Mango, Blueberry, Raspberry, Pomegranate, Avocado, Guava, Papaya	E vitamins in Fruit is very important as this vitamin protects the essential fats and cells from damage. E vitamins is great for maintaining healthy skin, it helps to heal wounds and reduces dry skin.
Vitamin K (phyllo-quinone)	Banana, Apple, Pear, Grapes, Pineapple, Blueberry, Cranberry, Strawberry, Raspberry, Avocado	K vitamins in Fruit are essential for blood-clotting. A lack of this vitamin can result in haemorrhaging. Body also produces this vitamin in the gut.

19.4 Antioxidants Present in Fruits

Fruits and vegetables are some of the best sources of antioxidants, minerals, and vitamins. However, since all fruits have antioxidants in various amounts, it's impossible to single out one that has the highest antioxidant content. Fruits may contain mineral antioxidants (copper, zinc), vitamins acting as antioxidants (vit. C), and other phytonutrients that act as antioxidants (resveratrol, anthocyanins, lycopene). It's best to consume a variety of fruits each day to get a healthy dose of various nutrients.

Table 19.3. Antioxidants and their Fruit Sources

Antioxidant	*Fruit Sources*
Resveratrol	Grape (mainly the seeded variety), blueberry
Anthocyanins	Blueberry, raspberry, blackberry, cherry, and muscadine grape
Lycopene	Tomato, watermelon, grapefruit, papaya, and guava
Lutein	Tomato, orange, kiwi, grapefruit
Isoflavones	Currant, raisin
Vitamin C	Orange, gooseberry, grapefruit, kiwi
Selenium	Banana, mango, avocado, orange, blueberry, blackberry
Manganese	Pineapple, acacia
Zinc	Avocado, pomegranate, blackberry, raspberry, peach, apricot
Copper	Avocado, guava, kiwi, blackberry, banana, pineapple, durian

Antioxidants are substances that protect various molecules in the body from being attacked by harmful free radicals. When free radicals accumulate, they can damage cell structure and DNA. We are constantly exposed to free radicals, whether it's pollutants in the air, water, and soil, cigarette smoke, heavy metals, or chemicals in certain foods. Although limiting the exposure to free radicals is crucial, being equipped to fight them through antioxidants is important too. A diet rich in antioxidants helps keep tissues and organs healthy, ensuring that they perform their functions well. Antioxidants can also help prevent health conditions such as diabetes, high blood pressure, heart diseases, stroke, liver diseases, kidney diseases, and even some cancers.

19.5 Nutritional Value of Fruits

Fruits and fruit juices exert a favourable effect on the growth rate of children. Chancy (Rao *et al.,* 1967) reported that a daily ration of oranges given with the normal meals resulted in an improvement in the increase of weight of under-nourished children. The studies revealed that growth rate of rats was more in milk diet with mango, in comparison to control milk diet.

19.5.1 Effects on Mineral Metabolism

Million and Rosenblath (Rao *et al.,* 1967) reported that supplements of grape fruit juice increased the calcium retention in children by approximately7 per cent. Scientists found that the ingestion of additional 100g of banana daily resulted in an increased retention of calcium and phosphorus in children. The supplementation of children diet with citrus fruits resulted in an increased retention of nitrogen, calcium, phosphorus and magnesium. Similar beneficial effects were observed with a daily supplement of 600to 700 ml of orange juice. Bioavailability of nutrients increases when fruits are consumed with milk.

19.5.2 Effect of Bioflavonoids

These are complex compounds closely associated with vit. C. These are found in several fruits and vegetables. All the bioflavonoids increase the effectiveness of vit. C and are recognized as potent antioxidants. Several flavonoids have been identified from fruits and vegetables and some of the important ones are quercetin, kaempferol, myricetin and luteolin. Quercetin has anti-carcinogenic activity. It also inhibits the growth of several types of cancer cells (Arya 2004).

19.5.3 Flavour Compounds Present in Fruits

The flavour compounds are a complex mixture of a large number of volatile compounds, contributing to overall sensory quality of fruits and vegetables. Horticultural crops contain some volatile flavour compounds in the intact tissues and some are produced enzymatically after the tissue is damaged or crushed. These compounds undergo further changes after heating. Flavour compounds of fruit and vegetables include non-volatile sugars, amino acids, fatty acids, organic acids and volatiles like aromatic hydrocarbons, aldehydes, acetates, ketones, alcohol, esters and sulphur compounds. Flavour compounds make the fruits more acceptable. Flavour is the interaction between taste, orthonasal and retronasal olfaction perceptions, commonly denominated as 'taste and aroma', which is one of the major quality traits of fruits and together with texture is responsible for repeated purchases of a given commodity (Beaulieu and Baldwin 2002). The aroma fraction of flavour can even influence the perception of other traits, as recorded for sweetness and sourness (Baldwin *et al.,* 1998). Furthermore, flavour, which is the interaction of taste and aroma, hence dependent on chemical traits, is strongly linked to the individual preferences of consumers and can be seen as the 'modern concept of quality' (Brückner 2008). Knowing the preferences of consumers and aiming to fulfil those expectations regarding the flavour of fruits, besides increasing the probability of producers to easily sell their commodities, they will also be linked to an expected improvement in nutritional uptake, as better-tasting fruits will likely replace less healthy snack foods (Klee 2010). New tools, namely those related to molecular techniques, allow the identification of genes responsible for biosynthesis of compounds and open new perspectives for the improvement of flavour, by cloning those genes, increasing that specific pathway or silencing the expression of a gene responsible for an undesired compound (Wyllie 2008).

19.6 Medicinal Value of Fruits

Striking therapeutic results have been obtained in a majority of patients with both primary and secondary hypertension by use of diet, mainly based on fruits, fruit juice and rice (Arya 2004). Lemons are a citrus fruit that people often use in traditional remedies because of their health benefits. Like other citrus fruits, they contain vit. C and other antioxidants. Antioxidants are essential for human health. These compounds mop up free radicals in the body that can damage the body's cells and lead to diseases, such as cancers. Researchers believeTrusted Source that the flavonoids in lemon and other citrus fruits have antibacterial, anticancer, and antidiabetic properties. Citrus fruits, including lemons, contain active components called phytochemicals that benefit health. Strawberries are a juicy red fruit with a high water content. The seeds provide plenty of dietary fiber per serving. Strawberries contain many healthful vitamins and minerals. They contain anthocyanins, which are flavonoids that can help boost heart health. The fiber and potassium in strawberries can also support a healthy heart. In a study people assigned female at birth who ate 3 or more servings per week of strawberries and blueberries – which are both known for their high anthocyanin content – had a lower risk of having a heart attack than those with a lower intake. Strawberries and other colorful berries also contain a flavonoid called quercetin. This is a natural anti-inflammatory compound. The flavonoids in grapefruits can help protect against some cancers, inflammation, and obesity. A review study suggests the compounds called furanocoumarins found in grapefruits can help protect against oxidative stress and tumors, and they may support healthy bones. Some research from this review suggests that grapefruit furanocoumarins may have anticancer properties, which may be especially effective against breast cancer, skin cancer, and leukemia. However, researchers still need to carry out more studies to confirm these properties.

19.7 Use of Fruit Milk Diets

A milk diet can take on a number of forms, but at its core, it is a diet in which only milk is consumed for an extended period of time – days or weeks – without any other type of food. In some cases, the milk diet consists of drinking just 4 pints of milk per day, which is equivalent to about 1,000 calories. Considering that your daily calorie intake should be roughly twice that high, this diet is obviously quite extreme. Weight loss is the ultimate goal for most people on a milk diet, but improving other aspects of your health is also possible.

19.8 Value Addition in Processed Fruits

Drinks made with pomegranate juice are very soothing and strengthening. An important aid to digestion, especially after a very heavy meal or one that deranges the doshas, is an after- dinner drink- an Ayurvedic version of the French *digestf.* Common *digestifs* are or warm water, a drink made from mung dal, the juice of sour fruit, sour rice gruel, milk, concentrated meat broth, asava (wine) and madya (spirits) (Sen 2015). In order to commercialize avocado on a large scale, it is important to develop food products derived from this fruit with prolonged shelf life long. This would also promote the creation of processing plants, which in turn would generate new jobs, and increase the profit of the farmers. Products like Arka avocado chutney, bread spread and spray dried powder are available in market. Arka Spray dried avocado powder developed and commercialized by ICAR-IIHR, Bengaluru has typical avocado flavour, stable for more than 3-6 months at room temperature, without change in taste, colour (pleasant greenish yellow) or texture at ambient, when stored in aluminium laminated pouches is obtained. Product will ensure year-round supply of value-added Spray Dried Avocado Powder, which is convenient and flexible to use and transport. Due to the availability of a large number of fruit crops in India and consumer preference for more palatable fruits of sweet taste; besides avocado has caught the imagination of the consumers for its nutrition security as a super fruit (Karunakaran *et al.,* 2023).

Presently, fresh juice of dragon fruit (*Hylocereus undatus*) is sold in Indian market. Many value added products can be produced on commercial scale with attractive colour without adding any synthetic colour as fruit contains bright purplish to pink colour pigment. Different products include ready to-serve beverage (RTS), squash, fermented beverage, jelly, osmotically dehydrated cubes or slices, frozen cubes or slices, ice cream, flavored milk and yoghurt can be prepared from the dragon fruit. Arka RTS beverage technology is developed and commercialized from ICARIIHR; Bengaluru can be helpful for startup business (Karunakaran *et al.,* 2023).

Nectar is a type of fruit beverage that contains at least 20 per cent fruit juice/ pulp and 15 per cent total soluble solids and also about 0.3 per cent acid. However, it is not diluted before serving. Guava (cv. Allahabad Safeda) fruits were harvested at the physiologically mature ripe stage, devoid of any blemishes, *i.e.* free from insect pests, diseases and mechanical damages; however its uniform fruits were quickly transported in CFB boxes to processing laboratory for extraction of pulp within 24 hours. The fruits were properly washed in running tap water, and after treatment the fruits were cut into halves using stainless steel knife and the seeds were removed and the juice was extracted using fruit pulper (Bajaj Process Pack Limited, Fruit Processing and Packaging Tech., Noida, Uttar Pradesh). Grape berries were selected having uniform bunch and free from insect pests and fungal diseases infection. The berries were separated from rachis and quickly transported in CFB boxes to processing laboratory. The fruits were properly washed in running tap water, and thereafter juice was extracted using fruit pulper (500-4000 Ib\In2 Johnston Automation Co. Kirti Nagar, New Delhi). The extracted juice was sieved using muslin cloth to remove the seeds and peel. However, extracted juices of both the fruits were blended with defined proportion of 60:40, 50:50 and 40:60 (Guava: coloured grape). The combination having 50:50 proportion was found to maintain good amount of nutrition as compared to other blend proportions (Singh *et al.,* 2023).

REFERENCES

Abu'l Fazal (1873)ibn Mubarak Allami. The Ain-I- Akbari, trans H. Blochmann (1974) New Delhi 1989

Achya K.T. (2002) A Historical dictionary of Indian food. New Delhi p.84

Arya A. (1996) Promising Food plants Indian Hort. 41(2): 10-12

Arya A, (2004) Tropical Fruits Diseases and Pests. Kalyani Pub. Ludhiana pp. 217.

Baldwin E, Scott J, Einstein M, Malundo T, Carr B, Shewfelt R, Tandon K. (1998) Relationship between sensory and instrumental analysis for tomato flavor. Journal of the American Society of Horticultural Science. 123: 906-915. DOI: 98A0967495

Beaulieu J, Baldwin E. (2002) Flavor and aroma of fresh-cut fruits and vegetables. In: Fresh-cut Fruit and Vegetables Science, Technology and Market. Boca Raton, FL, USA: CRC Press; 2002. pp. 391-425. ISBN: 9781420031874

Brückner B. (2008) Consumer acceptance of fruit and vegetables: The role of flavour and other quality attributes. In: Brückner B, Wyllie SG, editors. Fruit and Vegetable Flavour. Recent Advances and Future Prospects Woodhead Publishing Series in Food Science, Technology and Nutrition. New York, Washington DC, USA: Woodhead Publishing; 2008. pp. 11-17. ISBN: 9781845694296

Chen C, Chaudhary A, Mathys A. (2019) Dietary Change Scenarios and Implications for Environmental, Nutrition, Human Health and Economic Dimensions of Food Sustainability. Nutrients. 2019 Apr 16; 11(4): 856. doi: 10.3390/nu11040856. PMID: 30995719;

Dutt S. (2001) Ushering golden revolution in India. Indian Hort. 46 (3): 1

González-García S, Esteve-Llorens X, Moreira MT, Feijoo G. (2018) Carbon footprint and nutritional quality of different human dietary choices. Sci Total Environ. 2018 Dec 1 0; 644: 77-94. doi: 10.1016/j.scitotenv.2018.06.339. Epub 2018 Jul 4. PMID: 29981520.

Karunakaran G, Arivalagan M., Sakthivel T., Tripathi P.C., Kumar P.C., Shameena A, Muralidhara B.M. and Ruchitha T.(2023) Promising High Value Fruit Crops for Northeastern Region. Indian Horticulture 68 (6): 35-39

Klee HJ. (2010) Improving the flavor of fresh fruits: Genomics, biochemistry, and biotechnology. New Phytologist. 2010; 187: 44-56. DOI: 10.1111/j.1469-8137.2010.03281.x

Lal B. Arya A (1982) A soft rot of grapes caused by *Phomopsis viticola.* Indian Phytopath. 35(2): 261-264

Nithin. K.N. (2016). Production and Export Performance of Grapes from India: An Econometric Analysis. 143-146.

NHB. (2003) Horticulture Data Base. National Horticulture Board, Ministry of Agriculture, Gurgaon, Haryana.

Panwar R. Singh S.K., Singh C.P. Singh P.K. (2007) Mango fruit yield and quality improvement through fertigation along with Mulch. Indian. Journal of Agricultural Sciences 77 (10): 680-684

Rai R. N. Arya A. Tewari D.K. and Lal B. (1982)An unrecorded postharvest rot of guava (*Psidium guajava* L.) Geobios New Reports 1: 134

Rao N. Sastry L.V.L. and Parpia H.A.B. (1967)What nutrition fruits contain. Indian Hort. 11(4): 87-95

Robinson T. (2008) William Roxburgh: The founding father of Indian Botany. Pub. By Phillimore and Company England. Pp 286.

Roxburgh W. (1791) Letter from William Roxburgh to Sir Joseph Banks, 30 August 1791. Sutro Library EI 1: 47 and British library OIOC IOR, MSS/EUR/K 148,ff 243-7

Roy S.K. (2001)Strategic postharvest management of fruits and vegetables. Indian Hort. 45(4): 4-7

Sen C.T.(2015) A history of food in India. Speaking Tiger Pub. Pvt. Ltd. New Delhi pp.349.

Singh A.K., Bajpai A. and Ravishankar H. (2014) Seedless jamun fruits. Indian Horticulture 59: 9-10

Singh A.K., Sagar V.R., Singh S.K., Sethi S.and Shankar K. (2023) Anthocyanin enriched guava nectar: An alternative source of antioxidant biofortification Indian Horticulture 68 (6): 47-49.

Wyllie SG. (2008) Flavour quality of fruit and vegetables: Are we on the brink of major advances? In: Brückner B, Wyllie SG, editors. Fruit and Vegetable Flavour. Recent Advances and Future Prospects. Woodhead Publishing Series in Food Science, Technology and Nutrition. New York, Washington DC, USA: Woodhead Publishing; 2008. pp. 3-10. ISBN: 9781845694296

Chapter 20

Nutraceuticals for Good Health

A wholesome diet with plenty of fruits and vegetables along with the right combination of macronutrients is vital for optimum health. Although, in some cases, we don't always get enough of these major nutrients. Not getting enough nutrients can lead to oxidative stress which can then lead to other health issues. Research in nutrition is constantly developing with new discoveries and innovative technologies. One such advancement that bridges the gap between nutrition and medicine is nutraceuticals. Dietary supplements are those products that contain concentrated bioactive nutrients from a food source processed into a suitable dosage form. These supplements can contain one or more of the following: amino acids, vitamins, herbs or other botanicals, minerals, important metabolites, and certain enzymes. These are available in tablets, capsules, powders, and liquids, gummies, energy bars, and any other suitable forms. Nutraceuticals that fall under the umbrella of dietary supplements not only supplement diet but also promote health and prevent disease. The functional foods include whole foods and fortified, enriched, or enhanced foods that can improve health when consumed regularly as part of a varied diet. Functional food looks like food and is available as pasta, cereal, whole grains, yogurt, snacks, and more.

***Keywords**: Nutraceuticals, Foods, Probiotics, Fortified cereals, Processed foods.*

20.1 Introduction

The term "nutraceutical" is used to describe the medicinally or nutritionally functional foods. Nutraceuticals, which have also been called medical foods, designer foods, phytochemicals, functional foods and nutritional supplements, include such everyday products as "bio" yoghurts and fortified breakfast cereals, as well as vitamins, herbal remedies and even genetically modified foods and supplements. Many different terms and definitions are used in different countries, which can result in confusion.

The term "nutraceutical" was coined in 1989 by Stephen De Felice, founder and chairman of the Foundation for Innovation in Medicine. He defined a nutraceutical as a "food, or parts of a food, that provide medical or health benefits, including the prevention and treatment of disease" (Cencic and Chingwaru 2010). In Canada, a functional food has been defined as being "similar in appearance to conventional foods ... consumed as part of a usual diet" whereas a nutraceutical is "a product produced from foods but sold in pills, powders and other medicinal forms not generally associated with food".

In Britain, the Ministry of Agriculture, Fisheries and Food has developed a definition of a functional food as "a food that has a component incorporated into it to give it a specific medical or physiological benefit, other than purely nutritional benefit". Many of these new products that are being promoted to treat various disease states, find their origins in the plant kingdom.

20.2 What are Nutraceuticals?

These products include dietary supplements, diets, herbal products, genetically engineered foods, and vitamins. They contain a high concentration of bioactive compounds, derived from a natural source and have physiological benefits and aid in the prevention and treatment of disease. Nutraceuticals even include everyday foods like pre- and probiotics, fortified cereals, processed foods, and beverages.

Essentially, a nutraceutical is a substance that has a physiological benefit or provides protection from chronic disease. Unfortunately, the definition of nutraceuticals varies from country to country depending on how they are categorized and regulated. At the moment there is no clear internationally accepted definition of a nutraceutical. The idea of using food for both nutrition and medicinal purposes has been ingrained in many ancient cultures. In fact, the concept of nutraceuticals is nearly 3,000 years old! It began to catch on when Hippocrates, the father of modern medicine, recognized the relationship between food and health. Traditionally, the people of India and China consume various natural foods that are considered to be medicinal. Countries like Germany, France, and England were the first to consider one's diet more important than both exercise and hereditary factors in people's attempt to achieve good health (Mannion 1998).

Nutraceuticals can improve health, delay the aging process, prevent chronic diseases, increase life expectancy, or support the structure and functioning of the body (Jack 1995). They are also used in the prevention and treatment of mental health issues and disorders.

20.3 Marketing of Nutraceuticals?

Food labeling regulations do not allow food labels to carry health claims in many countries. This makes it hard for companies marketing nutraceutical to advertise the benefits of their products without a medicine licence, so they may decide either not to do any research or they may research a new product thoroughly and possibly obtain a patent. Unfortunately, many companies tend towards the former option due to either expense, the problems of obtaining a patent on a natural product, or the fact that they cannot put their claims on the label without a product licence, whether verified or not.

An example was the case of Cholestin, a cholesterol lowering supplement marketed by Pharmanex (Simi Valley, CA) in 1996. The product was based on an ancient Chinese remedy and proved very popular. When the pharmaceutical company Merck (White house Station, NJ) tested this successful "supplement", it found that the active ingredient was lovastatin, and the US Food and Drug Administration subsequently banned its incorporation in Cholestin. As can be seen from this example the difference between pharmaceutical and nutraceutical is sometimes difficult to define, depending only on regulatory issues.

In Britain, unlicensed herbal remedies and supplements are marketed without medical claims and as such are legally regarded as food supplements and are controlled by the Ministry of Agriculture, Fisheries and Food. In the past, as long as the products did not claim to have medicinal uses, the Medicines Control Agency (MCA), which is responsible for regulating medicinal products, has not required that the product satisfy medical legislation.[8] However, more emphasis is now being placed on the function of the marketed products as Britain moves into line with the rest of Europe.

It goes on: "Any substance or combination of substances which may be administered to human beings or animals with a view to making diagnosis or to restoring, correcting or modifying physiological functions in human beings or animals is likewise considered a medicinal product." This definition describes a product which has a medical function irrespective of whether these claims have been stated or not, which would include many supplements being taken for medical purposes (Aarts 1998).

British licensed medicines for human use are regulated according to the Medicines Act 1968 and the Medicines for Human Use (Marketing authorisations *etc.*) Regulations 1994 SI 1994:3144 (the regulations) (Anonymous 1994). In 1995, these regulations came into effect, implementing all the controls as listed in Directive 65/65EEC. According to these regulations there are controls over variation and renewal of UK licenses as well as labeling and package inserts. A five-year study of side effects of traditional remedies and food supplements was carried out between 1991 and 1995 in the medical toxicology unit at Guy's hospital, London (Shaw *et al.,* 1997). The authors concluded that the overall risk factor with these products was low and that they were generally considered to be safe. However, because many health care professionals considered these products as little more than placebo, side effects that were occurring were often not traced back to the supplements being used.

Many of these products are self-prescribed which may lead to inappropriate use, overdose, or interactions with other medicines. This highlights why the safety of these supplements must be ensured as many consumers are unaware of the difference between licensed and unlicensed products, and the differences between food legislation and medicines legislation. There are many inaccurate claims made for many supplements available, as well as much variation in the actual products, depending on storage, manufacturing process, or even batches of the same material. It is vital that these discrepancies be resolved, possibly with the creation of a third type of regulatory category for supplements (Shaw *et al.,* 1997).

In America, a new product may be eligible for regulatory status as a food, a dietary supplement, or as a medical food (Bidlack *et al.,* 1998). The US Congress passed the Dietary Supplement Health and Education Act (DSHEA) in 1994. This set up a new regulatory body for dietary supplements under the FDA. Unlike foods, dietary supplements are allowed to use "nutritional support statements". These may offer nutritional support in nutrient-deficiency disease, or state a description of the intended role, mechanism of action or effect on general health as a result of taking the product. For those dietary supplements for which such a statement is used the label must also state: "This statement has not been evaluated by the Food and Drug Administration. This product is not intended to diagnose, treat, cure or prevent any disease." In effect, because legally they are in a class by themselves, dietary supplements in the US can be marketed without the FDA being satisfied that they are safe. This makes it relatively easy for a manufacturer to market a product without investing the time and money required for its safety.

Of all three categories (food, dietary supplement or medical food), medical foods have the least formal regulatory controls. The US Congress in 1990 defined a medical food as "a food which is formulated to be consumed or administered eternally under the supervision of a physician and which is intended for the specific dietary management of a disease or condition for which distinctive nutritional requirements, based on recognized scientific principles, are established by medical evaluation". These products are usually promoted through health care professionals, including pharmacists, and in the UK this would make them prescription only or pharmacy medicines.

20.4 Implications for Pharmacy

Nutraceutical venture is becoming a popular topic in the pharmaceutical industry as it is difficult to develop new medicines fast and companies like, Du Pont, Abbott Laboratories and Warner Lambert, are turning to nutraceuticals (Brower 1998). This offers them a chance of a very large market. Datamonitor, a website that follows market trends, estimates the nutraceutical market at $17bn, and Dr Felice himself, speaking at a conference in 1998, put the figure at $250bn in America alone (Mannion 1998). Although these estimates vary widely, the nutraceuticals market is undoubtedly very large and growing. Most products reviewed are naturally occurring biochemicals which are effectively being used to supplement endogenous levels of these constituents that occur naturally in human metabolism, whether single component products such as glucosamine, or more complex products such as proanthocyanidin.

20.5 How Nutraceuticals Work ?

A healthy diet contributes to your health by providing the nutrients your body needs to repair itself, grow, and function well. When your diet does not supply enough of these essential nutrients and vitamins, nutraceuticals can act as a supplement. Nutrients are the chemical elements that make up food. Nutrients like proteins, carbohydrates, and fats provide energy while other nutrients like vitamins, water, electrolytes, and minerals are needed for a healthy metabolism.

Metabolism is a process that involves all chemical reactions that take place in the body to convert the food we eat into energy, which is then used to synthesize compounds needed by all cells of the body. Metabolism maintains the living state of all cells and organs. The pathways of metabolism depend on the nutrients the process of their breaks down. If all the nutrients are present in the right amount, our body, including all the organs and systems, function well. This means that your body can heal or repair properly and quickly. But if we don't have enough nutrients, then our body takes longer to repair or one might deal with harmful complications. When this happens, nutraceuticals can provide extra nutrients needed for optimum metabolic reaction and for your body to function properly.

20.6 Nutraceuticals as Dietary Supplements

Dietary supplements are those products that contain concentrated bioactive nutrients from a food source processed into a suitable dosage form. These supplements can contain one or more of the following: amino acids, vitamins, herbs or other botanicals, minerals, important metabolites, and certain enzymes. These are available in tablets, capsules, powders, and liquids, gummies, energy bars, and any other suitable forms. Nutraceuticals that fall under the umbrella of dietary supplements not only supplement diet but also promote health and prevent disease.

Functional foods are any food or food ingredient that provides a health benefit other than basic nutrition (Mazza 1998; Heasman and Mellentin 2001). The foods under this category include whole foods and fortified, enriched, or enhanced foods that can improve health when consumed regularly as part of a varied diet. Functional food looks like food and is available as pasta, cereal, whole grains, yogurt, snacks, and more.

Functional foods can be:

- ☆ **Traditional Functional Food:** These are nutrient-rich natural foods that deliver health benefits other than basic nutrition, such as omega-3 fatty acids in salmon and lycopene in tomatoes.
- ☆ **Non-Traditional Functional Food:** These are artificial foods prepared by adding bioactive components to help enhance health and well-being. Examples include fortified nutraceuticals such as juice with added calcium, cereal fortified with iron, flour with added folic acid. Modified functional foods can also include recombinant nutraceuticals, which

are energy-producing foods such as bread, yogurt, cheese, and vinegar produced via biotechnology techniques.

20.7 Certain Plant Based Nutraceuticals

One version of the weight-loss supplement OxyELITE Pro from USPlabs, LLC contained the compound 1,3- dimethylamylamine (DMAA) in addition to ingredients such as caffeine, *Bauhinia purpurea, Bacopa monniera, Cirsium oligophyllum*, and rauwolscine (Yohimbe) extract. Studies that were supported financially by USPlabs, LLC with a small number of healthy volunteers suggested that this supplement formulation could increase lipolysis, metabolic rate, heart rate and systolic blood pressure in the short term (2 h) as well as lead to small decreases in appetite, body weight, and BMI after intake for 8 weeks (McCarthy *et al.,* 2012). Accidental intake of supplements with DMAA, mainly in children, have caused relatively mild adverse effects such as tachycardia, nausea, and vomiting (Forrester 2013). However, serious cardiovascular events after DMAA intake have also been reported. Recreational use of DMAA in New Zealand has been associated with cases of cerebral hemorrhage (Gee *et al.,* 2010; 2012). Furthermore, there are three cases of cardiac arrest occurring during physical exercise in the military and in a gym following intake of DMAA-containing supplements (Karnatovskaia *et al.,* 2015). Two of the events eventually led to the deaths of US soldiers. While it cannot be proven that DMAA is the causative agent for all the adverse events, it is clear that DMAA does indeed have cardiovascular effects such as vasoconstriction and elevation of blood pressure (Swanson and Chen 1946; Bloomer *et al.,* 2011). Following receipt of 42 adverse events reports on products with DMAA, the FDA in April 2012 sent warning letters to 10 manufacturers of DMAA-containing supplements stating that since the safety of DMAA had not been documented, the products were adulterated, according to US law (USFDA 2012). In 2013, a series of severe hepatic liver disease was observed in individuals from Hawaii taking the weight-loss supplement OxyELITE Pro from USPlabs, LLC (Roytman *et al.,* 2014).

20.7.1 Arogyappacha for Health and Vitality

During the course of ethno-medico-botanical investigation of the 'Kani' tribe of Agastyar hills of Kerala, Pushpangadan (1988) has come across with a very interesting wild plant species known as *Trichopus zeylanicus* Gaerin, used by the 'Kani' tribe for getting instant stamina, health and vitality. The unripe fruits of 'Arogyappacha' (as 'Kani' tribe calls the plant) is eaten fresh to remain healthy and agile by 'Kani' tribe during their long tracking trips in he high mountainous forests of Agastyar hills. The word 'Arogyapacca' means the greener of health *i.e.* the one that gives very good health and vitality. It is claimed by 'Kani' men that one can live days together without food and still remain energetic and could perform even very rigorous physical work exercises by eating few fruits of this plant daily. The 'Kani' tribe also claims that if one eats the fruits of 'Arogyappacha' regularly he will remain always healthy agile, young and that no disease will afflict him. The 'Kani' tribe is normally very secretive about this plant.

But thanks to the chance incident that led to the discovery of this secret by the authors from them. The rapid recovery from fatigue experienced after consuming the fresh fruits of 'Arogyappacha' suggests its effects on neuro-physhological changes. The tonic effects this experienced is comparable to that of 'Ginseng'. This suggests that the other claims on the efficacy of 'Arogyappacha' as a protector of health by giving greater resistance to diseases and as elixir of life *etc.*, may also be sometimes turn out to be true. This little known wild plant used by 'Kani' tribe may be the drug stocked in the dispensary of mother Nature for combating the most dreadful killer disease of man namely Acquired Immuno Deficiency Syndrome shortly known as AIDS. From the preliminary observation made by the authors the 'Arogyappacha' may be investigated as a drug belonging to the 'Svathahita (that which helps to maintain positive health) group of drugs (Pushpangadan 1988).

20.7.2 Pomegranate for Cancer and Cardiovascular Diseases

Ellagic acid (EA) was first studied in the 1960s for its effects on blood pressure and clotting (Bhargava and Westfall 1969). Afterwards, many studies in cell cultures and animal models found that EA may slow the growth of some tumors caused by certain carcinogens. The dietary administration of ETs-containing foods such as strawberries and raspberries to rats has proved to inhibit events associated with both the initiation and promotion/progression of chemically-induced colon and oesophageal cancers (Harris *et al.,* 2001; Chen *et al.,* 2006). However, a more recent study showed the lack of effect on the number or size of adenomas in the small intestine of Apc-mutated Min mice upon administration of pure EA (1.5 g/kg), or cloudberry diets containing ETs (approx. 0.8 g) and EA (34 mg) (Paivarinta *et al.,* 2006). These contradictory results may be due to the fact that, the possible cancer chemopreventive effects of EA and related molecules may differ depending on the type of tumour, the animal model, *etc.*

Pomegranate juice is currently recognized as one of the most powerful *in vitro* antioxidant food. This remarkable activity has been associated to ETs, such as punicalagin, that are characteristic of this fruit (Gil *et al.,* 2000). The antioxidant activity and the punicalagin content have been suggested as the possible mediators of the different health effects reported for pomegranate juice. These effects include: protection against cardiovascular diseases (decrease in atherosclerosis risk factors such as hypertension, platelet aggregation, oxidative stress, and blood lipid profiles) (Aviram *et al.,* 2004; Rosenblat *et al.,* 2006), and cancer prevention (Pantuck *et al.,* 2006). The bioavailability and metabolism of ETs and EA are key issues that need to be resolved in order to understand the biological role of these phytochemicals and their *in vivo* effects. In general and due to their large molecular size, ETs are not absorbed (Cerda *et al.,* 2003a; 2005). However, small amounts of punicalagin were detected in the plasma of rats following long term administration with pomegranate ETs at high doses (Cerda *et al.,* 2003b). ETs are mostly hydrolysed to EA under the physiological conditions in the small intestine (Larrosa *et al.,* 2006).

20.7.3 Coconut Water for Good Health

Coconut (*Cocos nucifera* L.) water provides refreshment and nutrition to humans in a pure, additive-free manner. It earned global popularity and is known as a natural health drink (Prades *et al.,* 2012). Its lower calorie and sugar content have gradually positioned it as an alternative to sugary beverages and other traditional drinks (Medeiros and de Paiva 2012). Coconut water contains various electrolytes such as potassium, sodium, calcium, and magnesium, which assist the body in reestablishing hydration and electrolyte balance (Costa *et al.,* 2021). Therefore, whether facing scorching heat or intense workouts, coconut water's rapid electrolyte and hydration replenishment combats dehydration and fatigue (Zulaikhah 2019). The importance of coconut water extends beyond its function as a beverage. Coconut water finds its role as a food addictive in culinary endeavors, cocktail, and dessert (Rethinam and Krishnakumar 2022). Its fresh taste and subtle sweetness enhance the flavors of food and beverages. People utilize coconut water to craft several types of juices, alcoholic concoctions, pastries, imparting a distinctive coconut essence (Segura-Badilla *et al.,* 2020), with the Hainanese "Qing Bu Liang" standing out as an exceptional example (Xinmeng 2023).

Coconut water is comprised of sugars, minerals, vitamins, amino acids, enzymes, volatile aromatic compounds, and various biochemical compounds. The makeup of fresh coconut water is influenced by several factors such as the cultivation region, affecting soil, environmental conditions, and fertilizer usage, as well as the coconut variety, cultivar, and the harvesting stage's maturity (Burns *et al.,* 2020). Tender Coconut Water (TCW) is called the "fluid of life". It is a highly nutritious beverage derived from palm trees, serving as a natural isotonic drink with a similar osmotic pressures closely resembling our body's blood plasma (Zulaikhah 2019). The rich array of macro and micro nutrients in tender coconut water has the potential to reduce lipid levels and provide protective benefits to both the heart and liver (Shubhashree *et al.,* 2014). It has been used worldwide to treat oral rehydration, childhood illnesses, intestinal flu, and cholera (Tuyekar *et al.,* 2021). Coconut has been used in traditional Indian medicine for thousands of years to treat a variety of ailments, such as gonorrhea, bronchitis, fever and gingivitis (DebMandal and Mandal 2011). Ayurvedic medicine (ancient India) describes coconut as medicine that increases semen, promotes digestion, and clears the urinary path (Ediriweera 2003). In Sri Lanka, where coconut is a staple food, there are many references to its medicinal uses (Rethinam and Nanda Kumar 2001). Coconut water is a better choice than plain water or fruit drinks at relieving symptoms of diseases that cause dehydration (Prades *et al.,* 2012). It has a great impact to the health of human race and animals; the beginning of these research was started in 1942. This early research indicated that biologic value of coconut water was decided by proteins, amino acids. Also, the medical application of coconut water, for example, the feasibility of parenteral injection of coconut water and the applications in pediatrics (feeding infants). Coconut water has been widely used to treat infectious diseases, dysentery, influenza, and other diseases that cause dehydration over the past decades (Prades *et al.,* 2012). Most of the experiments were conducted using rats but those positive results indicated the future perspective clearly. Recently, *in-vivo* and *in-vitro* research showed that coconut water has the potential to lower blood sugar levels, similar to standard oral diabetes medications. Coconut water could have a significant impact on the treatment of diabetes (Joseph *et al.,* 2019). Nursing mothers who regularly drink coconut water can help baby's uptake calcium. Coconut water also contains vitamin-C, vit.-B and B9. The amount of vitamin-C in coconut water decreases as the coconut matures. Vit.-B9 is important for preventing anemia in pregnancy and mitochondrial toxicity caused by methanol metabolites (Jackson *et al.,* 2004). Furthermore, coconut water has antioxidant properties, and a mixture of coconut water and mauby bark syrup from Trinidad and Tobago has been shown to reduce hypertension in Humans (Alleyne *et al.,* 2005). Coconut water was used as the sperm extender of equine sperm (Brasileiro *et al.,* 2019) and rabbit sperm (Jimoh 2020), which ensures that a high pregnant percentage, and large, healthy litters. Coconut antimicrobial peptide-1, abbreviated as CnAMP1, is classified as an antimicrobial peptide, which is naturally present in the green coconut (Shubhashree *et al.,* 2014). Antimicrobial peptides are compact proteins that constitute a component of the innate immune system found in various organisms, including plants. Their primary function involves protecting the organism against microbial infections.

20.7.4 Delicious Drink: Thandai

To make thandai you make a paste of almonds, peppercorns, poppy seeds, cardamom, fennel seeds and watermelon seeds. This mixture of spices and nuts is then mixed with milk along with sugar, saffron and rose water. Once you have mixed in the spices-nut mixture with the milk, set it aside for around 1 hour so that the milk absorbs all the flavors. Because we eventually strain the mixture it's important for the milk to absorb the flavors before we do that. Thandai is supposed to be served chilled, it doesn't taste half as good when it's at room temperature.

Remember to use whole milk for this thandai recipe. Yes, you can very well make this drink with fat free milk too but it just tastes so much better with whole milk. The quantity of sugar and the amount of spices in this recipe can be adjust to taste. If you want stronger flavors in your thandai, add more of peppercorns that for little heat or more of the fennel seeds for that cooling effect. This recipe can be made in advance and refrigerated. So if you have a holi party planned, make thandai 1 to 2 days in advance and then just serve at your party.

Besides the fact, this thandai recipe is yum, this luscious thirst quencher is packed with some healthy stuff. It has ingredients listed below –

- ☆ **Provides energy** – melon seeds, almonds, cashews, and pistachios.
- ☆ **Boost Immunity** and **Helps fight infection** – spices like peppercorn, cardamom, cinnamon, and nutmeg.
- ☆ **It has a cooling effect on the body** – fennel seeds, saffron, and rose petals.

- ☆ **Detoxifies the body** – cardamom.
- ☆ **Aids in digestion too** – poppy seeds.

20.8 Nutraceuticals and Disease Cure

Over the past few years, nutraceuticals have become very popular. They're being used an alternative or supplemental treatment along with pharmaceuticals to help prevent and treat a wide range of diseases. They have attracted considerable interest because of their potential nutritional values, safety, affordability, and multiple therapeutic effects and are often seen as an attractive option to conventional treatments. Nutraceuticals can play an important role in the body's various biological processes, which help prevent various diseases and improve overall health and well-being. Nutraceuticals play an important role in preventing the onset of chronic diseases and reduce the complications involved. Evidence suggests they are used in the prevention and treatment of cardiovascular diseases, cancer, diabetes, obesity, and inflammatory-based diseases (Nasri *et al.,* 2014). ÿþPatients with eye disorders and conditions like age-related macular degeneration, glaucoma, and visual disorders can benefit from certain nutraceuticals (Hogg and Chakravarthy 2006). Nutraceuticals like lutein, zeaxanthin, vitamin C, and vitamin E reduce the risk of cataracts. Essential fatty acids like omega-3s are important for visual development and retinal function (Querques *et al.,* 2011).

20.8.1 Improve Immune Function

Immune booster nutraceuticals like green tea, blueberries, amino acids, and vitamin D are found to be useful in improving and strengthening immune health and thus help to prevent disease. Herbs and plant extracts like *Echinacea* and *Astragalus* help in the development and regenerations of stem cells. The anti-inflammatory effects of garlic can enhance the functioning of the immune system (Arreola *et al.,* 2015).

20.8.2 Improve Gut Health

Many nutraceuticals like prebiotics, probiotics, flavonoids, *Aloe vera*, resveratrol, and omega-3 and omega-6 acids can restore and stimulate the growth of good bacteria and reduce the effect of harmful bacteria in the gut. Nutraceuticals are shown to influence inflammation pathways and play an important role in reinforcing the body's natural gut defense mechanisms.

20.8.3 Support Reproductive Health

Many nutraceuticals play an important role in male and female reproductive health. Nutraceuticals help with male infertility and dysfunction, and prevent damage to sperm. Certain nutraceuticals can influence hormones and are also used to improve fertility, preconception, pregnancy outcomes in women, and provide support during menstruation.

20.8.4 Treat Inflammation

Nutraceuticals are showing promising results in the prevention and treatment of inflammation due to the presence of phytochemicals. They can reduce oxidative stress in chronic inflammatory diseases like asthma, osteoarthritis, rheumatoid arthritis, cancer, and autoimmune diseases.

20.8.5 Enhance Sports Medicine

Nutraceuticals are useful in sports medicine as they help athletes improve their performance and reduce oxidative stress and mitochondrial dysfunction, which occurs because of heavy exercise. Particularly, mitochondria-targeted nutraceuticals (MTNs) such as resveratrol and quercetin, have been shown to have antioxidant effects at a molecular level, which improves mitochondrial bioenergetics and leads to better exercise performance and recovery.

20.8.6 Nutraceuticals for Prolonged Life

Nutraceuticals provide preventative care to the body, which increases life expectancy as it reduces the risk of many diseases and helps the body recover quickly. Certain nutraceuticals like those present in citrus fruits and soybeans help renew healthier cells and improve gene expression. Some nutraceuticals can slow the aging process and thus promote longevity. Researches shows that nutrients and nutraceuticals play an important role in the prevention, management, and treatment of mental health disorders and psychological functioning. Certain nutrients and nutraceutical compounds including omega-3, vitamins B and D, magnesium, iron, zinc, curcumin, lycopene, and b-carotene are showing promising results in improving mood, stress, anxiety, depression, and cognitive functioning. Nutraceuticals with antioxidant properties help prevent neurodegenerative diseases, including Parkinson's and Alzheimer's disease.

20.9 Nutraceuticals: Adverse Effects

Compared to pharmaceuticals, nutraceuticals are potentially safe and can improve health naturally. However, certain nutraceuticals can have side effects such as allergic reactions caused by interactions with other nutraceuticals or therapeutic drugs (Singh 2016). A few nutraceuticals can also cause toxicity. In addition, many people tend not to disclose nutraceutical use to their physicians, which can result in adverse drug-supplement reactions (Ronis *et al.,* 2018). Both garlic and *Ginkgo biloba* use have been involved in several cases of excessive bleeding. For example, a 71 year-old man had persistent surgical bleeding that was attributed to indulgent garlic ingestion prior to the operation (Woodbury and Sniecinsk 2016). Furthermore, aged garlic extract inhibits platelet aggregation (Rahman *et al.,* 2016). Ginkgolide B, an active component of *Ginkgo biloba,* has been shown to inhibit platelet aggregating factor, and men and women taking *G. biloba* have suffered spontaneous bleeding (Bent *et al.,* 2005). Nutraceuticals have a powerful effect on your body, and it is very important to talk to your family doctor about your symptoms and nutraceutical use.

REFERENCES

Aarts TA. (1998) How long will the "Medical Food" window of opportunity remain open? J Nutr Function Med Foods 1: 45-57.

Anonymous (1994) The Medicines for Human Use (Marketing Authorisations *etc.*) Regulations. Statutory Instrument 3144.

Alleyne T, Roache S, Thomas C, Shirley A. (2005) The control of hypertension by use of coconut water and mauby: two tropical food drinks. West Indian Med J 54: 3–8

Arreola, R, Quintero-Fabián, S. Rocio Ivette, L., Flores Gutierrez, E., Reyes-Grajeda, J., Carrera-Quintanar, L., Ortuño, D. (2015) Immunomodulation and Anti-Inflammatory Effects of Garlic Compounds. Journal of Immunology Research. 2015. 1-13. 10.1155/2015/401630.

Aviram, M., Rosenblat, M., Gaitini, D., Nitecki, S., Hoffman, A., Dornfeld, L., Volkova, N., Presser, D., Attias, J., Liker, H., Hayek, T. (2004) Pomegranate juice consumption for 3 years by patients with carotid artery stenosis reduces common carotid intima-media thickness, blood pressure and LDL oxidation. Clin. Nutr. 23, 423–433

Bent S, Goldberg H, Padula A, Avins AL. (2005) Spontaneous bleeding associated with *Ginkgo biloba:* A case report and systematic review of the literature. J. Gen. Intern. Med 20: 657– 61

Bhargava, U.C., Westfall, B.A. (1969) Mechanism of blood pressure depression by ellagic acid. Proceed. Soc. Exp. Biol. Med. 132, 754–756.

Bidlack WR, Omaye ST, Meskin MS, Jahner D. (1998) Phytochemicals. Pennsylvania, US: Technomic Publishing Company Inc, 1998.

Bloomer RJ, Harvey IC, Farney TM, Bell ZW, Canale RE (2011) Effects of 1,3-dimethylamylamine and caffeine alone or in combination on heart rate and blood pressure in healthy men and women. Phys. Sportsmed 39: 111–20

Brasileiro LS, Segabinazzi LGTM, Menezes E, Salgueiro CC, Novello G, (2019) Coconut water as an extender component for cooled equine sperm. Journal of equine veterinary science 78: 69–73

Brower B. (1998) Nutraceuticals: Poised for a healthy slice of the market. Nat Biotechno l1998; 16: 728-33.

Burns DT, Johnston E-L, Walker MJ. (2020) Authenticity and the potability of coconut water-a critical review. Journal of AOAC International 103: 800–06

Cencic A, Chingwaru W (2010) The role of functional foods, nutraceuticals, and food supplements in intestinal health. Forum Nutr 2(6): 611–625

Cerda B., Llorach R., Cero´n J.J., Espý´n J.C., Toma´s-Barbera´n, F.A. (2003a) Evaluation of the bioavailability and metabolism in the rat of punicalagin, an antioxidant polyphenol from pomegranate juice. Eur. J. Nutr. 42: 18–28.

Cerda´ B., Cero´n J.J., Espý´n J.C., Toma´s-Barbera´n F.A. (2003b) The repeated oral administration of high doses of the pomegranate ellagitannin punicalagin to rats for 37 days is not toxic. J. Agric. Food Chem. 51, 3493–3501.

Cerda´ B., Toma´s-Barbera´n F., Espý´n J.C. (2005a) Metabolism of chemopreventive and antioxidant ellagitannins from strawberries, raspberries, walnuts and oak-aged wines in humans: identification of biomarkers and individual variability. J. Agric. Food Chem. 53, 227–

Chen, T., Hwang, H.J., Rose, M.E., Nines, R.G., Stoner, G.D., 2006. Chemopreventive properties of black raspberries in N-nitrosomethylbenzylamine-induced rat esophageal tumorigenesis: down-regulation of cyclooxygenase-2, inducible nitric oxide synthase, and c-Jun. Cancer Res. 66, 2853–2859.

Costa JR, Monteiro MJ, Tonon RV, Cabral LM, Pastrana L, Pintado ME. 2021. Fortification of coconut water with microencapsulated grape pomace extract towards a novel electrolyte beverage: Biological, sensorial and quality aspects. Future Foods 4: 100079

DebMandal M, Mandal S. 2011. Coconut (Cocos nucifera L. : Arecaceae): in health promotion and disease prevention. Asian Pacific Journal of Tropical Medicine 4: 241-47

Ediriweera E. 2003. Medicinal uses of coconut (Cocos nucifera L.). Cocoinfo Int 10: 11–21

Forrester MB. 2013. Exposures to 1,3-dimethylamylamine-containing products reported to Texas poison centers. *Hum. Exp.Toxicol* 32: 18–23

Jack DB. Keep taking the tomatoes – the exciting world of nutraceuticals. Mol Med Today 1995: 118-21.

Gee P, Jackson S, Easton J. 2010. Another bitter pill: a case of toxicity from DMAA party pills. N. Z. Med. J 123: 124–27

Gee P, Tallon C, Long N, Moore G, Boet R, Jackson S. (2012) Use of recreational drug 1,3 Dimethylamylamine (DMAA) associated with cerebral hemorrhage. Ann. Emerg. Med 60: 431–34

Gil, M.I., Tomas-Barbera´n, F.A., Hess-Pierce, B., Holcroft, D.M., Kader, A.A. (2000 Antioxidant activity of pomegranate juice and its relationship with phenolic composition and processing. J. Agric. Food Chem. 48, 4581–4589.

Harris, G.K., Gupta, A.M., Nines, R.G., Kresty, L.A., Habib, S.G., Frankel, W.L., LaPerle, K., Gallaher, D.D., Schwartz, S.J., Stoner, G.D. (2001) Effects of lyophilized black raspberries on azoxymetaneinduced colon cancer and 8-hydroxy-20 -deoxyguanosine levels in the Fischer 344 rat. Nutr. Cancer 40, 125–133.

Heasman M, Mellentin J (2001) The functional foods revolution. Healthy people. Healthy Earthscan Publications Ltd., London, pp 135–147

Hogg R, Chakravarthy U. (2006) Age-related eye disease study(Areds). In: Nutrition and the Eye. Elsevier; 2006: 205-210.

Jackson JC, Gordon A, Wizzard G, McCook K, Rolle R. (2004) Changes in chemical composition of coconut (*Cocos nucifera*) water during maturation of the fruit. Journal of the Science of Food and Agriculture 84: 1049–52

Jimoh O. (2020) Potential of coconut water to enhance fresh semen quality and fertility in rabbits. Tropical animal health and production 52: 249–55

Joseph D, George J, Mathews MM, Mathew F, Varghese B, Sunny B. (2019) A compilation on anti-diabetic profile of *Cocos nucifera*. Research Journal of Pharmacy and Technology 12: 3791–96

Larrosa, M., Toma´s-Barbera´n, F.A., Espý´n, J.C. (2006a.) The hydrolysable tannin punicalagin releases ellagic acid which induces apoptosis in human colon adenocarcinoma Caco-2 cells by using the mitochondrial pathway. J. Nutr. Biochem. 17, 611–625.

Karnatovskaia LV, Leoni JC, Freeman ML. (2015) Cardiac arrest in a 21-year-old man after ingestion of 1,3-DMAA-containing workout supplement. Clin. J. Sport Med 25: e23–25

Mannion M. (1998) Nutraceutical revolution continues at Foundation for Innovation in Medicine Conference. Am J Nat Med 1998; 5: 30-3.

Mazza G. (1998) Functional foods. Pennsylvania, US: Technomic Publishing Company Inc, 1998.

McCarthy CG, Canale RE, Alleman RJ Jr, Reed JP, Bloomer RJ. (2012) Biochemical and anthropometric effects of a weight loss dietary supplement in healthy men and women. Nutr. Metab. Insights 5: 13–22

Medeiros AC, de Paiva VdFL. (2012) Therapeutic use of coconut water. Journal of Surgical and Clinical Research 3: 83–91

Nasri, H.; Baradaran, A.; Shirzad, H.; Rafieian-Kopaei, M. (2014) New Concepts in Nutraceuticals as Alternative for Pharmaceuticals. Int. J. Prev. Med. *5:* 1487–1499.

Prades A, Dornier M, Diop N, Pain J-P.(2012) Coconut water uses, composition and properties: a review. Fruits 67: 87–107

Pantuck, A.J., Leppert, J.T., Zomorodian, N., Aronson, W., Hong, J., Barnard, R.J., Seeram, N., Liker, H., Wang, H., Elashoff, R., Heber, D., Aviram, M., Ignarro, L., Belldegrun, A. (2006) Phase II study of pomegranate juice for men with rising prostate-specific antigen following surgery or radiation for prostate cancer. Clin. Cancer Res. 12, 4018–4026.

Paivarinta, E., Pajari, A.M., Torronen, R., Mutanen, M. (2006) Ellagic acid and natural sources of ellagitannins as possible chemopreventive agents against intestinal tumorigenesis in the Min mouse. Nutr. Cancer 54, 79–83.

Pushpangadan P. (1988) Arogyappacha'(*Trichopus zeylanicus* Gaerin), the'ginseng'of kani tribes of agashyar hills (kerala) for ever green health and vitality. Anc. Sci. Life 8: 13–16.

Querques G, Forte R, Souied EH. (2011) Retina and omega-3. J Nutr Metab. 2011; 2011: 748361. doi: 10.1155/2011/748361. Epub 2011 Oct 31.

Rahman K, Lowe GM, Smith S. (2016) Aged garlic extract inhibits human platelet aggregation by altering intracellular signaling and platelet shape change. J Nutr 146: 410S–15S

Rethinam P, Nanda Kumar T. (2001) Tender coconut-an overview

Rethinam P, Krishnakumar V. (2022) Value Addition in Coconut Water. In Coconut Water: A Promising Natural Health Drink-Distribution, Processing and Nutritional Benefits: 287- 384: Springer. Number of 287-384 pp.

Ronis MJJ, Pedersen KB, Watt J. (2018) Adverse effects of nutraceuticals and dietary supplements. Annu Rev Pharmacol Toxicol. 2018; 58: 583-601.

Rosenblat M., Hayek T., Aviram M. (2006) Anti-oxidative effects of pomegranate juice (PJ) consumption by diabetic patients on serum and on macrophages. Atherosclerosis 187, 363–371

Roytman MM, Pörzgen P, Lee CL, Huddleston L, Kuo TT, Bryant-Greenwood P, Wong LL, Tsai N. (2014) Outbreak of severe hepatitis linked to weight-loss supplement OxyELITE Pro. Am. J. Gastroenterol 109: 1296–98

Shaw D, Leon C, Kolev S, Murray V. (1997)Traditional remedies and food supplements. A five- year toxicological study. Drug Safety 17: 342-56.

Segura-Badilla O, Lazcano-Hernández M, Kammar-García A, Vera López O, Aguilar-Alonso P, (2020) Use of coconut water (*Cocus nucifera* L.) for the development of a symbiotic functional drink. Heliyon 6

Singh RK. (2016) Nutraceuticals in reproductive and developmental disorders. In: Nutraceuticals. Elsevier; 123-134pp.

Swanson EE, Chen KK. (1946) Comparison of pressor action of aliphatic amines. *J. Pharmacol.* Exp. Ther 88: 10–13

Shubhashree M, Venkateshwarlu G, Doddamani S. (2014) Therapeutic and nutritional values of Narikelodaka (tender coconut water)-A review. Research Journal of Pharmacognosy and Phytochemistry 6: 195

Tuyekar S, Tawade BS, Singh KS, Wagh VS, Vidhate PK, (2021) An Overview on Coconut Water: As a Multipurpose Nutrition. Int. J. Pharm. Sci. Rev. Res 68: 63–70

US Food and Drug Administration (FDA). 2012. *FDA challenges marketing of DMAA products for lack of safety evidence* U.S. Food and Drug Administration, Silver Spring, MD: http: //www.fda.gov/NewsEvents/Newsroom/PressAnnouncements/ucm302133.htm

Woodbury A, Sniecinski R. (2016) Garlic-induced surgical bleeding: How much is too much? A A Case Rep 7: 266–69

Xinmeng L. (2023) A Study on the Linguistic Landscape of Catering Store Names in Haikou – Taking the Four Main Urban Districts of Haikou City as an Example. Modern Linguistics 11: 2025

Zulaikhah ST. (2019) Health benefits of tender coconut water (TCW). Int J Pharm Sci Res 10: 4: 74-80

Part B

Plants with Medicinal Value

Many herbs and teas offer harmless subtle ways to improve your health. Pay attention to what the evidence says about each herb's effectiveness as well as potential interactions or safety issues. Avoid using herbs for infants and children and for those who are pregnant and breastfeeding. As one of the oldest tree species, Ginkgo is also one of the oldest homeopathic plants and a key herb in Chinese medicine. The leaves are used to create capsules, tablets, and extracts, and when dried, can be consumed as a tea. the Ginkgo tree is considered a living fossil, with fossils dating from 270 million years ago. These trees can live up to 3,000 years. It's perhaps best-known for its ability to boost brain health. Studies say that Ginkgo can treat patients with mild to moderate dementia, and can slow cognition decline in dementia and Alzheimer's disease. Recent research is looking into a component that can help diabetes, and there continue to be more studies, including an animal study that says it might influence bone healing.

Chapter 21

A Herb of Health, Vigour and Vitality: Safed Musli (*Chlorophytum borivilianum* L.)

Safed musli (Chlorophytum borivilianum L.) is an indispensable medicinal plant and is considered as a 'white gold' or 'divya aushad' in Indian systems of medicine. In Ayurveda, C. borivilianum belongs to the group of "Vajikaran Rasayana" corroborated to its rejuvenating, aphrodisiac, natural sex tonic properties and effective in alleviating sexual disorders. It is largely used as ethnic medicine by local healers of indigenous communities of India. It is used to cure arthritis and increasing body immunity, curative for natal and postnatal problems, for rheumatism and joint pains, increase lactation in feeding mothers, effective as antimicrobial, anti-inflammatory, antitumor agent, also used in diarrhea, dysentery, gonorrhea, leucorrhea etc. It has spermatogenic property and is found useful in curing impotency, now it is considered as an alternative 'Viagra'. Its root contains steroidal and triterpenoidal saponins, sapogenins and fructans which act as therapeutic agents and play vital role in many therapeutic applications.

Keywords: *Herb, Health, Vigour, Vitality, Safed Musli, Chlorophytum borivilianum.*

21.1 Introduction

In Ayurveda, *C. borivilianum* belongs to the group of "Vajikaran Rasayana' corroborated to its rejuvenating, aphrodisiac, natural sex tonic properties and effective in alleviating sexual disorders. It is largely used as ethnic medicine by local healers of indigenous communities of India (Oudhia 2001). Traditionally, *C. borivilianum* is well known for treating male impotency in India. The multi therapeutic and nutritional importance of *C. borivilianum* is attributed to the rich source of phytochemicals particularly saponins. Recently, it has gained well established domestic (Indian) and international market for being herbal alternative of "Viagra" without any side effects. Under the trade name 'Nai Chetna', the state government of Gujrat, India, has launched a novel potency drug from *C. borivilianum*. Modern pharmacological studies of *C. borivilianum* have demonstrated a wide range of pharmacological activities, most importantly aphrodisiac, immunomodulatory and anticancer activities (Thakur *et al.,* 2009; Yadav and Chauhan 2022). The increased commercial exploitation of *C. borivilianum* and low productivity of this endangered plant has raised the concern over its conservation.

The use of plants for treatment of various ailments dates back to over 5000 years. A great source of ancient information is contained in the 'Vedas' and more specifically 'Yajur Veda' is the main source of such information. In these 'Vedas' the medicinal importance of many plants has been mentioned (Prakash 2001). The earliest monumental contribution on 'Ayur-veda' is the 'Samhitas' of 'Charak' and 'Sushrut' (1000-700 B.C.), which included 500 plants with their therapeutic properties. Although with the invasion of Greeks, Sagthians, Huns, Mughals and Europeans, the progress of Ayurveda declined, but the plant based drugs caught the attention of the west from the early colonial days. 'Ayurvedic Rasayanas' have been designated as a special class of herbs having multifaceted action inside the human body. Their activities range from improvement in mental acuity, keenness to enhancement of anabolism, maintenance of homeostasis, prevention of degenerative diseases, antioxidant activity and improvement in failing sexual functions (Kaushik 2005). 'Chyawanprash' is an excellent example of Rasayana which is a convergence of numerous therapeutic benefits and was formulated by 'Rishi Chyawan' after years of research put together, to keep the ailments at bay and look young and rejuvenated (Handa 1996). Safed musli is an important ingredient of this age old, time tested formulation (Thakur *et al.,* 2009).

21.2 Habitat and Cultivation

Ayurvedic texts have recommended *C. borivilianum* as the source of 'safed musli' raw drug. Its growth season is limited to middle of May to October, after that the aerial parts of the plant dry up and plant undergoes in dormancy. Roots of safed musli are fasciculate, sessile, cylindrical or ellipsoidal, 1-9 in number, number and size increases with the age of the plant. Leaves are radical up to 8 in number, narrow and spreading horizontally. Flowers are white born in clusters of three flowers. The seed are orbicular to coffee-bean-shaped.Safed musli is distributed in tropical and sub tropical climate of the world. In India, it is naturally distributed in the forest areas of Maharashtra, Gujarat, Rajasthan and Madhya Pradesh. It is inhabitant in wide range of habitats from the open, rocky places to shady and highly humus rich soil of the forest. Commercial cultivation has been started in Central India and has spread up to South India in recent past.

Safed musli prefer warm and humid climate during its crop growth period. Temperatures more than 35°C and less than 15°C are not suitable for its growth and yield. Areas receiving 50 to 150 cm rainfall between June end to October end are considered to be suitable for its successful cultivation. Safed musli can be cultivated on various types of soils from sandy loam, sandy clay, low in organic matter and acidic soils of Bihar, West Bengal to heavy black cotton soils of Maharashtra, Madhya Pradesh, Andhra Pradesh to red lateritic soils of Karnataka to calcareous soils of Gujarat and Rajasthan. However, crop prefers well drained loamy soils rich in organic matter for good fasciculated root development and high productivity. Soils having high pH and high calcium carbonate content are not suitable for the crop and show severe micro-nutrients deficiency symptoms especially for iron and cause chlorosis.

Land is prepared well by ploughing and harrowing to make soil pulverized and porous. Since water stagnation is harmful for the crop, therefore, proper drainage is to be created by providing proper slope during land preparation. At the time of land preparation care must be taken to make the field free from perennial weeds like *Cynodon dactylon* and *Cyprus rotundus* as the roots and stems of these weeds penetrate the fleshy roots of safed musli causing physical damage as well as invite fungal infection. Apply 15-30 tonnes of farm yard manure or 5 tonne vermicompost or poultry manure per hectare at the time of last ploughing. Planting is done during the month of June-July depending on the onset of monsoon. The best time for planting based on sprouting percentage of the dormant buds is during 25th June to 5th July.

Safed musli is propagated both by seedling raised from true seeds as well as vegetatively using fleshy roots having condensed stem disc and shoot buds. Under commercial cultivation, the dormant plants of the previous year are generally used as vegetative propagules which sprout as early as in the last week of April to second week of May. Planting material weighing 3-4 g root is ideal for highest root yield. The higher price of the vegetative planting material has forced to raise the crop through true seeds. However, due to low seed germination percentage, farmers use roots as propagating material.

Planting is usually done on raised beds or ridges or flat beds depending upon the soil texture and amount of rainfall. The places where soils are heavy, rainfall is more and drainage system is poor, the crop is grown on raised beds or ridges. While in low rainfall areas and in light textured soils, planting on gentle-slopped flat bed is advisable. Planting should be done preferably in rainy days when high humidity prevails in atmosphere for high survival rate. Single or double row planting is followed in ridges. Planting should be done at 30 cm row to row and 10 cm plant to plant spacing to accommodate 3.33 lakh plants per hectare. Mulching is advisable with organic matter such as dried leaves, straw *etc.*, in the nursery as well as in the main field after planting of vegetative propagules. It helps in conserving the soil moisture as well as to protect the sprouts from direct sun burn at initial stage and increase the root yield.

Two varieties 'Jawahar Safed Musli 405' and 'Rajvijay Safed Musli 414' were released from AICRP on MAP unit, Rajmata Vijayaraje Scindia Krishi Vishwa Vidyalaya, Mandsaur, Madhya Pradesh are available for cultivation.

Figure 21.1. Safed Musli (*Chlorophytum borivilianum* L.).

21.3 Medicinal Uses

Chlorophytum borivilianum is a traditional rare Indian medicinal herb which has many therapeutic applications in Ayurvedic, Unani, Homeopathic and Allopathic system of medicines. Its roots (tubers) are very famous for various therapeutic applications *viz.* adaptogenic, aphrodisiac, immunomodulatory and antidiabetic. It is used to cure physical illness and weakness, as an aphrodisiac agent and revitalizer, as general sex tonic, remedy for diabetes, arthritis and increasing body immunity, curative for natal and postnatal problems, for rheumatism and joint pains, increase lactation in feeding mothers, also used in diarrhea, dysentery, gonorrhea, leucorrhea *etc.* It has spermatogenic property and is found useful in curing impotency, now it is considered as an alternative 'Viagra'. Recent pharmacological studies on tubers of *C. borivilianum* has indicated antiviral, anticancer, antioxidant, antidiabetic, antistress, aphrodisiac, antimicrobial, hypolipidemic, hypocholesteremic, antiinflammatory, immunomodulatory activities (Loo *et al.,* 2005; Siddiqui 2005; Jamal 2005; Govindarajan *et al.,* 2005a; Visavadiya and Narsimhacharya 2007; Deore and Khadabadi 2009a). Steroidal saponins, polysaccharides, fructans and fructooligosaccharides, beta sitosterol, stigmasterol, hecogenin and mucilage are major constituents of *C. borivilianum* which act as therapeutic agents in many therapeutic applications.

The plant yields a flavonone glycoside, which is a powerful uterine stimulant, steroidal saponins which have muscle building properties and their structure is similar to male anabolic hormones testesterone. Roots of *Chlorophytum* contain 42 per cent carbohydrate, 80–89 per cent protein, 3–4 per cent fiber and 2–17 per cent saponin (Wagle *et al.,* 2000). Research studies on *Chlorophytum* conducted in India and elsewhere indicate that saponins (Jat and Bordia 1990) are responsible for medicinal properties. A saponin is a category of phytonutrients (plant nutrients) found abundantly in many beans, and other plants such as ginseng, alfalfa, yucca, aloe, quinoa seed and also in Safed Musli. Saponins in Safed Musli are known to ensure the defense of the plant against microbial or fungal attack. Several drugs owe their anti-inflammatory and antiedemic properties to saponins. Inulin type 2, 1 linked fructans have been reported by Sreevidya *et al.* (2006) and Kothari (2004). In their study, presence of mannans of pure type has also been determined in a separate set of experiments. Apart from biologically effective steroidal and triterpenoidal saponins, sapogenins and fructans having prebiotic importance (Devon and Scott 1976), the other phytoconstituents reported from the plant are high quantities of simple sugars mainly sucrose, glucose, fructose, galactose, mannose and xylose (Sreevidya *et al.,* 2006). Proteins, phenolics, triterpenoids, gallo-tannins and mucilage are also reported from *C. borivilianum* (Thakur and Dixit 2005). The simple sugars like fructose, glucose, galactose, sucrose, xylose and mannose are also found in the Musli roots. The medicinal properties may be due to the saponin and polysaccharide content in the Safed musli roots (Rath and Panja 2013).

21.3.1 Uses as an Aphrodisiac

Chlorophytum borivilianum has been traditionally acclaimed and advocated for its aphrodisiac activity (Oudhia 2001). In a recent study, ethanolic extract of roots as well as sapogenins isolated from the roots were studied for effect on sexual behavior and spermatogenesis in albino rats. Treatment had pronounced anabolic and spermatogenic effect in treated animals, evidenced by weight gains of body and reproductive organs. Administration of extracts markedly affected sexual behavior of animals reflected in reduction of mount ejaculation, post ejaculatory and intromission latency. An increase in mount frequency and attractability towards female was observed (Thakur 2005).

It is also called Herbal Viagra by many! It also helps in increasing sexual time and helps in enhancing performance. It helps improve male weakness, helps in cases of impotency or erectile dysfunction due to the presence of Rho-Kinase II. It helps increase sperm count and reduces frigidity in women. Provide vaginal lubrication, cures premature ejaculation and boost urge for sex in both males and females. Safed musli is very useful for treating oligospermia and improves count, volume, liquefaction time and motility. It also improves serum testosterone levels and improves testicular functions. Safed musli extract is studied for these benefits and according to scientific study, it is significantly beneficial as compared to placebo. In ayurveda, Safed musli is well-known for its spermatogenic potential. However, it is used in conjunction with other herbs (Singh *et al.,* 2012).

21.3.2 Uses as an Antioxidant

C. borivilianum has many additional beneficial effects including properties of adaptogen, immunomodulation and anti-aging and is also a rich source of zinc and iron. Zinc increases the activity of natural killer cells and boost the production of antibodies in response to infection. As zinc is required for the production of the antioxidant enzyme superoxide dismutase, which neutralizes potentially damaging free radicals. The mineral also aids in the production of testosterone and boost sperm production. Physiological supplementation of zinc in ageing and in age related degenerative diseases corrects immune defects, reduces infection relapse and prevents ageing. Zinc is also an antioxidant mineral that is important in protecting sperm against free radical attack. Iron helps in the formation of hemoglobin, which is present in the red blood cells, and myoglobin, a molecule that transports oxygen in muscles. It also prevents iron deficiency anemia, as well as stimulates the immune system. It is also used to improve muscular and athletic performance, and prevents fatigue. The roots are reported to contain 42 per cent of carbohydrates, 8–9 per cent of proteins, 3–4 per cent fibres, and 2–17 per cent of saponins (Bordia *et al.,* 1995).

21.3.3 Uses for Good Health and Bodybuilding

It is effective for improving male performance, inducing spermatogenesis and providing strength to the muscles. It would not be wrong to call Safed musli as a domicile of the medical properties like. anti-inflammatory properties, antioxidant properties, antacid properties, aphrodisiac properties, androgenic properties and many more. The use of Safed musli has shown good effects on the body keeping good for bodybuilder and athletes, help fight fatigue and physical weakness, help gain more healthy weight which makes it ideal for underweight people, ideal for bodybuilders as they help muscle gain (Khanam *et al.,* 2013)

21.4 Uses as Nutraceutical

Safed musli has some important phytochemical constituents which make it very popular and useful for a nutritional dietary regime in body builders and sports person. The important constituent of *C. borivilianum* is Inulin type fructans. Researchers of India and Vienna have reported that presence of nutraceutical components *viz.* fructans and fructo-oligosaccharides in *C. borivilianum* (Sreevidya *et al.,* 2006). Fructans improve calcium absorption in bone and provide important fiber and roughage to the body, thus providing a perfect blend for a healthy diet. Fructans are also very low in calorie and help in prevention of diabetes induced problems. Inulin supplementation also fulfills the urgent energy requirements to overcome muscle fatigue during body-building processes. *C. borivilianum* root contains considerable quantity of zinc and iron, (3.75 mg/100g and iron as 4.65 mg/100g). Zinc is responsible for the antioxidant property of *Chlorophytum* (Jaykumar *et al.,* 2008).

C. borivilianum has many additional beneficial effects including properties of adaptogen, immunomodulation and anti-aging and is also a rich source of zinc and iron. Zinc increases the activity of natural killer cells and boost the production of antibodies in response to infection. As zinc is required for the production of the antioxidant enzyme superoxide dismutase, which neutralizes potentially damaging free radicals. The mineral also aids in the production of testosterone and boost sperm production. Physiological supplementation of zinc in ageing and in age related degenerative diseases corrects immune defects, reduces infection relapse and prevents ageing. Zinc is also an antioxidant mineral that is important in protecting sperm against free radical attack. Iron helps in the formation of hemoglobin, which is present in the red blood cells, and myoglobin, a molecule that transports oxygen in muscles. It also prevents iron deficiency anemia, as well as stimulates the immune system. It is also used to improve muscular and athletic performance, and prevents fatigue. The roots are reported to contain 42 per cent of carbohydrates, 8–9 per cent of proteins, 3–4 per cent fibers, and 2–17 per cent of saponins (Bordia *et al.*, 1995).

There are number of beneficial phyto-constituents or phytochemical substances that have muscle building benefits which are present in abundance in *C. borivilianum*. The major proportions are steroidal saponins (Kaushik 2005). Steroidal saponins of *C. borivilianum* have a steroidal chemical structure which is very similar to the chemical structure of male anabolic hormone testosterone. The steroidal saponins are rapidly absorbed in the body and are converted to male hormone testosterone which assists in building up muscles and gaining masculine power (Tandon and Shukla 1992). Beta sitosterol, stigmasterol and hecogenin are very important naturally occurring chemical compounds that resemble in structure to testosterone and help in body building process. *C. borivilianum* has a prominent anabolic effect, and when taken as a dietary supplement at a dose of 5-10 gm daily (dry root powder) or 50-100 mg of extract prior to exercise, the drug can be beneficial in improving the muscle strength building process.

21.5 Conclusion

It helps to relieve stress and depression, reduces anxiety, improves sleeping disorder, and It has mind calming property. Safed musli is hugely known as a sex tonic and Ayurvedic aphrodisiac compound by many. They are ideal for men especially and can be used by women also. Safed musli provides nourishment to the malnourished body and it increases the body weight. According to Ayurveda, nocturnal emission (night fall or wet dreams) occurs due to excess Pitta associated with Vata. Dreams, thinking and visualization of physical relations or even watching TV with such scenes can trigger this condition. Safed musli acts on both Pitta and Vata. If patient feel weakness, low backache, and loss of strength or energy after nocturnal emission, then Safed musli powder along with Shakkar (sugarcane brown sugar) should be used for a few weeks. This remedy also helps reducing frequency of nocturnal emissions and revitalizes the body. No doubt, it provides strength to penile tissue, improves hardness, and helps sustaining longer erections, but it has less stimulant action and erectogenic effect as compared to *Anacyclus pyrethrum* (Akarkara). Its main function is to provide the strength, acts on testes, improves hormone profile, and induces spermatogenesis (Mujeeb *et al.,* 2009). Safed musli has some important phytochemical constituents which make it very popular and useful for a nutritional dietary regime in body builders and sports persons.

It has been envisaged that efforts should be made to standardize, encourage and popularize the cultivation of *C. borivilianum* as a commercial crop. The analysis of previous pharmacological investigations suggested lack of substantial scientific evidences in various studies and do not stand the test of critical assessment. Due to high economic value, *C. borivilianum* has also encountered a problem of adulteration with closely resembling medicinally inferior species. The studies available on toxicity, safety and quality of Safed musli are inadequate for providing information on commercial utilization. Thus, the present review summarizes comprehensive information on Safed musli and possible scope for future research to fill the existing lacunae on its different aspects of the study.

REFERENCES

Bordia, P.C.; Joshi, A. and Simlot, M.M. (1995) "Safed Musli" in: Advances in Horticulture: Medicinal and Aromatic Plants, (Chadha, K.L. and Gupta, R., Ed.), Malhotra Publishing House: New Delhi, pp. 440-449.

Deore, S.L. and Khadabadi, S.S. (2009a) A study of hypocholesteremic effects of *Chlorophytum borivilianum* tubers in rats. Pharmacologyonline, **1**, 398-403.

Devon, T.K. and Scott, A.I (1976) Handbook of Naturally Occurring Compounds, Academic Press, New York, Vol. I, p. 498.

Govindarajan, R.; Sreevidya, N.; Vijayakumar, M.; Thakur, M.; Dixit, V.K.; Mehrotra, S. and Pushpangadan, P. (2005a) Standardization and determination of antioxidant activity of *Chlorophytum borivilianum*. Nat. Prod. Sci., 11(3), 165-169.

Handa, S.S. (1996) Rasayana Drugs. in Supplement to Cultivation and Utilization of Medicinal Plants, (Handa, S.S. and Kaul, M.K. Ed.) Regional Research Laboratory, CSIR Jammu-Tawi, pp. 509- 513.

Jamal, M. A. (2005) Effects of Safed musli on cell kinetics and apoptosis in human breast cancer cell lines. International conference on promotion and development of botanicals with international co-ordination exploring quality, safety, efficacy and regulation organized by School of natural product study, Jadavpur university, Kolkata, India, p. 26.

Jaykumar, B. Sagar, S.K. and Sinnathurai, G. (2008) Neutraceutical compositions and methods of use of Safed Musli extracts. Patent No. WO/2008/020451 A2.

Jat, R.D. and Bordia, P.C. (1990) Propagation studies in Safed Musli (*Chlorophytum* species) Proc. of National symposium on Adv. In Plant Sci.: Current Status and Emerging Challenges. Deptt. of Botany, Sukhadia Univ., Udaipur, 46.

Kaushik, N. (2005) Saponins of *Chlorophytum* species. Phytochem. Rev. 4: 191-196.

Khanam Z, Singh O, Singh R, Bhat IUH (2013) Safed musli (*Chlorophytum borivilianum*): A review of its botany, ethnopharmacology and phytochemistry. J Ethnopharmacol. 150 (2): 421-441.

Kothari, S.K. (2004) Safed Moosli (*Chlorophytum borivilianum*) revisited. J. Med. Aromat. Plant Sci., 26: 60-63.

Loo, J.V.; Clune, Y.; Bennett, M. and Collins, J.K. (2005) The specificity of interaction with intestinal bacterial fermentation by prebiotics determines their physiological efficacy. Br. J. Nutr. (Suppl. 1), 91-98.

Mujeeb M, Khan SA, Ali M, Mall A, Ahmad A (2009) Antidiabetic activity of the aqueous extract of *Chlorophytum borivilianum* L. in Streptozotocin induced-hyperglycemic rats–a preliminary study. J Pharm Res. 2009, 2: 5-53.

Oudhia, P. (2001) Problems perceived by Safed Moosli (*Chlorophytum borivilianum*) growers of Chattisgarh (India) region: A study. J. Med. Aromat. Plants Sci., DOI 22/4A and 23/1A, 396-399.

Prakash, V. (2001) Indian medicinal plants; current status. in Himalayan Medicinal Plants Potential and Prospects, (Samanth, S.S.; Dhar, U.and Palni, L.M.S., Ed.). Gyan Odaya Prakashan Publisher, Nainital, pp. 45-63.

Rath SK, Panja AK. (2013) Clinical evaluation of root tubers of Shweta Musali (*Chlorophytum borivilianum* L.) and its effect on semen and testosterone. Ayu 2013; 34(3): 273-5

Siddiqui, Y. M. (2005) Potent antiviral effect of Safed musli on poliovirus replication, International conference on promotion and development of botanicals with international co-ordination exploring quality, safety, efficacy and regulation organized by School of natural product study, Jadavpur University, Kolkata, India, p. 25.

Singh D, Pokhriyal B, Joshi YM, Kadam V. (2012) Phytopharmacological aspects of *Chlorophytum borivilianum* (safed musli): A review. Int J Res Pharm Chem 2(3): 853-9.

Sreevidya, N.; Govindarajan, R.; Vijayakumar, M.; Thakur, M.; Dixit, V.K.; Mehrotraand, S. and Madhusudan, K.P. (2006) Action of fructo-oligo polysaccharide fraction of *Chlorophytum borivilianum* against streptozotocin induced oxidative stress. Planta Med., 72: 1421-1424.

Tandon, M. and Shukla, Y.N. (1992) Sapogenins from *Asparagus adscendens* and *Chlorophytum arundinaceum*. J. Indian Chem. Soc., 69: 893.

Thakur, M. (2005) Phytochemical and Pharmacological studies on *Chlorophytum borivilianum* Sant. and F., M. Pharm Thesis, Department of Pharmaceutical Sciences, Dr. H.S. Gour University, Sagar.

Thakur, M. and Dixit, V.K. (2005) Fructan; The polymer with unexplored potential. Ind. Pharm., 4 (40): 7-12.

Thakur G.S., Bag M., Sanodiya B.S., Debnath M., Zacharia A., Bhadauriya P., Prasad GBKS and Bisen, P.S. (2009) *Chlorophytum borivilianum*: A White Gold for Biopharmaceuticals and Nutraceutical. Current Pharmaceutical Biotechnology, 10: 650-666

Visavadiya, N.P. and Narsimhacharya, A.V.R.L. (2007) Ameliorative effect of *Chlorophytum borivilianum* root on lipid metabolism in hyperlipaemic rats. Clin. Exp. Pharmacol. Physiol. 34(3): 244 - 249.

Wagle, A.; Kelkar, G. D. and Heble, M. R. (2000) in : Biotechnology: Secondary Metabolites, Oxford and IBH Publishing Co, Pvt. Ltd., New Delhi, p 219-220.

Yadav M.K. and Chauhan R.S.(2022) Phytochemical analysis of *Chlorophytum borivilianum* (safed musli) dried root powder extract. J. Exp. Zool. India 25, 1485-1487. DocID: https: //connectjournals.com/03895.2022.25.1485

Chapter 22

A Natural Wonder Tree: Babul (*Vachellia nilotica* (L.) P.J.H. Hurter and Mabb.)

The Babul tree, scientifically known as Vachellia nilotica, is a resilient and drought-resistant plant species that belongs to the Fabaceae family. This evergreen tree is native to the African and Asian regions and is renowned for its ability to thrive in arid and challenging environments. Throughout its range, the Babul tree goes by several names, including "gum arabic tree," "Egyptian thorn," and "babool." This diversity in nomenclature reflects its global presence and cultural significance. The history of the Babul tree is deeply intertwined with human civilization. It has been used for various purposes for thousands of years, dating back to ancient Egypt and India. Babul trees can reach impressive heights of up to 20 meters, with a broad, umbrella-like canopy that provides ample shade in the scorching sun. One of the defining features of the Babul tree is its thorns, which serve as a natural defense mechanism against herbivores. These thorns are not only protective but also have unique medicinal properties. The tree's small, feathery leaves are a vital source of food for various herbivorous animals, contributing to the overall ecological balance. In regions where water is scarce, the Babul tree plays a crucial role in preventing soil erosion and maintaining groundwater levels. Babul trees provide shelter and sustenance to a wide array of wildlife, including birds, insects, and mammals, making them a cornerstone of local ecosystems.

Keywords: *Natural wonder, Arid tree, Babul, Vachellia nilotica, Gum Arabic, Oral health.*

22.1 Introduction

Modern agricultural practices have recognized the importance of Babul tree

Vachellia syn. *Acacia nilotica* in agroforestry, it promotes sustainable land use and biodiversity conservation. Agroforestry is a sustainable farming practice that integrates trees like the Babul tree into agricultural systems to promote biodiversity and sustainable land use. Babul tree provide shelter and food for various wildlife species, enhancing the biodiversity of their ecosystems. The Babul tree's bark, leaves, and gum contain compounds with anti-inflammatory and antibacterial properties, making it valuable in traditional medicine. Gum arabic, extracted from the Babul tree's resin, is used in the food and pharmaceutical industries for its emulsifying and stabilizing properties (Houerou 1980). The genus name *Acacia* derives from Ancient Greek: κακßα, which may come from the Ancient Greek word for its characteristic thorns, κέὸ, *ákis*, "thorn", or may have been borrowed from Pre-Greek. The specific epithet *nilotica* was probably given by Linnaeus from this tree's originally known range along the Nile river. In Australia the tree is known as a prickly acacia, despite usurping Dioscorides' two millennia-old etymology, the Australian species classified as *Acacia* in Australia do not have thorns.inoculating the seed with improved rhizobia strains can increase early growth (Luna 1996), and salt tolerant *Rhizobium* isolates have also been tested which can nodulate and fix nitrogen in saline soils (Lal and Khanna 1994). There are a number of reports of vegetative propagation (Garg *et al.,* 1996; Mathur and Chandra 1983), although regeneration by seed is the preferred and easier option.

22.2 Bark: Nature's Pharmacy

The bark of the Babul tree is a rich source of medicinal compounds that have been used in traditional medicine for generations.The Babul tree, scientifically known as *Vachellia nilotica*, is not just a botanical wonder; it's also a treasure trove of medicinal properties that have been harnessed for centuries by various cultures across Africa and Asia. Among its most notable attributes is its potent anti-inflammatory properties. This quality makes it a valuable remedy for a range of inflammatory conditions, from minor skin irritations to more serious ailments like arthritis. The application of Babul bark paste or infusion can provide relief from pain and swelling, offering a natural alternative to pharmaceutical anti-inflammatories. Additionally, the bark contains compounds with antibacterial and antimicrobial properties. These properties make it effective in treating various infections, both internal and external. In traditional practices, a decoction of Babul bark has been used to alleviate symptoms of respiratory infections and gastrointestinal disorders. The bark of the babul tree is used to tan leather. A by-product from felling is the bark, which has high levels of tannin (12-20 per cent) used for tanning leathers in India and the pods of *A. nilotica* subsp. *nilotica* have been used for over 6000 years in Egypt for tanning. *A. nilotica* subsp. *adstringens* is used both for tanning and as a dye in Nigeria.

22.3 Leaves: A Green Wellness Treasure

The leaves of the Babul tree are equally medicinal, holding therapeutic promise in their green abundance. Rich in tannins, saponins, and alkaloids, these leaves offer a spectrum of health benefits. One prominent use of Babul leaves is in oral care. Chewing on Babul leaves or using them to make a natural toothpaste has been a time-honored practice in many cultures. The natural astringent properties of the leaves help strengthen gums and teeth, prevent cavities, and combat bad breath. It's no wonder that Babul leaves are often referred to as "nature's toothbrush." Beyond oral health, Babul leaf extracts have been explored for their potential in managing diabetes. Research suggests that the compounds found in the leaves may help regulate blood sugar levels, offering hope to those grappling with this chronic condition. The young leaves and flowers of the babul tree are edible and can be used in cooking. The leaves and pods of the babul tree are used as fodder for livestock. Charak Samhita mentions that consuming buds of babul with lime in form of tablets, when chewed relieves from sour throat (Saxena 2023).

22.4 Gum Arabic: A Versatile Elixir

Derived from the resin of the Babul tree, gum arabic is a prized natural substance known for its versatility. While not a direct product of the tree's medicinal attributes, gum arabic's use in various pharmaceutical formulations and traditional remedies cannot be overlooked. One of its key roles is as a binding and emulsifying agent in the pharmaceutical industry. Medicinal formulations often require stable suspensions, and gum arabic's natural emulsifying properties make it a valuable ingredient in creating such suspensions. Furthermore, its presence can enhance the delivery and absorption of certain medications. In traditional medicine, gum arabic has also found applications as a soothing agent for coughs, asthma (Chauhan 2020) and sore throats. Its demulcent properties help relieve irritation in the throat, providing relief to those suffering from respiratory ailments. T*he versatile elixir that is gum arabic, the Babul tree continues to be a source of healing and wellness, bridging the ancient wisdom of traditional medicine with the demands of modern pharmaceutical and healthcare practices.* Charak Samhita mentions that 2 to 4 g of gum consuming early morning and late in the evening helps in curing dysentery (Saxena 2023).

22.4.1 Gum Arabic: Nature's Goldmine

The Babul tree, renowned for its multifaceted contributions to both nature and humanity, unveils yet another treasure from its repertoire: gum arabic. This natural exudate, extracted from the tree's resin, is nothing short of a precious resource, cherished for its remarkable emulsifying and stabilizing properties. In this exploration, we will delve into the depths of gum arabic, uncovering its vital role in the food and pharmaceutical industries and understanding why it is hailed as a true gem. The gum of the babul tree is used in a variety of products, including food, cosmetics, and pharmaceuticals. Gum is a viable income sources for thousands of forest dwellers, especially tribals in India. Exploiting Babul tree for gum tapping will help farmers to strengthen their socioeconomic conditions as well as to help conserve environment and biodiversity too (Raj 2015). As per Abtew *et al.* (2014) gum and resin income contributes to 14 per cent –23 per cent of the small-scale producers' household income in the dryland of Ethiopia and Sudan. Therefore, it represents important sources of livelihoods of the local community, employment and foreign exchange earnings (Chikamai and Casadei 2005). In addition to the wider industrial applications in developed countries, the products are consumed locally as traditional medicines, foods and beverages, and are used in religious and cultural rituals (Chikamai *et al.,* 2009).

22.4.2 The Essence of Gum Arabic

- ✰ Gum arabic, also known as acacia gum, is a water-soluble substance derived from the hardened sap of the Babul tree.
- ✰ Its origins trace back to the arid regions of Africa and Asia, where the Babul tree thrives.
- ✰ What sets gum arabic apart is its unique composition, characterized by a complex blend of polysaccharides and glycoproteins.
- ✰ This composition is responsible for its exceptional properties and versatility.

22.4.3 Emulsifying Powerhouse

- ✰ In the food industry, achieving and maintaining the desired texture and consistency of various products can be a formidable challenge.
- ✰ This is where gum arabic takes center stage as a natural emulsifier.
- ✰ Its ability to create stable emulsions, where typically immiscible substances like oil and water coexist harmoniously, is nothing short of remarkable.
- ✰ Gum arabic's emulsifying prowess finds extensive use in the production of a wide range of food products.
- ✰ Salad dressings, beverages, confectioneries, and even ice cream owe their creamy and uniform textures to this natural wonder.
- ✰ By stabilizing these emulsions, gum arabic ensures that products not only look and taste appealing but also have an extended shelf life.

22.4.4 Stability Booster in Pharmaceuticals

- ✰ Beyond the culinary realm, gum arabic plays a pivotal role in the pharmaceutical industry. Used for treatment of Asthma (Chauhan 2020).
- ✰ Medicinal formulations often contain a diverse array of active ingredients that must remain in a stable state until they reach the end-user.
- ✰ Gum arabic steps in as a reliable stabilizing agent, preventing the separation or degradation of these components.
- ✰ In the realm of oral medications, gum arabic assists in the creation of suspensions that evenly distribute active pharmaceutical ingredients.

- ☆ This is particularly crucial for pediatric and geriatric formulations where ease of administration and precise dosing are imperative.
- ☆ Moreover, gum arabic's presence can improve the palatability of liquid medications, making them more acceptable to patients, especially children.
- ☆ Its ability to mask the bitterness of certain drugs makes it an invaluable ally in ensuring patient compliance.

22.4.5 A Sustainable and Renewable Resource

- ☆ One of the remarkable aspects of gum arabic is its sustainability.
- ☆ The Babul tree's resin can be harvested without harming the tree, making it a renewable resource.
- ☆ This sustainability aligns with the growing global awareness of eco-friendly and socially responsible sourcing practices, a trend that the food and pharmaceutical industries increasingly embrace.

22.5 Traditional Practices

The Babul tree, a natural marvel in its own right, extends its influence far beyond the realms of nature and science. It weaves a rich tapestry of cultural significance, touching the lives of people across diverse communities.

In this exploration, we delve into the profound connection between the Babul tree and traditional practices, uncovering how its products have been integral to traditional medicine, oral care, and religious rituals, nurturing a cultural heritage that spans generations.

22.5.1. Medicine from the Past

For countless generations, the Babul tree has been a cherished source of remedies and healing in many cultures. Its bark, leaves, and gum are prized for their medicinal properties, offering solace to those seeking natural solutions to various ailments.

- ☆ In traditional medicine, the bark of the Babul tree shines as a natural panacea.
- ☆ Its anti-inflammatory and antibacterial properties make it a trusted remedy for a multitude of health issues, ranging from skin irritations to digestive disorders.
- ☆ Its application is often as simple as creating a paste or infusion, reflecting the wisdom of traditional healers who have harnessed its benefits for centuries.
- ☆ The leaves of the Babul tree, known for their astringent qualities, find their place in herbal medicine, particularly in the treatment of oral health issues.

- ☆ Chewing Babul leaves or using them to craft toothpaste has been a common practice, not only for maintaining oral hygiene but also for alleviating dental problems.
- ☆ This tradition continues to thrive as a testament to the enduring efficacy of nature's solutions.

22.5.2 Nurturing Oral Traditions

Oral care holds a special place in the traditions of many cultures, and the Babul tree plays a significant role in this aspect. Its leaves, with their natural cleansing and strengthening properties, serve as a "green toothbrush" in many households.

- ☆ Chewing these leaves not only promotes healthy teeth and gums but also imparts a sense of cultural continuity.
- ☆ Moreover, the use of Babul-based oral care products extends beyond the practical realm.
- ☆ It serves as a bridge between generations, where elders pass down the knowledge of traditional practices to younger members of the community.
- ☆ This oral tradition fosters a sense of cultural identity and unity, reinforcing the importance of preserving age-old customs.

22.5.3 A Spiritual Connection

Religious rituals often draw upon the symbolism and sanctity of natural elements, and the Babul tree is no exception.

- ☆ In various cultures, the Babul tree is revered for its connection to spirituality. Its presence in religious rituals signifies purity, resilience, and the enduring cycle of life.
- ☆ Offerings made from Babul tree products hold special significance, whether it's the use of its leaves in sacred ceremonies or the inclusion of its gum resin as an offering to deities.
- ☆ The tree's thorns, representing protection and strength, are also incorporated into rituals as symbolic elements, emphasizing its spiritual importance.

22.6 Medicinal Uses of Babul

The Babul Tree is used to treat a wide range of ailments, (Singh *et al.,* 2009) including:

22.6.1 Digestive Problems

It has been used to treat digestive problems for centuries in traditional medicine systems such as Ayurveda. It is believed to have astringent, anti-inflammatory, and antimicrobial properties, which make it effective in relieving diarrhea, dysentery, and indigestion.

A) Diarrhea

- ☆ Diarrhea is a condition characterized by loose or watery stools. It can be caused by a variety of factors, including bacterial and viral infections, food poisoning, and certain medications. The babul tree is effective in treating diarrhea caused by all of these factors.
- ☆ To use babul for diarrhea, a decoction can be made from the leaves or bark of the tree. The decoction is then consumed orally. The babul can also be used in the form of a powder, which can be mixed with water or yogurt and consumed.

B) Dysentery

- ☆ Dysentery is a type of diarrhea that is accompanied by blood and mucus in the stool. It is usually caused by a bacterial infection. The babul is effective in treating dysentery by killing the bacteria that cause the infection.
- ☆ To use the babul for dysentery, the same decoction or powder that is used for diarrhea can be consumed. The babul can also be used in the form of a paste, which can be applied topically to the rectum to relieve inflammation and pain.

C) Indigestion

- ☆ Indigestion is a condition characterized by discomfort in the stomach after eating. It can be caused by a variety of factors, including overeating, eating fatty or spicy foods, and certain medical conditions. The babul is effective in relieving indigestion by improving digestion and reducing inflammation in the stomach.
- ☆ To use the babul for indigestion, a decoction or powder can be made from the leaves or bark of the tree and consumed orally. The babul can also be used in the form of a candy, which is made by boiling the leaves of the tree in sugar syrup.

22.6.2 Respiratory Problems

The babul (*Acacia arabica*), is used to treat respiratory problems such as coughs and sore throats for centuries in traditional medicine systems such as Ayurveda. It is believed to have expectorant, anti-inflammatory, and antimicrobial properties, which make it effective in relieving these symptoms.

A cough is a reflex action that helps to clear the airways of irritants and mucus. It can be caused by a variety of factors, including viral infections, allergies, and asthma. The babul is effective in relieving coughs caused by all of these factors. To use the babul for coughs, a decoction can be made from the leaves or bark of the tree. The decoction is then consumed orally. The babul can also be used in the form of a syrup, which is made by boiling the leaves of the tree in sugar syrup.

A sore throat is a condition characterized by pain and inflammation in the throat. It can be caused by a variety of factors, including viral infections, bacterial infections, and allergies. The babul is effective in relieving sore throats caused by all of these factors. To use the babul for sore throats, a gargle can be made from the leaves or bark of the tree. The gargle is then used to rinse the throat several times a day. The babul can also be used in the form of a lozenge, which is made by mixing the powder of the leaves or bark of the tree with honey.

22.6.3 Skin Conditions

The babul, tree is used to treat skin conditions such as eczema and fungal infections for centuries in traditional medicine systems such as Ayurveda. It is believed to have anti-inflammatory, antiseptic, and antifungal properties, which make it effective in relieving these symptoms.

Eczema is a chronic skin condition characterized by dry, itchy, and inflamed skin. It can be caused by a variety of factors, including genetics, allergies, and irritants. The babul is effective in relieving the symptoms of eczema by reducing inflammation and itching. To use the babul for eczema, a paste can be made from the leaves or bark of the tree. The paste is then applied to the affected area of the skin. The babul can also be used in the form of a bath, by adding a decoction of the leaves or bark of the tree to the bathwater.

Fungal infections are caused by fungi, which are microscopic organisms that can grow on the skin, hair, and nails. The babul is effective in treating fungal infections by killing the fungi that cause the infection. To use the babul for fungal infections, a paste can be made from the leaves or bark of the tree. The paste is then applied to the affected area of the skin. The babul can also be used in the form of a wash, by adding a decoction of the leaves or bark of the tree to water and washing the affected area.

22.6.4. Wounds and Burns

The babul tree, has been used to treat wounds and burns for centuries. It is believed to have astringent, anti-inflammatory, and antimicrobial properties, which make it effective in promoting wound healing and preventing infection. Wounds can be caused by a variety of factors, such as accidents, surgery, and certain medical conditions. The babul is effective in healing wounds by reducing inflammation and promoting the growth of new tissue. To use the babul for wounds, a paste can be made from the leaves or bark of the tree. The paste is then applied to the wound. The babul can also be used in the form of a wash, by adding a decoction of the leaves or bark of the tree to water and washing the wound.

Burns are injuries to the skin caused by heat, chemicals, or electricity. The babul is effective in treating burns by reducing inflammation and pain. It also helps to promote wound healing and prevent infection. To use the babul for burns, a paste can be made from the leaves or bark of the tree. The paste is then applied to the burn. The babul can also be used in the form of a bath, by adding a decoction of the leaves or bark of the tree to the bathwater.

22.6.5 Urinary Problems

The babul tree, also known as Acacia arabica, has been used to treat urinary problems, such as urinary tract infections (UTIs), for centuries in traditional medicine systems such as Ayurveda. It is believed to have antimicrobial, anti-inflammatory, and diuretic properties, which make it effective in relieving the symptoms of UTIs and preventing the infection from spreading. UTIs are most commonly caused by bacteria, such as Escherichia coli (*E. coli*). The symptoms of UTIs include pain or burning when urinating, frequent urination, and cloudy or bloody urine.

The babul tree can help with UTIs in a number of ways.

- ✰ First, its antimicrobial properties help to kill the bacteria that cause the infection.
- ✰ Second, its anti-inflammatory properties help to reduce inflammation in the urinary tract.
- ✰ Third, its diuretic properties help to flush out toxins and bacteria from the urinary tract.

There are a number of ways to use the babul tree for UTIs.

- ✰ One way is to drink a decoction made from the leaves or bark of the tree.
- ✰ Another way is to use the babul tree in the form of a powder, which can be mixed with water or yogurt and consumed.
- ✰ The babul tree can also be used in the form of a sitz bath, by adding a decoction of the leaves or bark of the tree to the bathwater and sitting in the bath for 10-15 minutes.

22.6.6 Mouth Ulcers

It is effective in relieving the pain and discomfort of mouth ulcers and promoting healing. The babul tree can help with mouth ulcers in a number of ways.

- ✰ First, its anti-inflammatory properties help to reduce inflammation in the mouth, which can reduce the pain and discomfort of mouth ulcers.
- ✰ Second, its antibacterial properties help to kill the bacteria that can cause mouth ulcers.
- ✰ Third, its analgesic properties help to relieve pain and discomfort.

Different ways to use the babul tree for mouth ulcers.

- ✰ One way is to chew on a fresh leaf from the babul tree.
- ✰ Another way is to rinse the mouth with a decoction made from the leaves or bark of the tree.

- ✰ The babul tree can also be used in the form of a paste, which can be applied directly to the mouth ulcers.

22.6.7. Malaria

The babul tree, has been used to treat malaria for centuries in traditional medicine systems such as Ayurveda. Some studies have shown that the babul tree contains compounds that can kill the malaria parasite, *Plasmodium falciparum*. Another study, published in the journal "Journal of Ethnopharmacology", found that the babul tree extract was effective in reducing the severity of malaria in mice. More research is needed to confirm the efficacy and safety of the babul tree for malaria in humans. There are a number of ways to use the babul tree for malaria.

- ✰ One way is to drink a decoction made from the leaves or bark of the tree.
- ✰ Another way is to use the babul tree in the form of a powder, which can be mixed with water or yogurt and consumed.
- ✰ The babul tree can also be used in the form of a paste, which can be applied to the skin.

22.7 Certain other Uses

Conservationists and environmentalists are increasingly emphasizing the importance of preserving babul tree populations to combat desertification and promote biodiversity. It is used to treat diabetes (Gilani *et al.,* 1999).The babul tree is also used for a variety of other purposes, including food, (Houerou 1980) fodder (Carter 1998), gum, fuelwood and timber (Rao and Purkayastha (1972), Goldsmith and Carter (1981), Tewari and Rajput (1987) and Troup and Joshi (1983). The plant was used to make tooth brush, as it helps to maintain oral health. The wood of the babul tree is strong and durable, and is used for a variety of purposes, such as making furniture, tools, and carts. The babul tree is also beneficial for the environment. It is a nitrogen-fixing plant, which means that it can help to improve soil fertility. It is also a drought-tolerant plant, and can help to prevent soil erosion. The tree is reported to be a good host for the lac insect in Gujarat and adjoining areas of Rajasthan (India), although in other areas of India it has been disappointing (Troup and Joshi 1983). It is also a good host for sandalwood (Padmanabha *et al.,* 1988).

The dyeing of Indian chintzes was an important use of plant dyes, and Roxburgh appears to have been aware of this from one of his earliest extant letters to Sir Joseph Banks, sending him some caducay galls. These came from the tree he knew as *Mimosa arabica* (*Acacia nilotica* ssp. *indica*) and from which a yellow dye was made (Anonymous 1782). The tree is widespread in India and grew commonly on the Coast, but the galls from which the dye was extracted occurred on plants growing further inland. Parasitic insect feed on leaves and produced galls having bright yellow dye (Robinson 2008).

22.8 Conclusion

In villages on road side and in arid conditions the most common tree is Babul. In the grand tapestry of nature, the Babul tree stands as a testament to adaptability, resilience, and its indispensable role in sustaining life. From its thorny branches to the healing properties of its bark, this remarkable tree continues to amaze and inspire. As we move forward, let us cherish and protect this gift from nature for generations to come.

REFERENCES

Abtew AA, Pretzsch J, Secco L, Mohamod TE. (2014). Contribution of small-scale gum and resin commercialization to local livelihood and rural economic development in the Dry-lands of eastern Africa. Forests, 5: 952-977.

Anonymous (1782) Letter from William Roxburgh to Sir Joseph Banks, 11 Nov. 1782 British Library. Add MSS 33977, ff181-2

Carter, J.O. (1998) *Acacia nilotica*: a tree legume out of control.*Forage tree legumes in tropical agriculture*, [ed. by Gutteridge, R.C., Shelton, H.M.]. St. Lucia, Queensland, Australia: The Tropical Grassland Society of Australia.

Chauhan K. (2020) Role of Ethnobotany on Indian Society: A Review.Journal of Arts, Culture, Philosophy, Religion, Language and Literature 4 (2): 109-111

Chikamai B, Casadei E. (2005). Production and marketing of gum resins; frankincense, myrrh and opoponax; Food and Agriculture Organization of the United Nations: Nairobi, Kenya, pp 1–97.

Chikamai B, Tchatat M, Tieguhong JC, Ndoye O. (2009). Forest management for non-wood forest products and services in Sub-Saharan Africa. Discovery and Innovation, 21: 50–59.

Garg, L., Bhandari, N.N., Rani, V., Bhojwani, S.S. (1996) Somatic embryogenesis and regeneration of triploid plants in endosperm cultures of *Acacia nilotica*. *Plant Cell Reports*, 15(11) 855-858.

Gilani AH, Shaheen F, Zaman M, Janbaz KH, Shah BH, Akhtar MS (1999). Studies on antihypertensive and antispasmodic activities of methanol extract of *Acacia nilotica* pods. Phytother. Res., 13: 665- 669

Goldsmith, B., Carter, D.T. (1981) The indigenous timbers of Zimbabwe. In: *Zimbabwe Bulletin of Forestry Research*, 9 x + 406 pp.

Houérou, H.N. le, (1980) Chemical composition and nutritive value of browse in tropical West Africa. In: *Browse in Africa*, [ed. by Houérou, H.N. le]. Addis Ababa, Ethiopia: International Livestock Centre for Africa. 261-289.

Lal, B., Khanna, S. (1994) Selection of salt-tolerant *Rhizobium* isolates of *Acacia nilotica*.*World Journal of Microbiology and Biotechnology*, 10(6) 637-639.

Luna, R.K. (1996) *Plantation trees.*Dehra Dun, India: International Book Distributors. xii + 975 pp.

Mathur, I., Chandra, N. (1983) Induced regeneration in stem explants of *Acacia nilotica*. *Current Science*, 52(18) 882-883.

Padmanabha, H.S.A., Nagaveni, H.C., Rai, S.N. (1988) Influence of host plants on growth of sandal. *Myforest*, 24(2) 154-160.

Raj A. (2015). Gum exudation in Acacia nilotica: effects of temperature and relative humidity. In Proceedings of the National Expo on Assemblage of Innovative ideas/work of post graduate agricultural research scholars, Agricultural College and Research Institute, Madurai (Tamil Nadu), pp 151

Rao, R.K., Purkayastha, S.K. (1972) *A. arabica* Willd. *Indian Woods, their identification, properties and uses. Vol III. Leguminosae to Combretaceae*. Delhi, India: Manager of Publications. 20-22, 208.

Robinson T (2008) William Roxburgh: The founding father of Indian Botany. Pub. By Phillimore and Company England. Pp 286.

Saxena O.P. (Nidar) (2023) Charak Samhita in Hindi by DPB Pub. Delhi pp.332

Sharma M. and Kumar A. (2013) Ethnobotanical uses of medicinal plants: a review. International Journal of Life Science and Pharma Research. 3(2): 52-57

Singh BN, Singh BR, Sarma BK, Singh HB (2009). Potential chemoprevention of N- nitrosodiethylamine-induced hepatocarcinogenesis by polyphenolics from *Acacia nilotica* bark. Chem-Biol. Interact., 181: 20-28

Tewari, M.C., Rajput, S.S. (1987) *Acacia nilotica* - utilisation aspects.*Journal of the Timber Development Association of India*, 33(2) 28-32.

Troup, R.S., Joshi, H.B. (1983) *The silviculture of Indian trees. Vol IV. Leguminosae*. Delhi, India: Controller of Publications.

Chapter 23

Phytochemical Composition and Pharmacological Potential of Ardusi (*Adhatoda zeylanica* (Linn.) Nees)

Innumerable natural products with inherent potential biological activity procured from plants have played a significant role in improving the human health since the dawn of civilization. As more than 50 per cent of modern clinical drugs originate from natural products. Therefore based on latter, the approach to new drug discovery and its development in the pharmaceutical industry has emerged significantly. A. vasica is a good source of pyrroloquinazoline alkaloids such as vasicine, vasicol, adhatonine, vasicinone, vasicinol, vasicinolone etc., which are the chief constituents of different extracts of the plant. These compounds have shown many biological and pharmacological activities viz. anti-malarial, anti-inflammatory, antioxidant, antidiabetic, anti-bacterial, anti-cancer etc. Further, Adhatoda leaves have been used extensively in Ayurvedic medicine for the treatment of diarrhea, dysentery, tuberculosis, skin diseases, vomiting and leprosy etc. Herbal medicine is still the mainstay of about 75–80 per cent of the world population, mainly in the developing countries, for primary health care because of better cultural acceptability, better compatibility with the human body and lesser side effects. However, the last few years have seen a major increase in their use in the developed world. As far as contemporary drugs are concerned they must be further characterized after their pharmacological screening by studying the pharmacokinetic and pharmacodynamic properties, including toxicity. Therefore, the present communication enumerate the phytochemical composition and pharmacological potential of A. vasica.

Keywords: *Adhatoda vasica, Secondary metabolites, Pyrroloquinazoline alkaloids, Biological activities.*

23.1 Introduction

Adhatoda vasica is popularly known as Basak in Bangla, Malabar Nut in English, Adosa in Hindi and Vasaka in Sanskrit (Figure 23.1). It has been used in preparation of herbal medicines for the past 2000 years. It is a medicinal plant native to Asia, widely used in Siddha Medicine, Ayurvedic and Unani system of medicine. The plant's range includes Sri Lanka, Nepal, India, Pakistan, Indonesia, Malaysia, and China, as well as Panama where it is thought to have been introduced (Ahmad and Javed 2007). This evergreen perennial shrub has leathery leaves. The flowers are dense having large bracts of whitish pink/purple colored. The herb is often grown as a hedge and its leaves and twigs are utilized as green-manure. The whole plant or its roots, leaves, bark and flowers are used in various herbal preparations. They are oppositely arranged, smooth-edged, and borne on short petioles (Dymock *et al.,* 1890; Jain 1965).

Figure 23.1. Flowering Plant of *Adhatoda vasica.*

It has been revealed from the literature that the quinazoline based alkaloids have been the major constituents present in the different parts of *A. vasica*, which have been mainly responsible for their wide range of pharmacological potential. Research performed has revealed that the alkaloids, vasicine and vasicinone (quinazoline ring derivatives) present in the leaves, possess respiratory stimulant activity whereas its roots contain vasicinolone, vasicol and peganine. Vasicine (1,2,3,9–tetrahydropyrrolo [2,1-b] quinazolin-3-ol, C11H12N2O) also called Peganine (Gupta *et al.,* 1977). Other chemical constituents of this plant include vasicinone (3–hydroxy–2,3- dihydropyrrolo[2,1–b]quinazolin–9(1H)–one, C11H10N2O2) (isolated from leaves, stem and roots) and deoxyvasicinone (derived from leaves) *etc.* Recent investigations on vasicine showed bronchodilatory activity comparable to theophylline, both *in vitro* and *in vivo*. Both the alkaloids in combination showed pronounced bronchodilatory activity. Deoxyvasicinone possesses antimicrobial, anti-inflammatory and antidepressant activities. It is a primary medicinal plant for the respiratory tract ailments in the treatment of cough, bronchitis, asthma and symptoms of common cold (Iyengar *et al.,* 1994; Khan and Yadava 2010). The medicinal uses of Adulsa leaves are attributed to its antitussive, antimicrobial and anti-inflammatory properties. Its leaf juice is the most common home remedy for cough, respiratory diseases and bleeding disorders. It is also a highly effective natural medicine for respiratory infections (Hussain and Hore 2007). In ayurveda, Adulsa juice is also called Adusa Swarasa, Adulsa Swarasa and Vasa Svarasa.

Constituents of the plant have been exposed to have anti-stress effects, which might be occasioned partly by an endocrine and partly by an immunomodulatory mechanism of action. This plant has medicinal uses, mainly antispasmodic, fever reducer, anti-inflammatory, anti-bleeding, bronchodilator, anti-diabetic, disinfectant, anti-jaundice and oxytocic. It is antiperiodic, astringent, diuretic, purgative and is also used as an expectorant in addition to liquefy sputum. The leaves, flowers and roots of this plant are used as herbal drug against tubercular activities cancer and possess anti-helmintic properties (Baral *et al.,* 2019).

23.2 Phytochemical Composition

Water, ethanolic and petroleum ether extracts of *Adhatoda* leaves enriched with alkaloids, phenols, tannins and reducing sugars, main constituent vasicine *etc.* have been found to possess antibacterial activity (singularly or in combination) against *S. epidermidis, S. aureus B. subtilis, E. faecalis, E. coli, P aeroginosa, P. vulgaris, K. pneumoniae* and *C. Albicans. A. vasica* possesses abortifacient and uterotonic properties (Kirtikar and Basu 1975; Karthikeyan *et al.,* 2009). The leaves are mostly used in the treatment of respiratory ailments in Ayurveda. The two main alkaloids, vasicine and vasicinone present in the leaves, possess respiratory stimulant activity. Vasicine, at low concentrations induces bronchodilation and relaxation of the tracheal muscle. It also has immense potential as an anti-ulcer agent which is used to treat or ameliorate peptic ulcer or irritation of the gastrointestinal track of great therapeutic relevance (Atta-Ur-Rahman *et al.,* 1997; Sultana *et al.,* 2005; Ayyanar and Ignacimuthu 2008).

A wide range of phytochemical constituents have been isolated from this plant having activities like antitussive, abortifacient, antimicrobial, insecticidal, hepatic protection, cardiovascular protection, anticholinesterase, antioxidant, antiinflammatory and other important activities (Chakrabarty and Brantner 2001). Some important bioactive compounds have been reported in various part of *A. vasica* are essential oils and quinazoline alkaloids. So this plant can form one of the best options for developing novel compounds having medicinal value. Two new aliphatic hydroxyketones, isolated from the aerial parts of *A. vasica* have been characterized as 37- hydroxyhexatetracont-1-en-15-one and 37- hydroxyhentetracont an-19-one, respectively, on the basis of spectral data and chemical studies. 2',4- Dihydroxychalcone 4-glucoside has been identified in the flowers of *A. vasica*. A novel alkaloid and a galactoside isolated from the roots of *A. vasica* have been characterized as 9-acetamido-3,4- dihydropyrido-(3,4-b)-indole and O-ethyl-α-D- galactoside respectively by chemical and spectroscopic methods. In addition sitosterol β-D-glucoside, D-galactose and deoxyvasicin one have also been isolated from the roots of this plant (Jain and Sharma 1982). Two new pyrroloquinazoline alkaloids, *viz.* 1,2,3,9- tetrahydropyrrolo (2,1-b)-quinazolin-9-one-3Rhydroxy-3(2'-dimethylamino phenyl (desmethoxyaniflorine) and 7-methoxy-3R-hydroxy-1,2,3,9- tetrahydropyrrolo-[2,1-b]- quinazolin-9-one (7-methoxyvasicinone), together with several known compounds were isolated from the leaves of *A. vasica*. Their structures were established by spectroscopic and X-ray diffraction analyses.

The chemical compounds found in *A. vasica* plant includes essential oils, fats, resins, sugar, gum, amino acids, proteins and vitamins 'C' *etc.* (Reddy *et al.,* 1988). The phytochemical analysis show that phenols, tannins, alkaloids, anthraquinone, saponins, flavonoids and reducing sugars were found in the leaves of plant (Maurya and Singh 2010). But the pharmacologically most studied chemical component in plant is vasicine, a bitter quinazoline alkaloid, the novel alkaloid isolated leaves and characterized as 1, 2, 3, 9-tetrahydro- 5-methoxypyrrol [2, 1-b] quinazolin-3-ol. It is present in the leaves, roots and flowers. Besides vasicine, the leaves contain several alkaloids (Vasicinone, Vasicinol, Adhatodine, Adhatonine, Adhvasinone, Anisotine and Hydroxypeganine), betaine, steroids and alkanes (Lahiri and Pradhan 1964; Suthar *et al.,* 2009).

23.3 Pharmacological Activities

The pharmacological activities of vasicine and vasicinone are well known (Huq *et al.,* 1967). Recent investigations on vasicine showed bronchodilatory activity comparable to theophylline, both *in vitro* and *in vivo* (Amin and Metha 1959; Atal 1980). Both the alkaloids in combination showed pronounced bronchodilatory activity. Vasicine also exhibits strong respiratory stimulant activity. There has also been a report of thrombopoetic (platelet increasing) activity with vasicine. Uterine stimulant activity and moderate hypotensive activity of the alkaloids have been observed. Vasicine is metabolized to vasicinone and analysis of plant leaf extract showed that it contained 0.85 per cent vasicine and 0.027 per cent vasicinone. Sitosterol, β-glucoside-galactose and deoxyvasicine have been isolated from the roots of the plant. 2'-4- dihydroxychalcone- 4-glucoside has also been recognized in the flowers. Flowers of *A. vasica* mainly contain kaempferol and quercetin. A new triterpenoid, 3-hydroxy-D-friedoolean-5-ene, along with the known compounds, epitaraxerol and peganidine have been isolated from the aerial parts of *A. vasica*. The leaves also contain a very small amount of an essential oil and a crystalline acid. An analysis published in India in 1956 showed that the seeds as containing 25.8 per cent of deep yellow oil composed of glycerides of arachidic 3.1 per cent, be- henic 11.2 per cent, lignoceric 10.7 per cent, cerotic 5 per cent, oleic 49.9 per cent and linoleic acids 12.3 per cent and β-sitosterol (2:6 per cent). Elemental analysis using atomic absorption spectrophotometry revealed the presence of major (K, Na, Ca and Mg) and trace (Zn, Cu, Cr, Ni, Co, Cd, Pb, Mn and Fe) elements in *A. vasica*. It has alkaloids containing pyrroquinazoline and ring derivatives like vasicine, vasicol, vasicinone along with other miner constituents. Vasicine is a major bioactive alkaloid of vasica which contain pyrroquinazoline ring (Thappa *et al.,* 1996).

23.4 Traditional Uses

Adhatoda vasica, an ayurvedic medicinal plant has been used in various chest and respiratory track infection *viz.*- whooping cough, chronic bronchitis, asthma and has been employed as sedative expectorant in the treatment of excessive phlegm (mucus with bacteria, debris, and sloughed-off inflammatory cells) and menorrhagia (abnormal blood clotting, disruption of normal hormonal regulation of periods, or disorders of the endometrial lining of the uterus) in Sri Lanka. It is also used for the treatment of bleeding piles, impotence and sexual disorders (Bhaduri *et al.,* 1968). antispasmodic and anthelmintic drug. Glycodin, which is a famous product used for the cure of bronchitis has been a product extracted from the leaves of *A. vasica*. Further, the extracts have been found to be quite effective against tuberculosis. Various parts of the plant have been used in Indian traditional medicine for the treatment of asthma, joint pain, lumber pain, sprains, cold, cough, eczema, malaria, rheumatism, swelling and venereal diseases. In India paste of the fresh root applied on abdomen and vagina minutes before childbirth facilitates easy delivery. The extract of roots of *A. vasica* has commonly been used by rural population against diabetes, cough and certain liver disorders (Jain and Sharma 1982), whereas its paste has been used for curing tuberculosis, diphtheria, malarial fever, leucorrhoea, eye diseases, acute nightfall (paste mixed with sugar) in Sitapur district, Uttar Pradesh, India. Further, the root's decoction has been used for gonorrhoea and as an expectorant, antispasmodial/anthelmintic agent.

The various preparations of leaves has been used for curing bleeding, hemorrhage, skin diseases, wounds, headache and leprosy (chronic infection) while their infusion or the solvent extract has been observed to be an excellent agent for the destruction of white ants, flies, and mosquitoes. The fresh juice of leaves mixed with honey and ginger juice cures all types of acute cough, chronic bronchitis, breathlessness and liquefies sputum and asthma. The extract obtained by decoction of its leaves along with fruit of *Phyllanthus emblica*, mixed with honey has been effective against asthma. The crushed fresh leaves of the plant have been used to treat snake bites (India and Sri lanka). Macerated extract from leaves/Yellow leaves/smoke from dry leaves acts against cough and phlegm blockage during cold (Bangladesh) and an infusion protects from headache (Myanmar and Pakistan). The leaf powder boiled in sesame oil stops bleeding, earaches as well as pus from ears and jaundice. Leaf juice has been used as the best medicine to enhance platelet count during dengue like viral fevers, postpartum hemorrhage, urinary trouble, acidity and belching and its mixture with jaggery reduces excessive menstrual flow get cured. Externally warmed leaves have been used for rheumatic pains and dislocation of joint, stomach catarrh with constipation, rheumatism, gout, fever and urinary stone. Juice made from the bark and leaves help against vomiting. The flowers have been known to possess expectorant and anti-asthmatic, antiseptic properties and have been used against of opthalmia, cold, phthisis, asthma, bronchitis, cough, antispasmodic, high fever and gonorrhoea. Also the flowers improve blood circulation and hectic heat of blood. Fruits of the *A. vasica* are four seeded capsules. In Pakistan, fruits of the plant are used for cold, antispasmodic and bronchitis. The fruit of the plant are also used for diarrhea, dysentery, fever and as laxative.

23.5 Conclusion

A wide range of phytochemical constituents have been isolated from this plant having activities like antitussive, abortifacient, antimicrobial, insecticidal, hepatic protection, cardiovascular protection, anticholinesterase, antioxidant, anti-inflammatory and other important activities. *A. vasica* has emerged as imperative medicinal herb well known for its applications in different traditional medicinal system such as Ayurveda, Siddha and Unani *etc.*

REFERENCES

Ahmad, S.A.and Javed, S. (2007) Exploring the Economic value of underutilized plant species in Ayubia National Park. Pak. J. Bot., 39: 1435-1442.

Amin, A.H. and Metha, D.R. (1959) A bronchodilator alkaloid (Vasicinone) from *Adhatoda vasica* Nees. Nature, 184: 1317.

Atal C.K. (1980) Chemistry and Pharmacology of vasicine: A new oxytocin and abortifacient. Indian Drugs, 15: 15-18.

Atta-Ur-Rahman, Nighat, S., Farzana, A., Nighat, M. and Iqbal, C. (1997) Phytochemical studies on *Adhatoda vasica* Nees. Natural Product Research, 10(4): 249 – 256.

Ayyanar M, Ignacimuthu S (2008) Medicinal uses and pharmacological Actions of five commonly used Indian Medicinal plants: A mini-review. Iranian J. Pharm. Therapeut., 7: 107-114.

Baral P K., Roy S., Sultana S. (2019). A review article on *Adhatoda vasica* Nees: A potential source of bioactive compounds. Indian Journal of Pharmaceutical Sciences; 70; nb. (1): 36 – 42

Bhaduri B, Ghose CR, Bose AN, Moza BK, Basu UP, (1968) Antifertility activity of some medicinal plants. Indian J. Exp. Biol., 6: 252-253.

Chakrabarty, A. and Brantner, A.H. (2001) Study of alkaloids from *Adhatoda vasica* Nees on their anti-inflammatory activity. Phytother. Res., 15: 532-534.

Dymock, W. Waeden, C.J.H. and Hooper, D. (1890) A history of the principal drugs of vegetable origin. London. Pharmacographia Indica, 50–54.

Gupta, O.P., Sharma, M.L., Ghattak, B.J.R. and Atal, C.K. (1977) Pharmacological investigation of vasicine and vasinone- The alkaloids of Adhatoda vasica. Indian J. Med. Res., 66: 680–691.

Huq, M.E., Ikram, M. and Warsi, S.A. (1967) Chemical composition of *Adhatoda vasica* D - Vasicine Vasicinine Beta Sito Sterol Tri Triacontane Alkaloids ii. Pakistan Journal of Scientific and Industrial Research, 10: 224-5.

Hussain, S. and Hore, D.K. (2007) Collection and conservation of major medicinal plants of Darjeeling and Sikkim, Himalayas. Indian J. Trad. Know. 6: 352-357.

Iyengar, M.A, Jambaiah, K.M., Kamath, M.S. and Rao, G.O. (1994) Studies on antiasthma Kada: A proprietary herbal combination. Indian Drugs, 31: 183-186.

Jain, S.K. (1965) Medicinal plant lore of the tribals of Bastar. Economic Botany, 19: 236–250.

Jain, M.P. and Sharma, V.K. (1982) Phytochemical investigation of roots of *Adhatoda vasica*. Planta Med., 46: 250-252.

Khan, M.H. and Yadava, P.S. (2010) Herbal remedies of asthma in Thoubal District of North East India. Indian J. Nat. Prod. Resourc, 1: 80-84.

Karthikeyan A, Shanthi V, Nagasathya A. (2009) Preliminary Phytochemical and antibacterial screening of crude extract of the leaf of *Adhatoda vasica* (L). Int. J. Green Pharm., 3: 78- 80.

Kirtikar, K.R., Basu, B.D. (1975) Indian Medicinal plants (second Ed.) Bishen Singh Mahendra Pal Singh, Delhi, 3: 1899-1902.

Lahiri P.K, Prahdan S.N. (1964) Pharmacological investigation of Vasicinol- an alkaloid from *Adhatoda vasica* Nees. Indian J. Exp. Biol., 2: 219-223.

Maurya S, Singh D. (2010) Quantitative analysis of total phenolic content in *Adhatoda vasica* Nees extracts. Int. J. Pharm. Tech. Res., 2: 2403-2406.

Reddy, M.B., Reddy, K.R. and Reddy, M.N. (1988) A survey of medicinal Plants of Chenchu tribes of Andhra Pradesh, India. Ind. Int. J. Crude Drug Res., 26: 189-196.

Sultana, N., Muhammad Aijaz A., Yousuf A., Nighat A. (2005) Phytochemical studies on *Adhatoda vasica*. Pakistan Journal of Scientific and Industrial Research; 48; nb. (3); 180 – 183.

Suthar, A.C., Katkar, K.V., Patil, P.S., Hamarapurkar, P.D., Mridula, G., Naik, V.R., Mundada, G.R. and Chauhan, V.S. (2009) Quantitative estimation of vasicine and vasicinone in *Adhatoda vasica* by HPTLC. J. Pharm. Res., 2: 1893-1899.

Thappa, R. K., Agarwal, S. G., Dhar, K. L., Gupta, V. K., Goswami, K. N. (1996). Two pyrroloquinazolines from *Adhatoda vasica*, Phytochemistry; 42; nb. (5): 1485 –1488.

Chapter 24

A Spice for Overall Health Welfare: Fennel (*Foeniculum vulgare* Mill.)

Saunf or Fennel seeds are an edible and healthy seeds used for many years in cooking as well as medicinal purposes. It is also called as motisaunf in India. It belongs to the Apiaceae or Umbelliferae family and Foeniculum genus, cultivated in many countries all over the world. India is the top producer of the saunf however it is cultivated in average amount in Iran, Mexico, China, Canada, Syria, Bulgaria, United States, Australia and etc. Fennel seeds are an edible and healthy seeds used for many years in cooking as well as medicinal purposes. India is the top producer of the saunf however it is cultivated in average amount in countries like Iran, Mexico, China, Canada, Syria, Bulgaria, United States, Australia and etc. Fennel juice is an extremely nutritional beverage. It contains vitamin A, B, C, E, and beta-carotene. It is a rich source of minerals like magnesium, iron, cobalt, potassium, phosphorus, sodium, silicon, and zinc. It also provides high levels antioxidants, carotenoids, and flavonoids. It contains anethole, a phytonutrient that provides fennel juice its strong aroma. The high levels of antioxidants neutralize free radicals in the body, preventing cell damage.

***Keywords**: Spice. Health, Fennel, Foeniculum vulgare, Anethole, Saunf*

24.1 Introduction

In India, it is an old and most common trend of offering some fennel seeds with mishri (or sugar coated saunf) after a meal to chew. It is mostly carried out in hotels, restaurants and parties in order to prevent after eating problems like gas, indigestion, acid reflux, abdominal cramps, flatulence, and constipation. DO not be confused with the fennel seeds (motisaunf with green or yellowish color, used for cooking purpose) and aniseeds (patli or chhoti saunf with green color, used for raw eating purpose after meal). Fennel is an aromatic herb belonging to the parsley family. It is used as a spice taste that is similar to anise. It is an essential ingredient in the Mediterranean cuisine (Blamey and Grey-Wilson 1989). It is native to the South European region but is consumed well in all of the European regions. It is a two dark green leaves and yellow flowers. Fennel seeds are not only tasty but are also very healthy owing to their nutritional content. Fennel bulb is also eaten as a vegetable in some parts of the Mediterranean region. The genus name *Foeniculum* is derived from latin word 'foenum' means hay, foeniculum means little hay because dry leaves of fennel resemble fine dried hay. In Europe, it was commonly called fenkel or fennil, which gradually became fennel in English (Weiss 2002). It bears strong medicinal value (Rathore *et al.,* 2013); it is also cited in various herbal medicines prescribed in Ayurveda the ancient science of India and also in Yunani formulation practiced in Middle East countries. Fennel is active ingredient of traditional Chinese medicine, particularly for rheumatism, cold pain and stomach disorders (He and Huang 2011; Sahu 2012).

Apart from being used for adding flavor, it is also used for various health benefits that are derived from its anti-oxidants. These anti-oxidants are essential to curb unwanted free radical reactions in the body. This phenomenon also prevents many diseases that can get serious at a later stage. Some of the antioxidants listed in these seeds are kaempferol and quercetin that are known to prevent degenerative reactions (Blamey and Grey-Wilson 1989; Irma 1997; Anonymous 2012; GRIN 2023).

24.2 Major Varieties

Variety selection depends primarily on its adaptation to the soil and climatic conditions and preferably should have resistance/tolerance to pests and diseases prevailing in that region. There are many varieties released for cultivation to different areas. A brief description of each variety is given here (Solanki *et al.,* 2017).

- ☆ **Ajmer Fennel -1** :This variety was developed at NRCSS, Ajmer, suitable for growing both as early crop and as rabi crop, The plants are erect and tall, bearing large size umbels. The seeds are attractive bold, medium sized fragrant with volatile oil content up to 1.6 per cent. The variety is from late maturity group and produce average yield of 1950 kg/ha during rabi crop.
- ☆ **RF-101 :** It was developed by SKN College of Agriculture (RAU) Jobner, through recurrent half sib-selection from a local collection of Sohela (Tonk) and identified for release in 1995. The plants are tall erected, with stout stem. It bears large umbel with long bold grain. It mature in 150-160 days and gives an average yield of 1550 kg/ha
- ☆ **RF-125:** It was developed by SKN College of Agriculture (RAU) Jobner, through recurrent half sib selection from an exotic collection, EC-243380 from Italy and identified for release in 1997. The plants are early, short

stature with compact umbels and long bold. It matures in 110-130 days and gives an average yield of 1730 kg/ha.

- ☆ **Pilwai Fennel -35 (PF-35:** It was developed by Gujarat Agriculture University though selection from Pilwai local germplasm in 1973. It is suitable for Gujarat and Rajasthan. The plants are tall and spreading with medium sized hairless and green seeds. It is moderately tolerant to sugary disease, leaf spot and leaf blight. It contains 1.9 per cent essential oil. It mature 215-225 days and gives an average yield of 1289 kg/ha.
- ☆ **Gujarat Fennel-1:** It was developed at Spice Research Centre (Gujarat Agriculture University), Jagudan, through pure line selection from a local germplasm (VC-14-3-3) and released in 1985. The plants are tall, bushy type suited for kharif, rabi and drought, umbels bigger in size, more number of umbel, seeds bold, long dark green
- ☆ colour, do not shatter, contain 1.6 per cent essential oil. It is moderately tolerant to sugary disease and leaf spot. It is suitable for early sowing and is reasonably tolerant to drought. It is long duration variety (220-230 days) yielding 1700 kg/ha. It is suitable for all the fennel growing areas of Gujarat and also suitable for other parts of India.
- ☆ **Gujarat Fennel- 2 :** It was developed at Spice Research Centre (Gujarat Agriculture University), Jagudan, through selection from a local material and therefore yield superiority is more than 20 per cent over local variety. It is a tall and long duration variety. This variety produced an average yield of 2400 kg/ha.
- ☆ **Hisar Swarup:** It was developed at Haryana Agricultural University, Hisar. This is a selection from indigenous germplasm and has a yield potential of 1700 kg/ha. Plant has an average height of 132 cm, spreading type late in maturity, resistant to lodging and shattering of grain fruits are long, bold and has oil content of 1.6 per cent.
- ☆ **Co-1:** It was developed by Tamil Nadu Agriculture University, Coimbatore through reselection from PF-35 and released in 1985. The plants are medium with profuse branching. It matures in 210-220 days and gives an average yield of 567 kg/ha. It is suited for intercropping with other crops. It can thrive in saline and alkaline soil, hilly areas, drought prone and water logged conditions.

24.3 Pharmacology and Pharmacognosy

Apart from being used for adding flavor (Dalal 2021), it is also used for various health benefits that are derived from its anti-oxidants. These anti-oxidants are essential to curb unwanted free radical reactions in the body. This phenomenon also prevents many diseases that can get serious at a later stage. Some of the antioxidants listed in these seeds are kaempferol and quercetin that are known to prevent degenerative reactions. Fennel seeds contain a good amount of dietary fiber that is essential for the body. The fiber content helps improve digestion in the body, by facilitating timely breaking down of food molecules that make energy reactions possible. The carcinogenic diseases should be treated at the right stage so that their effect can be curbed or eliminated in the future. This is made possible by the anti-oxidant capabilities of the fennel seeds. Another effect of prevention of free radical reactions by fennel seeds – Neurological diseases are helpfully kept at bay. Cholesterol absorption in the arteries is prevented by the action of the fiber binding to the bile salts. Fiber present in fennel seeds decreases reabsorption of cholesterol, thereby preventing any heart-related diseases.

A number of volatile oil compounds like myrcene, fenchone, chavicol, cineole are present in these seeds that assist in digestive, carminative, and antioxidant reactions in the body. Production of Red Blood Cells is enhanced by the consumption of Fennel Seeds. Copper, which is required by the body for the production of red blood cells, is present in these seeds. The growth and development of the body are geared up. Zinc, found in the fennel seeds, promotes regular growth and development. These days, when people face tensions in their day to day lives, it is very important to maintain the blood pressure and keep the heart rate balanced. Fennel Seeds help a lot in this process. The potassium content helps in maintaining the blood pressure and hence the heart rate is also maintained.

Fennel juice has a calming effect on the intestine. The anti-spasmodic properties of fennel juice can reduce intestinal cramps, bloating, and flatulence. The essential oils present in fennel juice stimulate the secretion of digestive juice, reducing stomach inflammation. It also improves the absorption of nutrients from the foods we eat (Ono *et al.,* 1995). Fennel juice is widely used as an antacid. The roughage in fennel juice balances the digestive tract, promoting regular bowel movements. Fennel also reduces bad bacteria and increases good bacteria in the gut. The high concentration of iron and histamine in fennel juice makes it a good natural remedy for anemia. Regular consumption of fennel juice increases the production of hemoglobin in the body, reversing anemia. It also prevents the formation of blood clots in the body. Fennel contains several essential nutrients, which helps to improve vision. The antioxidants and amino acids in fennel juice rejuvenate the eyes, preventing macular degeneration. Fennel juice, when applied to the eye, can reduce eye irritations and fatigue. The flavonoids present in fennel juice are very useful in preventing oxidative stress, which helps to prevent macular degeneration. The diuretic properties of fennel juice help to remove the toxic substance from the body through frequent urination. This helps to reduce inflammation that causes rheumatism and swelling. It also prevents the formation of kidney stones. Fennel juice contains phyto-estrogen, a chemical that is similar to the hormone estrogen in structure. Thus, fennel juice is very useful for women suffering from menopause and pre-menstrual problems. It also contains folate, which prevents birth defects in unborn babies. Fennel juice eases menstruation problems by regulating hormonal actions in the body. Lactating mothers can regularly drink fennel juice to increase milk production.

Fennel juice is believed to release endorphins into the bloodstreams, which relieve depression associated with certain diseases and creates a mood of euphoria. Potassium in fennel juice increases brain function and the cognitive abilities of the brain. In addition to this, fennel juice is also a vasodilator, which increases oxygen supply to the brain and delays the onset of dementia. Fennel essential oil is an aperitif. It helps boost appetite. So if you think you aren't feeling hungry when it is required, fennel essential oil is what you should turn to! Spasms are violent and involuntary contractions of the organs, respiratory tract, intestines, muscles, and nerves. It can lead to chronic cough and acute pain and inflammation. Severe cases of spasms can lead to fainting and even death. Fennel essential oil is an active antispasmodic agent, which relaxes the intestines, muscles and nerves. It provides instant relief from spasms. Fennel essential oil contains many chemicals, which exhibit antiseptic and bactericidal characteristics. Fennel oil helps prevent wounds from turning septic. Fennel essential oil prevents tetanus and other post-surgical infections as well. Fennel essential oil is an active carminative agent. It helps release gas from intestines and provides relief from stomach problems like pain, restlessness, heartburn and indigestion. It also inhibits the formation of gas and helps treat excessive flatulence.

Fennel bulb exhibits strong anti-inflammatory properties. Whether you talk of inflammation that is internal or external, fennel bulb cures both. This magic food contains anethole. Anethole serves as a very strong anti-inflammatory flavonoid. It easily cures inflammation caused in humans without causing any side-effects. Fennel bulb offers natural cooling and neutralizing benefits. Thus, it is used to treat the problem of acidity in the duodenum. It makes intestines active and eases the process of absorption. Hence, the toxins get separated from useful minerals. The separated toxins are easily flushed out of the body. Hence, the normal pH level is restored which results in relief from acidity. Vitamin C is a natural immunity boosting vitamin for humans. Fennel bulb is highly rich in vitamin C. Hence; its regular intake strengthens the immune system. It keeps the body safe from minor infections naturally. It also increases the body's capacity to fight back major infections. This is because – stronger immune system means active WBCs. As soon as some infection enters the body, the WBCs engulf all the disease-causing bacteria and viruses and destroy them. Fennel bulb is a rich source of cineole, which is an active agent that potentially clears nasal blockages. Thus, it is a boon for people suffering from congestion, asthma, and bronchitis. Cineole efficiently clears and controls the formation and accumulation of phlegm in the chest and nasal passage. Hence, it eases respiration. Fennel bulb is a rich source of potassium. Potassium is an active component of cranial nerves. It is also an active compound that maintains the electrolytic balance of the brain. Moreover, potassium is also a memory charger. In short, fennel bulb ensures proper functioning of the human brain by regulating the levels of potassium. It sharpens the memory and improves focus and concentration.

Fennel bulb can regulate hormonal changes in the human body. Its mild analgesic benefits work like a boon to control and cure menstrual pain in women.

24.4 Use in Skin Care

We all want to have healthy skin and hair but the mere thought of money and time involved in a beauty regime is enough to put us off. But what if I told you that you can stay beautiful with the help of a tiny thing called the fennel seed? Fennel seeds not only impart a special flavor to those gastronomic delights brewing in your kitchen, but can also be utilized for achieving a flawless skin and healthy hair. Fennel seeds can be beneficial for your skin and hair. Fennel seeds are inexpensive and readily available. Loaded with minerals like iron, copper, selenium, potassium, magnesium, zinc and even vitamins like Vitamin C, Vitamin A, and Vitamin E, fennel seeds are also known to be a rich storehouse of dietary fibers, active compounds, and antioxidants. Fennel seeds offer amazing effects for the skin. Not only fennel seeds are antiseptic by nature, but also are loaded with antioxidants. As a result, they help prevent acne, cell damage, while keeping the skin toned. Using fennel seeds as a part of your skin care will help you delay the onset of lines, dark spots and wrinkles.When you consume fennel seeds in one form or the other, you get glowing and flawless skin in return. Wondering about the skin care benefits of consuming fennel seeds?

When consumed, fennel imparts a cooling effect to the skin. It soothes inflammation and skin irritation. Fennel tea, when consumed on a daily basis, helps in improving the overall texture of the skin. It treats acne and makes your skin look youthful. Fennel seeds, when consumed daily, provide the body with minerals like iron, zinc, selenium and calcium. These minerals are effective in maintaining the oxygen balance, thus imparting healthy skin. Fennel seeds serve your skin by providing various vitamins, which in turn provides younger looking skin. So, whether you consume it in the form of herbal tea or as a mouth-freshener after usual meals, ensure that you include a considerable portion of fennel seeds in your regular diet alternately, you may use fennel while garnishing some of your favorite salad recipes.

24.5 Use of Saunf in Home Remedies

Saunf is an aromatic and sweet flavored seed acts as a mouth freshener and helps in removing bad breath, relief from toothache and gum problems. Eating saunf seeds provides relief from the common cold and cough problems. It is rich source of essential nutrients, fibers, anti-oxidants, vitamins and minerals which promotes courage and strength if consumed regularly. It is a powerful anti-oxidant as it contains some natural antioxidant compound called kaemferol and quercetin which helps in combating with harmful free radicals thus prevents from various degenerative diseases like cancer, aging, neurological diseases and *etc.* Regular intake of saunf helps in getting relief from the constipation and other digestive problems as it contains high level of dietary fibers which absorbs more water and makes stool soft and ease passage. Dietary fibers found in it helps in lowering the serum cholesterol level as it gets binded with bile salts and reduces its absorption in the colon. Its antioxidant nature protects colon and its mucus membrane from the cancer by inhibiting the growth and development of the cancerous cells. Fennel seed is rich source of essential oil compounds like limonene, anisic aldehyde, myrcene, pinene, fenchone, chavicol, anethole, cineole and *etc.* which make it able to prevent from various health disorders. It is rich source of copper which is highly required by the body to produce red blood cells thus it protects from anemia and other problems caused due to decreased amount of hemoglobin. It is rich source of zinc mineral which acts as a co-factor for many enzymes regulating the growth and development of body, sperm production, nucleic acid synthesis and *etc.* Fennel seed is rich source of potassium electrolyte which is very necessary for the balance of cells and body fluid as well as controlling heart rate and blood pressure. It is also a rich source of manganese mineral which is the important cofactor for various powerful antioxidant enzymes. Its antibacterial nature protects from the infection of various harmful bacteria including *Bacillus subtilis*, and fungi like *Cladosporium cladosporiodes*, *Aspergillus niger etc.* It provides relief from the problems of stomach, intestine, colon, irritable bowel syndrome, spleen, kidney and *etc.* It helps in curing fever by increasing perspiration. Washing mouth with the warm saunf water is very helpful in removing bad breath.

Charak Samhita mentions eating saunf after frying in cleared butter or ghee, along with mishri helps in curing dysentery (Saxena 2023).

Massaging the muscles on lower abdominal area with the mixture of saunf oil, whole fatty milk, geranium oil and clary sage oil helps in reducing menstrual pain, cramps and PMS disorders. It contains good level of phytoestrogens improving food digestion by soothing the irritable gastric glands. Fennel seed is a good blood purifier helps in purifying blood if a handful of raw saunf is eaten two times a day, morning and evening. It helps in curing heat stroke if saunf water (soaked into water for whole night) is taken by adding a pinch of salt. It helps in promoting proper bowel movements if one teaspoon of the roasted saunf is taken after meals. It is diuretic in nature helps in removing toxic materials from the body as well as provides relief from the body swelling and joints pain during rheumatism and arthritis. It stimulates the secretion of digestive enzymes and juices as well as reduces inflammation of digestive tract thus promote more absorption of the essential nutrients from the food while passing through the alimentary canal. Regular intake of it helps in recovering digestive system functioning after getting long radiation or chemotherapy treatments. Saunf water acts as a vasodilator thus keeps mind and body stress free and calm by enhancing oxygenated blood flow to the brain and all through the body thus keeps away from depression, dementia and other memory loss problem. Fennel seed is a most common mouth freshener used in the betel leaf to make it delicious and flavored to chew. Fennel seed is a natural home remedy to suppress appetite and go for weight maintenance. It has been used for years by the women during their fasting periods to reduce hunger. It is a fat digester and hunger suppressor thus helps in decreasing weight among obese and overweight people if saunf tea or saunf water is taken in the early morning with empty stomach. It helps in preventing from the eye disorders (optic nerves weakness, glaucoma and night blindness) and improving eye sight if carrot juice is taken with half tsf powdered saunf regularly. Regular consumption of saunf helps in maintaining clear vision. Fennel seed promotes blood vessels dilation in the eye by reducing the eye pressure thus prevents from glaucoma (caused due to increased eye pressure), retinopathy (vision loss in diabetics), vision impairment diseases and *etc.*

It is good source of all the minerals required for bone growth such as calcium, phosphorus, magnesium and *etc.* thus helps in improving bone strength and development. Fennel seed naturally slows down process of bone destroying cells production in the body as well as reduces occurrence of age related bone problems like osteoporosis. Applying the mixture of saunf water (boil two tsf of saunf into a glass of water for at least one hour at low heat) and whey (or dahi water) to the face for 15 to 20 min then wash, helps in getting fair look with smooth skin. Whether it is applied directly to the skin or eaten, helps in getting acne and spots free skin as it has antiseptic and antioxidant properties.

It slows down the aging process by preventing cell damage and reducing the effects of free radicals thus keeps skin flawless without wrinkles and fine lines. Applying fennel seed water to the skin helps in getting toned, flexible and soft skin as it hydrates the skin by enhancing blood circulation to the face.

Regular use of fennel seeds tea (drinking or applying to hair) helps in getting strong and lengthy hair as it prevents unusual hair loss and premature hair graying. It prevents from hair fall by providing relief from itchy scalp and dandruff. Saunf has emmenagogue property which helps in regulating the regular menstrual flow among women. Its availability with essential phytoestrogens helps in curing menstrual disorders such as premenstrual syndrome, early menopause, unusual breast enlargement and *etc.* It prevents from the cardiovascular diseases, hormone or estrogen dependent cancer, menopausal disorders, hot flashes, dysmenorrheal (painful menstruation) and *etc.* Fennel seed keeps mind very calm and stress free by reducing the level of anxiety, depression, stimulating proper menstruation, lowering blood pressure, and *etc.* It helps in relieving morning sickness, weakness, fatigue and other symptoms during pregnancy. Regular use of fennel seeds helps in improving milk production among women on breast feeding or nursing mothers because of its galactagogue property.

Drinking saunf water with little honey (boil saunf into water and add honey) helps in increasing the breast milk production. It is aphrodisiac in nature helps in enhancing libido as well as maintaining sexual power in both men and women. It is a better home remedy to enhance breast at home than other surgical breast enlargement procedures. Some of its benefits for increasing breast size are mentioned below: It is loaded with the natural hormones which involve in the developing and growing breast tissues to make breasts large and firm without any side effects. Saunf is used in manufacturing many creams and pills having estrogenic effects which are used by most of the women for years to get their enhanced breast. Its natural phytoestrogens availability prevents abnormal growth of the breast tissues thus protects from breast cancer. Fennel seeds availability with high amount of flavonoids helps in promoting the production of estrogen hormone which in turn promotes breast enlargement. Massaging breast with saunf oil for 10-15 min each (100-300 circular rubs for each) once a day in the morning or before going to bed helps in enhancing breast size. This procedure needs time and dedication but result oriented. It is very beneficial for the newborn babies as it helps in relieving from the colic pain as well as improving digestion. Massaging joints with the fennel seed oil or saunf oil helps in relieving joints pain (caused due to arthritis) and cellulite. It also provides relieve from the respiratory problems like cough, bronchitis and *etc.* It contains essential oil required for the health of liver thus protects from the liver disorders. Drinking jeera and saunf water on daily basis (boil jeera and saunf seeds powder into a glass of water till it remains half) helps in getting wrinkles and acne free skin as well as helps in good digestion by keeping acidity away. Eating one tsf of powdered mixture of saunf, almonds and mishri at every night before going to bed with a glass of warm milk for one and half months helps in improving eyesight by curing all the eye disorders. Avoid drinking water after having this mixture. Taking this mixture with milk also helps in enhancing the memory and concentration level by curing headache and eye pain.

Fennel seeds tea is a natural detoxifier (when used regularly) helps in releasing body toxins by detoxifying the body as well as prevents from water retention in the body. It can be prepared at home by boiling one or two tsf saunf seeds or leaves into a cup of water till it remains half and add some sweetener (sugar, honey or jaggery) and milk according to the need and taste. Taking saunf tea on daily basis helps in relieving gastrointestinal disorders including gas, acid reflux, bloating, loss of appetite, indigestion, acidity and *etc.* Generally it is used by the people during summer season as a home remedy to get relief from the summer heat. It improves digestion by curing intestinal worms problem among kids. It helps in regulating menses by easing menstrual spasms. Regular intake of saunf tea aids in proper digestion and assists in weight loss. It helps in keeping body fit and youth look. It helps in curing bloating, gas, indigestion and digestive tract inflammation which helps in decreasing belly fat and thus weight loss. Fennel leaves are used to reduce the effect of snake bite and neutralize the vegetable poison like mushroom *etc.* Inhaling the steam of boiled fennel leaves helps in treating respiratory infections of bronchitis and asthma.

24.6 Nutritional Benefits of Saunf/Fennel Seeds per 100g

Energy	345 Kcal	Carbohydrate	52.29 g
Protein	15.80 g	Fat	14.87 g
Dietary Fibers	39.8 g		

Vitamins			
Thiamin (B1)	0.408 mg	Riboflavin (B2)	0.353 mg
Niacin (B3)	6.050 mg	Pyridoxine (B6)	0.470 mg
Vitamin A	135 IU	Vitamin C	21 mg
Electrolytes			
Sodium	88 mg	Potassium	1694 mg
Minerals			
Calcium	1196 mg	Copper	1.067 mg
Iron	18.54 mg	Magnesium	385 mg
Manganese	6.533 mg	Phosphorus	487 mg
Zinc	3.70 mg		

Source: USDA Nutrient Database.

24.7 Conclusion

Fennel seeds or Saunf are aromatic and sweet taste spice used in kalaunji masala in India for cooking vegetables as well as various medicinal and home remedy to cure variety of health disorders. Because of its lots of medicinal properties (like appetizer, carminative, anti-spasmodic, anti-emetic, purgative *etc.*) it is widely used as a home remedy to treat various health disorders Pitta, vata, cough, gas, bloating and other stomach and respiratory problems.

REFERENCES

Anonymous (2012) "Fennel Pollen: Culinary Fairy Dust". The Wall Street Journal. 23 March 2012.

Blamey, M. and Grey-Wilson, C. (1989) Flora of Britain and Northern Europe. ISBN 0-340-40170-2

Dalal. T. (2021) Fennel seeds, Indian recipes using. http: //www.tarladalal.com/glossary-fennel-seeds-410i.

GRIN (2023) USDA-ARS Germplasm Resources Information Network (GRIN) Germplasm Resources Information Network: *Foeniculum vulgare.*

He W. and Huang B. (2011) A review of chemistry and bioactivities of a medicinal spice: *Foeniculum vulgare*. Journal of Medicinal Plants Research Vol. 5(16), pp. 3595- 3600, http: //www.academicjournals.org/JMPR ISSN 1996-0875

Irma R. (1997). Joy of Cooking. New York: Simon and Schuster Inc. p. 375. ISBN 0- 684-81870-1.

Ono M, Ito Y, Kinjo J, Yahara S, Nohara T and Niiho Y (1995) Four new glycosides of stilbenetrimer from *Foeniculi fructus* (fruit of *Foeniculum vulgare* Miller). Chemical and pharmaceutical bulletin, 43(5): 868–871.

Rathore, S.S., S.N. Saxena and B. Singh. (2013) Potential health benefits of major seed spices. Int. J. Seed. Spices, 3(2): 1-12

Sahu D. (2012). "The power of five seeds". The Times Of India.

Saxena O.P. (Nidar) (2023) Charak Samhita in Hindi by DPB Pub. Delhi pp.88

Solanki R.K., Verma P., Meena R.S., Meena NK., B Singh B. (2017) Fennel In:

Breeding of Horticultural Crops Vol. 1 -Part A : Spices (2017) : 409-424. Editors : V. A Parthasarathy, C. Aswath, K. Nirmal Babu and R. S Kumar Today and Tomorrow's Printers and Publishers, New Delhi -

Weiss EA (2002) Spices Crops. CABI, Wallingford, 356–60.

Chapter 25

Most Versatile Therapeutically Enriched Herb: Guduchi (*Tinospora cordifolia* (Willd.) Mier ex Hook f. and Thomson)

Tinospora cordifolia named as "Guduchi" in Sanskrit belonging to family Menispermaceae is a genetically diverse, large, deciduous climbing shrub with greenish yellow typical flowers, found at higher altitude. In racemes or racemose panicles, the male flowers are clustered and female are solitary. The flowering season expands over summers and winters. A variety of active components derived from the plant like alkaloids, steroids, diterpenoid lactones, aliphatics, and glycosides have been isolated from the different parts of the plant body, including root, stem, and whole plant. Recently, the plant is of great interest to researchers across the globe because of its reported medicinal properties like anti-diabetic, anti-periodic, anti-spasmodic, anti-inflammatory, anti-arthritic, anti-oxidant, anti-allergic, anti-stress, anti-leprotic, anti-malarial, hepatoprotective, immunomodulatory and anti-neoplastic activities. In this review, we focus our attention to: (i) the reported genetic diversity in the Plant (ii) biological roles reported in humans and animals and active components from the plant. (iii) biological roles reported in humans and animals.

Keywords: *Therapeutic use, Fever, Herb, Guduchi, Tinospora cordifolia, Steroid.*

25.1 Introduction

The use of plants for curing human ailments is an ancient practice in which there has been a recent revival of interest. Ethnobotanical studies has resulted into finding various uses of plants for human health. Various traditional uses are tested for medical use for treating various ailments from curing diabetes, and heart problems. Plant like *Tinospora cordifolia* provide immunity to the body. *T. cordifolia* commonly named as "Guduchi" in Sanskrit belonging to family Menispermaceae is a genetically diverse, large, deciduous climbing shrub with greenish yellow typical flowers, found at higher altitude (Anonymous 2001; Parthipan 2011; Rana *et al.,* 2012). A myriad of biologically active compounds, including alkaloids, diterpenoid lactones, glycosides, steroids, sesquiterpenoid, phenolics, aliphatic compounds, and polysaccharides have been isolated from different parts of the plant body (Upadhyay *et al.,* 2010). These compounds have been reported to have different biological roles in disease conditions thus enabling potential application in clinical research. *Tinospora* extract is extensively used in various herbal preparations for the treatment of different diseases and as anti-spasmodic, anti-microbial, anti-osteoporotic, anti-inflammatory, anti-arthritic, and anti-diabetic properties (Sharma *et al.,* 2012).

25.2 *Tinospora cordifolia*: A Genetically Diverse Plant

Reports on studies of morphological and physiological characters of the plant, including plant length, stem diameter, growth habit, floral morphology, flower color, stomatal density, trichomal density, lenticels density, petiole length, plant biomass, and other characteristics of the plant and diversity in the genetic components identified by markers have indicated the diversity in the medicinal plant which has profound importance for efficient and effective management of plant genetic resources. Reports using markers for random amplified polymorphic DNA, and inter-simple sequence repeat primers (Rout 2006; Upadhyay *et al.,* 2010; Rana *et al.,* 2012) have pointed toward the genetic variation within the population. However, reports on conservation strategies and propagation of the germplasm are few.

25.3 Biological Roles

The major biological property of *Tinospora cordifolia* includes:

25.3.1 Immunomodulatory Property

The immuomodulatory property of *Tinospora* is well documented. Active compounds 11-hydroxymustakone, N-methyl-2-pyrrolidone, N-formylannonain, cordifolioside A, magnoflorine, tinocordiside and syringin (Sharma *et al.,* 2012) has been reported to have potential immunomodulatory and cytotoxic effects (Kapil and Sharma 1997). They have been reported to function by boosting the phagocytic activity of macrophages, production of reactive oxygen species (ROS) in human neutrophil cells (More and Pai 2012), enhancement in nitric oxide (NO) production by stimulation of splenocytes and macrophages indicative of anti-tumor effects (Upadhyaya *et al.,* 2011). Aqueous *Tinospora* extracts has been also reported to influence the cytokine production, mitogenicity, stimulation and activation of immune effector cells (Upadhyaya *et al.,* 2011). In mice, *T. cordifolia* extracts has been shown to result in up-regulation of IL-6 cytokine, resulting in acute reactions to injury, inflammation, activation of cytotoxic T cells, and B cell differentiation (Sudhakaran *et al.,* 2006). Active compounds in aqueous extracts like alkaloids, di-terpenoid lactones, glycosides, steroids, sesquiterpenoid, phenolics, aliphatic compounds or polysaccharides in experimental rat model have been reported for their cytotoxic action. Dry stem crude extracts of *T. cordifolia* with a polyclonal B cell mitogen, G1-4A on binding to macrophages have been reported to enhance immune response in mice by inducing secretion of IL-1, together with activation of macrophages. Reports on *T. cordifolia* in prevention of oxidative damage also exist (Raghu *et al.,* 2009). The (1,4)-alpha-d-glucan, derived *T. cordifolia* have been shown to activate human lymphocytes with downstream synthesis of the pro- and anti-inflammatory cytokines, *in vitro* (Koppada *et al.,* 2009). Synergistic effects of compounds in the immunomodulatory activity of *T. cordifolia* are reported (Sharma *et al.,* 2012).

25.3.2 Anti-diabetes Property

The stem of *Tinospora cordifolia* is widely used in the therapy of diabetes by regulating the blood glucose (Sangeetha *et al.,* 2011) in traditional folk medicine of India. It has been reported to mediate its anti-diabetic potential through mitigating oxidative stress (OS), promoting insulin secretion and also by inhibiting gluconeogenesis and glycogenolysis, thereby regulating blood glucose (Sangeetha *et al.,* 2011). Alkaloids, tannins, cardiac glycosides, flavonoids, saponins, and steroids as the major phytoconstituents (Zinjarde *et al.,* 2011) of *T. cordifolia* have been reported to play an anti-diabetic role.

The isoquinoline alkaloid rich fraction from stem, including, palmatine, jatrorrhizine, and magnoflorine have been reported for insulin-mimicking and insulin-releasing effect both *in vitro* and *in vivo* (Patel and Mishra 2011). Oral treatments of root extracts have been reported to regulate blood glucose levels, enhance insulin secretion and suppress OS markers. Initiation and restoration of cellular defence anti-oxidant markers including superoxide dismutase (SOD), glutathione peroxidase (GPx) and glutathione (GSH), inhibition of glucose 6-phosphatase and fructose 1, 6-diphosphatase, restoration of glycogen content in liver was reported in *in vitro* studies (Patel and Mishra 2011). The crude stem ethyl acetate, dichloromethane (DCM), chloroforms and hexane extracts of *T. cordifolia* inhibited the enzyme's salivary and pancreatic amylase and glucosidase (Chougale *et al.,* 2009) thus increasing the post-prandial glucose level and finds potential application in treatment of diabetes mellitus.

The root extract has been reported to decrease the levels of glycosylated hemoglobin, plasma thiobarbituric acid reactive substances, hydroperoxides, ceruloplasmin and vitamin E diabetic rats (Umamaheswari and Prince Mainzen 2007). Oral administration of *T. cordifolia* extract in "Ilogen-Excel" formulation (Ayurvedic herbal formulation) composed of eight medicinal plants including *Curcuma longa, Strychnos potatorum, Salacia oblonga, T. cordifolia, Vetiveria zizanioides, Coscinium fenestratum, Andrographis paniculata,* and *Mimosa pudica* is reported to reduce GSH and vitamin C (Umamaheswari and Prince Mainzen 2007). in blood and urine glucose and lipids in the serum and tissues in alloxan diabetic rats with a subsequent decrease in body weight (Prince *et al.,* 2003). Decreased concentration of GSH, GPx, and SOD, catalase activity is reported in heart and brain of diabetic rats (Prince *et al.,* 2004). *T. cardifolia* root extract (TCE) has been reported to cause an increase in body weight, total hemoglobin and hepatic hexokinase (Stanely *et al.,* 2000) and lowering hepatic glucose-6-phosphatase, serum acid phosphatase (ACP), alkaline phosphatase (ALP), and lactate dehydrogenase (LDH) in diabetic rats thus having hypoglycemic and hypolipidaemic effect. The protective effects of TCE were reported in presence of higher levels of anti-oxidant molecules and enzymes. TCE has been shown to significantly counterbalance the diabetes-associated OS in the maternal liver by lowering the levels of malondialdehyde and ROS and the increased levels of GSH and total thiols (Shivananjappa and Muralidhara 2012).

25.3.3 Anti-toxic Effects

Tinospora cordifolia extracts have been reported to scavenge free radicals generated during aflatoxicosis (Gupta and Sharma 2011). It exhibited protective effects by lowering thiobarbituric acid reactive substances (TBARS) levels and enhancing the GSH, ascorbic acid, protein, and the activities of anti-oxidant enzymes *viz.,* SOD, CAT, GPx, Glutathione S-transferase (GST) and glutathione reductase (GR) in kidney. Alkaloids such as a choline, tinosporin, isocolumbin, palmatine, tetrahydropalmatine, and magnoflorine from *T. cordifolia* showed protection against aflatoxin-induced nephrotoxicity (Gupta and Sharma 2011). *T. cordifolia* stem and leaves extract has shown hepatoprotective effect in Swiss albino male mice against lead nitrate induced toxicity (Sharma and Pandey 2010a). Oral administration of plant extracts prevented the occurrence of lead nitrate induced liver damage (Sharma and Pandey 2010b). Decreased level of SOD, CAT and increased level of aspartate aminotransferase (AST), alanine aminotransferase (ALT), ALP, and ACP were observed in mice suffering from lead toxicity. Synergistic administration of aqueous extract of stem and leaf along with the lead nitrate increased the activities of SOD and CAT and decreased the levels of AST, ALT, ALP, and ACP enzymes (Sharma and Pandey 2010b). Protective role of aqueous extract of stem and leaves of *T. cordifolia* overcoming the toxic effects of lead is shown as its effects on the hematological values (Sharma and Pandey 2010a). Cyclophosphamide (CP) an anti-cancer drug has been reported to reduce the GSH content in both bladder and liver and lowered levels of cytokines Inerferon-γ and IL-2 an increased levels of pro-inflammatory cytokine TNF-α. This effect could be reversed on *T. cordifolia* treatment indicating the role of *T. cordifolia* in overcoming CP induced toxicities in cancer treatment (Hamsa and Kuttan 2012).

25.3.4 Anti-arthritic, Anti-osteoporotic Effects

Single or synergistic formulations of *Tinospora cordifolia* with *Zingiber officinale* has been used in rheumatoid arthritis treatment in traditional medicine

(Chopra *et al.,* 2012). *T. cordifolia* has been reported to affect the proliferation, differentiation and mineralization of bone like matrix on osteoblast model systems *in vitro* and hence finds potential application as an anti-osteoporotic agent. Ecdysteroids isolated from the plant have been reported of protein anabolic and anti-osteoporotic effects in mammals. Beta-Ecdysone (Ecd) from *T. cordifolia* extracts have been reported to induce a significant increase in the thickness of joint cartilage, induce the osteogenic differentiation in mouse mesenchymal stem cells (Gao *et al.,* 2008) and to relieve osteoporosis in osteoporotic animal models (Gao *et al.,* 2008). Further 20-OH-β-Ecd isolated from *T. cordifolia* has been reported of its anti-osteoporotic effects (Abiramasundari *et al.,* 2012). thus highlighting the role of *T. cordifolia* in the treatment of osteoporosis and osteoarthritis (Kapur *et al.,* 2010).

25.3.5 Anti-HIV Effects

TCE has been shown to demonstrate a decrease in the recurrent resistance of HIV virus thus improving the therapeutic outcome (Kalikar *et al.,* 2008). Anti-HIV effects of TCE was revealed by reduction in eosinophil count, stimulation of B lymphocytes, macrophages and polymorphonuclear leucocytes and hemoglobin percentage thus, revealing its promising role of application in management of the disease (Kalikar *et al.,* 2008; Akhtar 2010).

25.3.6 Anti-cancer Effects

The anti-cancer effects of *T. acordifolia* are mostly studied in animal models. TCE have been shown to have a radioprotective role by significantly increase in body weight, tissue weight, testes-body weight ratio and tubular diameter and inhibit the harmful effects of sub-lethal gamma radiation on testes in male Swiss albino mice. In pre-irradiating mice, TCE significantly affected radiation induced rise in lipid peroxidation and resulted in the decline of GSH concentration in testes (Sharma 2011). Pre-treatment of HeLa cells by TCE have been shown to decrease the cell viability, increase LDH and decrease in GSH S-transferase activity (Rao and Rao 2010). Dihydrotestosterone (DHT) in TCE has been reported to stimulate the growth and proliferation of Human LNCaP cells (which are androgen-sensitive human prostate adenocarcinoma cells). Androgenic compounds in TCE act via androgen receptor (Kapur *et al.,* 2009). Newly isolated compounds like (5R, 10R)-4R, 8R-dihydroxy-2S, 3R: 15, 16-diepoxycleroda-13 (16), 17, 12S: 18,1S-dilactone (ECD), a diterpenoid from *T. cordifolia* has been reported for its chemopreventive potential in diethylnitrosamine (DEN) induced hepatocellular carcinoma (HCC) in rats by decreasing anti-oxidant activities via SOD, CAT and detoxification enzymes like GSH, GPx and subsequent increase in the activities of the hepatic markers ((Serum glutamic oxaloacetic transaminase)SGOT, (Serum Glutamic Pyruvate Transaminase) SGPT, LDH) and decreased serum transaminase level thus confirming its anti-tumor effects and promising application as a potent chemo preventive drug for HCC (Dhanasekaran 2009).

The radiosensitizing activity of DCM extract of *T. cordifolia* has been reported in Ehrlich ascites carcinoma (EAC) mice enabling tumor-free survival via depletion of

GSH and glutathione-S-transferase by elevated levels of lipid peroxidation and DNA damage to tumor cells (Rao *et al.,* 2008). TCE hexane fraction has been shown to block the G1 phase in EAC mice and cause apoptosis by the formation of apoptotic bodies, nuclear condensation, activation of caspase-3, decreased cell number and ascites volume, increased expression of pro-apoptotic gene, *Bax*, and decreased expression of anti-apoptotic gene, *Bcl-2* (Thippeswamy and Salimath 2007). TCE could induce a reduction of papillomas, tumor yield, tumor burden, and tumor weight while increase phase II detoxifying enzymes (Chaudhary *et al.,* 2008) in skin carcinoma animal models. The effect of a hydroalcoholic (80 per cent ethanol: 20 per cent distilled water) extract of aerial roots of *T. cordifolia* on Swiss albino mice (Singh *et al.,* 2006) revealed a significant increase in acid-soluble sulfhydryl (-SH), cytochrome *P* (450) contents, and enzyme activities of cytochrome *P* (450) reductase, cytochrome b5 reductase, GST, DT-diaphorase (DTD), SOD, catalase, GPX, and GR activity in the liver highlighting the chemopreventive role of *T. cordifolia* against carcinogenicity (Singh *et al.,* 2006).

In vivo anti-angiogenic activity of TCE in B16-F10 melanoma was detected by increased levels of pro-inflammatory cytokines, including IL-1 β, IL-6, TNF-α, granulocyte monocyte-colony stimulating factor (GM-CSF) and the vascular endothelial cell growth factor (VEGF), increased production of anti-angiogenic agents IL^{-2} and tissue inhibitor of metalloprotease-1 (TIMP-1) in the B16-F10 extract-treated animals (Leyon and Kuttan 2004a). The polysaccharide fraction from *T. cordifolia* was found to be very effective in reducing the metastatic potential of B16-F10 melanoma cells. Markers of neoplastic development were reduced significantly in the treated animals compared with the untreated control animals (Leyon and Kuttan 2004b).

25.3.6 Anti-microbial Activity

The methanol extracts of *T. cordifolia* have been reported to have potential against microbial infections (Narayanan *et al.,* 2011). The anti-bacterial activity of *T. cordifolia* extracts has been assayed against *Escherichia coli, Staphylococcus aureus, Klebsiella pneumoniae, Proteus vulgaris, Salmonella typhi, Shigella flexneri, Salmonella paratyphi, Salmonella typhimurium, Pseudomonas aeruginosa, Enterobacter aerogene*, and *Serratia marcesenses* (Gram-positive bacteria) (Jeyachandran *et al.,* 2003; Tambekar *et al.,* 2009; Narayanan *et al.,* 2011). In mice models, TCE has been reported to function in bacterial clearance and improved phagocytic and intracellular bactericidal capacities of neutrophils (Thatte *et al.,* 1992). TCE has been reported of immuno stimulant properties on macrophages (Sengupta *et al.,* 2011). Intra-mammary infusion of hydro-methanolic extracts of *T. cordifolia* treatment showed enhanced phagocytic activity of polymorpho nuclear cells in bovine subclinical mastitis (Purandare and Supe 2007).

25.3.7 Anti-oxidant Activity

The anti-oxidant capacity of methanolic extracts of stem of *Tinospora* administered orally increased the erythrocytes membrane lipid peroxide and catalase activity. It also decreased the activities of SOD, GPx in alloxan-induced diabetic rats (Prince *et al.,* 2003; Sivakumar and Rajan 2010). *Tinospora* extracts possess possible inhibitors of aldose reductase and anti-oxidant agents (Gacche and Dhole 2011) thereby reducing chemotoxicity induced by free radicals (Rawal *et al.,* 2004). TCE has been reported of its strong free radical scavenging properties against superoxide anion (O_2^-), hydroxyl radicals (OH), NO radical, and peroxynitrite anion ($ONOO^-$) (Rawal *et al.,* 2004). The extract was also found to reduce the toxic side effects of CP in mice by the free radical formation (Mathew and Kuttan 1997). *Tinospora* lowers the levels of malondialdehyde and ROS and the higher levels of GSH and total thiols. The protective effects of *T. cordifolia* could be observed even in the fetal milieu, with higher levels of anti-oxidant molecules and enzymes (Shivananjappa and Muralidhara 2012).

Tinospora cordifolia has the ability to scavenge free radicals generated during aflatoxicosis. *T. cordifolia* showed protection against aflatoxin-induced nephrotoxicity due to the presence of alkaloids such as a choline, tinosporin, isocolumbin, palmatine, tetrahydropalmatine, and magnoflorine (Gupta and Sharma 2011). A significant increase in the concentration of TBARS in brain along with a decrease in heart has been observed in diabetic rats (Prince *et al.,* 2004). It also enhanced formation of SOD, GPx, and GSH in liver. Treatment with *T. cordifolia* also inhibited glucose 6-phosphatase and fructose 1, 6-diphosphatase; and restored glycogen content in liver. *T. cordifolia* has been shown to regulate blood glucose (Sangeetha *et al.,* 2011). (5R, 10R)-4R, 8R-dihydroxy-2S, 3R: 15, 16-diepoxycleroda-13 (16), 17, 12S: 18,1S-dilactone (ECD), a diterpenoid from *T. cordifolia* has been shown to possess chemo-preventive potential in DEN induced HCC rats. Treatment of ECD in both preventive and curative DEN induced animals increased the level of anti-oxidants and detoxification enzymes (Dhanasekaran *et al.,* 2009). An aqueous extract of *T.cordifolia* has a radio-protective enhancing the survival of mice against a sub-lethal dose of gamma radiation (Kapur *et al.,* 2010).

Tinospora cordifolia also has the components that decrease the recurrent resistance of HIV virus to antiretroviral therapy (ART) and improve the outcome of the therapy (Gupta *et al.,* 2010). The effect of a hydroalcoholic (80 per cent ethanol: 20 per cent distilled water) extract of aerial roots of *T. cordifolia* on carcinogen/ drug metabolizing phase-I and phase-II enzymes, anti-oxidant enzymes, GSH content, LDH and lipid peroxidation has been shown in liver of Swiss albino mice. The enhanced GSH level and enzyme activities involved in xenobiotic metabolism and maintaining anti-oxidant status of cells are suggestive of a chemo-preventive efficacy of *T. cordifolia* (Singh *et al.,* 2006). *T. cordifolia* has been reported to contain an alpha-glucosidase inhibitor, characterized as saponarin (apigenin-6-C-glucosyl-7-O-glucoside). The leaf extract had appreciable anti-oxidant and hydroxyl radical scavenging activities (Sengupta *et al.,* 2009). Pepticare, a herbomineral formulation of the Ayurveda medicine consisting of the herbal drugs: *Glycyrrhiza glabra, Emblica officinalis* and *T. cordifolia*, has anti-ulcer and anti-oxidant activity in rats (Bafna and Balaraman 2005).

25.4 Uses in other Diseases

A dose dependent reduction in size and in lipid peroxide levels of serum and heart tissue were observed with the prior treatment of *T. cordifolia* (Rao *et al.,* 2005). Charak Samhita mentioned that use of dry ginger and *Tinospora* root powder and atis in equal amounts and made in form of a quath, when consumed 25 ml. twice for 15 days cures sprue or sever dysentery(Saxena 2023p. 217).The activation of macrophages by cytotoxic T cells leads to increase in GM-CSF which leads to leucocytosis and improved neutrophil function (Thatte *et al.,* 1994). Octacosanol isolated from *T. cordifolia* inhibits proliferation of endothelial cells and Ehrlich ascites tumor cells, inhibits neovascularization induced by angiogenic factors in chick chorioallantoic membrane and rat cornea *in vivo* angiogenesis assays and also inhibits secretion of ascites fluid in the growing tumor cells *in vivo* (Thippeswamy *et al.,* 2008) by inhibiting activity of matrix metalloproteinases (MMPs) and translocation of transcription factor nuclear factor-kappa-B (NF-κB) to nucleus (Thippeswamy *et al.,* 2008). Oral administration of 70 per cent methanolic extract of *T. cordifolia* stem reduces sperm motility and density, lowering of serum testosterone, protein, sialic acid, glycogen contents, and depletion of vesicular fructose of testes leading to reduction of male fertility in rats (Gupta and Sharma 2003). The *in vivo* administration of alcoholic extract of *T. cordifolia* has been reported to increase bone marrow derived macrophages (BMDM) in bearing Dalton's lymphoma (DL) (Singh *et al.,* 2006). The polyherbal preparations Caps HT2 of *T. cordifolia,* could reduce plasma recalcification time and enhanced the release of lipoprotein lipase enzyme. Other polyherbal HP-1 has hepatocurative and anti-oxidant effects (Jeyachandran *et al.,* 2003). Charak Samhita mentioned that use of *Tinospora* root powder in form of a quath, when consumed twice cures leucorrhoea (Saxena 2023p. 818).

25.5 Conclusions

Tinospora cordifolia has an importance in traditional ayurvedic medicine used for ages in the treatment of fever, jaundice, chronic diarrhea, cancer, dysentery, bone fracture, pain, asthuma, skin disease, poisonous insect, snake bite, eye disorders (Parthipan *et al.,* 2011). Recent reports have shown the compounds and their biological roles in *T. cordifolia* extract. Such properties (Tambekar *et al.,* 2009) may be exploited for production of new formulations, which may be better and promising over conventional one. Although genetically diverse and reports of application of tissue culture based propagation of *Tinospora* exist, effective conservation strategies of the germplasm for such an economically important medicinal plant with many biological role remains yet to be accomplished.

A plant with as diverse a role as *T. cordifolia* is a versatile resource for all forms of life. There are reports as already discussed that the plant extracts have active compounds in the form of alkaloids, glycosides, lactones and steroids. All these active compounds have immunomodulatory and physiological roles of different types, thereby demonstrating the diverse versatility of the plant. Studies need to be conducted with aspects how the active compounds actually interact with the living systems and affects the structure-function relationships. Crystal structures of the membrane bound receptors and the activation of the downstream signaling cascades and the changes in the immediate environment of the site of action can lead us into identification of novel perspectives into our understanding of nature. The search into the vivacious sources of nature can also lead us into differential interactions among the evolutionary related groups of organisms. The future scope of the review remains in exploiting the biochemical and signaling pathways of the active components of *Tinospora* thus, enabling effective disease targeting. With so much to offer to the scientific world of medicine, the plant *Tinospora* truly acts as an incredible source.

REFERENCES

Abiramasundari G, Sumalatha KR, Sreepriya M. (2012) Effects of *Tinospora cordifolia* (Menispermaceae) on the proliferation, osteogenic differentiation and mineralization of osteoblast model systems *in vitro*. J Ethnopharmacol. 141: 474–80.

Anonymous (2001) The Ayurvedic Pharmacopoeia of India. Part I. 1st ed. Vol. 1. New Delhi: Department Of AYUSH, Ministry of Health and FW; 2001. pp. 53–5.

Akhtar S. (2010) Use of *Tinospora cordifolia* in HIV infection. Indian J Pharmacol. 42: 57.

Bafna PA, Balaraman R. (2005) Anti-ulcer and anti-oxidant activity of pepticare, a herbomineral formulation. Phytomedicine. 12: 264–70.

Chaudhary R, Jahan S, Goyal PK. (2008) Chemopreventive potential of an Indian medicinal plant (*Tinospora cordifolia*) on skin carcinogenesis in mice. J Environ Pathol Toxicol Oncol. 27: 233–43.

Chougale AD, Ghadyale VA, Panaskar SN, Arvindekar AU. (2009) Alpha glucosidase inhibition by stem extract of *Tinospora cordifolia*. J Enzyme Inhib Med Chem. 24: 998–1001.

Chopra A, Saluja M, Tillu G, Venugopalan A, Narsimulu G, Handa R, (2012) Comparable efficacy of standardized ayurveda formulation and hydroxychloroquine sulfate (HCQS) in the treatment of rheumatoid arthritis (RA): A randomized investigator- blind controlled study. Clin Rheumatol. 31: 259–69.

Dhanasekaran M, Baskar AA, Ignacimuthu S, Agastian P, Duraipandiyan V. (2009) Chemopreventive potential of Epoxy clerodanediterpene from *Tinospora cordifolia* against diethylnitrosamine-induced hepatocellular carcinoma. Invest New Drugs. 27: 347–55.

Gacche RN, Dhole NA. (2011) Profile of aldose reductase inhibition, anti-cataract and free radical scavenging activity of selected medicinal plants: An attempt to standardize the botanicals for amelioration of diabetes complications. Food ChemToxicol. 49: 1806–13.

Gao L, Cai G, Shi X. (2008) Beta-ecdysterone induces osteogenic differentiation in mouse mesenchymal stem cells and relieves osteoporosis. Biol Pharm Bull. 31: 2245–9.

Gupta R, Sharma V. (2011) Ameliorative effects of *Tinospora cordifolia* root extract on histopathological and biochemical changes induced by aflatoxin-b (1) in mice kidney. Toxicol Int. 18: 94–8.

Gupta RS, Sharma A. (2003) Antifertility effect of *Tinospora cordifolia* (Willd.) stem extract in male rats. Indian J Exp Biol. 41: 885–9.

Gupta GD, Sujatha N, Dhanik A, Rai NP. (2010) Clinical evaluation of shilajatu rasayana in patients with HIV Infection. Ayu. 31: 28–32.

Hamsa TP, Kuttan G. (2012) *Tinospora cordifolia* ameliorates urotoxic effect of cyclophosphamide by modulating GSH and cytokine levels. Exp Toxicol Pathol. 64: 307–14.

Jeyachandran R, Xavier TF, Anand SP. (2003) Antibacterial activity of stem extracts of *Tinospora cordifolia* (Willd) Hook. f and Thomson. AncSci Life. 23: 40–3.

Kalikar MV, Thawani VR, Varadpande UK, Sontakke SD, Singh RP, Khiyani RK. (2008) Immunomodulatory effect of *Tinospora cordifolia* extract in human immuno-deficiency virus positive patients. Indian J Pharmacol. 2008; 40: 107–10.

Kapur P, Wuttke W, Jarry H, Seidlova-Wuttke D. (2010) Beneficial effects of beta-Ecdysone on the joint, epiphyseal cartilage tissue and trabecular bone in ovari ectomized rats. Phytomedicine. 17: 350–5.

Koppada R, Norozian FM, Torbati D, Kalomiris S, Ramachandran C, Totapally BR. (2009) Physiological effects of a novel immune stimulator drug, (1,4)-α-D-glucan, in rats. Basic Clin Pharmacol Toxicol. 105: 217–21.

Leyon PV, Kuttan G. (2004a)Effect of *Tinospora cordifolia* on the cytokine profile of angiogenesis-induced animals. Int Immunopharmacol. 4: 1569–75.

Leyon PV, Kuttan G. (2004b) Inhibitory effect of a polysaccharide from *Tinospora cordifolia* on experimental metastasis. J Ethnopharmacol. 90: 233–7.

Mathew S, Kuttan G. (1997) Antioxidant activity of *Tinospora cordifolia* and its usefulness in the amelioration of cyclophosphamide induced toxicity. J ExpClin Cancer Res. 16: 407–11.

More P, Pai K. (2012) *In vitro* NADH-oxidase, NADPH-oxidase and myeloperoxidase activity of macrophages after *Tinospora cordifolia* (guduchi) treatment. Immuno pharmacol Immunotoxicol. 34: 368–72.

Narayanan AS, Raja SS, Ponmurugan K, Kandekar SC, Natarajaseenivasan K, Maripandi A, (2011) Antibacterial activity of selected medicinal plants against multiple antibiotic resistant uropathogens: A study from Kolli Hills, Tamil Nadu, India. Benef Microbes. 2: 235–43.

Parthipan M, Aravindhan V, Rajendran A. (2011) Medico-botanical study of Yercaud hills in the eastern Ghats of Tamil Nadu, India. AncSci Life. 30: 104–9.

Patel MB, Mishra S. (2011) Hypoglycemic activity of alkaloidal fraction of *Tinospora cordifolia*. Phytomedicine. 18: 1045–52.

Prince S. Mainzen P, Menon VP. (2001) Antioxidant action of *Tinospora cordifolia* root extract in alloxan diabetic rats. Phytother Res. 15: 213–8.

Prince S. Mainzen P, Menon VP. (2003) Hypoglycaemic and hypolipidaemic action of alcohol extract of *Tinospora cordifolia* roots in chemical induced diabetes in rats. Phytother Res. 17: 410–3.

Prince P S, Kamalakkannan N, Menon VP. (2004) Restoration of antioxidants by ethanolic *Tinospora cordifolia* in alloxan-induced diabetic Wistar rats. Acta Pol Pharm. 61: 283–7.

Purandare H, Supe A. (2007) Immunomodulatory role of *Tinospora cordifolia* as an adjuvant in surgical treatment of diabetic foot ulcers: A prospective randomized controlled study. Indian J Med Sci. 61: 347–55.

Raghu R, Sharma D, Ramakrishnan R, Khanam S, Chintalwar G J, Sainis K B. (2009) Molecular events in the activation of B cells and macrophages by a non-microbial TLR4 agonist, G1-4A from *Tinospora cordifolia*. Immunol Lett. 123: 60–71.

Rana V, Thakur K, Sood R, Sharma V, Sharma TR. (2012) Genetic diversity analysis of *Tinospora cordifolia* germplasm collected from northwestern Himalayan region of India. J Genet. 91: 99–103.

Rao PR, Kumar VK, Viswanath RK, Subbaraju GV. (2005) Cardioprotective activity of alcoholic extract of *Tinospora cordifolia* in ischemia-reperfusion induced myocardial infarction in rats. Biol Pharm Bull. 28: 2319–22.

Rao SK, Rao PS. (2010) Alteration in the radiosensitivity of HeLa cells by dichloromethane extract of guduchi (*Tinospora cordifolia*) Integr Cancer Ther. 9: 378–84.

Rao SK, Rao PS, Rao BN. (2008) Preliminary investigation of the radiosensitizing activity of guduchi (*Tinospora cordifolia*) in tumor-bearing mice. Phytother Res. 22: 1482–9.

Rawal AK, Muddeshwar MG, Biswas SK. (2004) *Rubia cordifolia*, *Fagonia cretica* Linn and *Tinospora cordifolia* exert neuroprotection by modulating the antioxidant system in rat hippocampal slices subjected to oxygen glucose deprivation. BMC Complement Altern Med. 4: 11.

Rout GR. (2006) Identification of *Tinospora cordifolia* (Willd.) Miers ex Hook F and Thomas using RAPD markers. Z Naturforsch C. 61: 118–22.

Sangeetha MK, Raghavendran Balaji HR, Gayathri V, Vasanthi HR. (2011) *Tinospora cordifolia* attenuates oxidative stress and distorted carbohydrate metabolism in experimentally induced type 2 diabetes in rats. J Nat Med. 65: 544–50.

Saxena O.P. (Nidar) (2023) Charak Samhita in Hindi by DPB Pub. Delhi pp.818

Sharma U, Bala M, Kumar N, Singh B, Munshi RK, Bhalerao S. (2012) Immunomodulatory active compounds from *Tinospora cordifolia*. J Ethnopharmacol. 141: 918–26.

Sengupta S, Mukherjee A, Goswami R, Basu S. (2009) Hypoglycemic activity of the antioxidant saponarin, characterized as alpha-glucosidase inhibitor present in *Tinospora cordifolia*. J Enzyme Inhib Med Chem. 24: 684–90.

Singh SM, Singh N, Shrivastava P. (2006) Effect of alcoholic extract of ayurvedic herb *Tinospora cordifolia* on the proliferation and myeloid differentiation of bone marrow precursor cells in a tumor-bearing host. Fitoterapia. 77: 1–11.

Stanely P, Prince M, Menon VP. (2000) Hypoglycaemic and other related actions of *Tinospora cordifolia* roots in alloxan-induced diabetic rats. J Ethnopharmacol. 70: 9–15.

Sharma P, Parmar J, Sharma P, Verma P, Goyal PK. (2011) Radiation-Induced Testicular Injury and Its Amelioration by *Tinospora cordifolia* (An Indian Medicinal Plant) Extract. Evid Based Complement Alternat Med. 2011: 643847.

Sharma V, Pandey D. (2010a) Beneficial effects of *Tinospora cordifolia* on blood profiles in male mice exposed to lead. Toxicol Int. 17: 8–11.

Sharma V, Pandey D. (2010b) Protective role of *Tinospora cordifolia* against lead-induced hepatotoxicity. Toxicol Int. 17: 12–7.

Shivananjappa MM, Muralidhara (2012) Abrogation of maternal and fetal oxidative stress in the streptozotocin-induced diabetic rat by dietary supplements of *Tinospora cordifolia*. Nutrition. 2012; 28: 581–7.

Sengupta M, Sharma GD, Chakraborty B. (2011) Effect of aqueous extract of *Tinospora cordifolia* on functions of peritoneal macrophages isolated from CCl_4 intoxicated male albino mice. BMC Complement Altern Med. 11: 102.

Sivakumar V, Rajan MS. (2010) Antioxidant Effect of *Tinospora cordifolia* Extract in Alloxan-induced Diabetic Rats. Indian J Pharm Sci. 72: 795–8.

Singh RP, Banerjee S, Kumar PV, Raveesha KA, Rao AR. (2006) *Tinospora cordifolia* induces enzymes of carcinogen/drug metabolism and antioxidant system, and inhibits lipid peroxidation in mice. Phytomedicine. 13: 74–84.

Sudhakaran DS, Srirekha P, Devasree LD, Premsingh S, Michael RD. (2006) Immunostimulatory effect of *Tinospora cordifolia* Miers leaf extract in *Oreochromis mossambicus*. Indian J Exp Biol. 44: 726–32.

Tambekar DH, Khante BS, Chandak BR, Titare AS, Boralkar SS, Aghadte SN. (2009) Screening of antibacterial potentials of some medicinal plants from Melghat forest in India. Afr J Tradit Complement Altern Med. 6: 228–32.

Taylor JC, Rapport L, Lockwood GB. (2003) Octacosanol in human health. Nutrition. 19: 192–5.

Thatte UM, Rao SG, Dahanukar SA. (1994) *Tinospora cordifolia* induces colony stimulating activity in serum. J Postgrad Med. 40: 202–3.

Thatte UM, Kulkarni MR, Dahanukar SA. (1992) Immunotherapeutic modification of *Escherichia coli* peritonitis and bacteremia by *Tinospora cordifolia*. J Postgrad Med. 38: 13–5.

Thippeswamy G, Sheela ML, Salimath BP. (2008) Octacosanol isolated from *Tinospora cordifolia* down regulates VEGF gene expression by inhibiting nuclear translocation of NF-<kappa>B and its DNA binding activity. Eur J Pharmacol. 588: 141–50.

Thippeswamy G, Salimath BP. (2007) Induction of caspase-3 activated DNase mediated apoptosis by hexane fraction of *Tinospora cordifolia* in EAT cells. Environ Toxicol Pharmacol. 23: 212–20.

Upadhyay AK, Kumar K, Kumar A, Mishra HS. (2010) *Tinospora cordifolia* (Willd.) Hook. f. and Thoms. (Guduchi)-validation of the Ayurvedic pharmacology through experimental and clinical studies. Int J Ayurveda Res. 1: 112–21.

Upadhyaya R, PR, Sharma V, Anita KV. (2011) Assessment of the multifaceted immunomodulatory potential of the aqueous extract of *Tinospora cordifolia*. Res J Chem Sci. 1: 71–9.

Umamaheswari S, Prince Mainzen PS. (2007) Antihyperglycaemic effect of 'Ilogen-Excel', an ayurvedic herbal formulation in streptozotocin-induced diabetes mellitus. Acta Pol Pharm. 64: 53–61.

Zinjarde PS, Bhargava SY, Kumar AR. (2011) Potent α-amylase inhibitory activity of Indian Ayurvedic medicinal plants. BMC Complement Altern Med. 11: 5.

Chapter 26

Phytoconstituents and Pharmacological Properties of Red Bead Tree (*Adenanthera pavonina* L.)

Adenanthera pavonina L. is an important medicinal tree and its barks are used in Experimental studies of this tree have shown reported to demonstrate numerous pharmacological activities such as antidiarrheal, anti-inflammatory, antinociceptive, antioxidant, anticancer, anti-inflammatory, analgesic activities, antihypertensive effect, antifungal, hepatoprotective, renal protective, CNS depressant, anticonvulsant, hyperlipidemic, antibacterial and BP-reducing activities of the bark, leaf, and seed extracts and their isolated compounds have been investigated. Seeds are anticephalgic and also used for the treatment of paralysis and blood pressure.A. pavonina constituted a number of phytochemicals, which disclose its uses for various therapeutic purposes. Previous phytochemical investigation reported the presence of many bioactive compounds like robinetin, chalcone, butin and flavanolampelopsin, stigmasterol glucosides, oleanolic acid, echinocystic acid, and sapogenins from the leaves and seeds of the tree. The phytochemical screening revealed various secondary metabolites with diverse biological effects, including flavonoids, alkaloids, saponins, tannins, steroids, triterpenoids, polyphenols, anthraquinones, coumarins, glycosides and polysaccharides. Seven compounds viz. ethyl 3,3-dimethyl-13-hydroxytridecanoate, stigmasta-5,22-dien-3β-ol, tert. butyltridecanoate, 6-α-hydroxy stigmast-20(21)-en-3-one(4)of dichloromethane extract and 18-(22 , 32 -dihydroxyphenyl)nonadec-17-en-2-ol, 1-(N-propyl amino)-2-henecosanone, and stigmast–5(6), 20(21)-diene-3-on were isolated from the barks of A. pavonina. Five new and two known compounds have been isolated and characterized from the bark of A. pavonina. The isolated compounds could be a potential template for the synthesis and development of new lead compounds with interesting pharmacological properties. The oil was found to be rich in neutral lipids and low in polar lipids. The neutral lipids consisted mainly of triaglycerols. This review summarizes the incredible phytochemical investigation, traditional uses and therapeutic potential of A. pavonina that needs to be explored, opening new perspectives for future researches.

Keywords: *Chalcone, Flavonoids, Alkaloids, Saponins, Steroids, Triterpenoids, CNS depressant, Anticephalgic, Red Bead Tree, Adenanthera pavonina.*

26.1 Introduction

Red Bead Tree (*Adenanthera pavonina*), a very fast-growing, medium-sized to large deciduous, perennial free-flowering tree of the legume family (Fabaceae). with somewhat irregularly rounded crown, is found growing in forests and forest edges, woodland, coastal areas, roadsides, dry open forest and disturbed areas, evergreen forests, seasonally dry forests, open savannahs and agricultural land. The tree is native to the Asian continent and mostly found in Africa, Pacific and Caribbean regions. It is also indigenous to India and Bangladesh particularly in the South-eastern region.

Overall a beautiful perennial, deciduous tree The ground seeds are used to treat boils and inflammation, as well as an industrial lubricant oil.

As a food source the young leaves can be cooked and eaten.

26.2 Phytochemical Constituents

Previous phytochemical investigation reported the presence of many bioactive compounds like robinetin, chalcone, butin and flavanolampelopsin, stigmasterol glucosides, oleanolic acid, echinocystic acid, and sapogenins from the leaves and seeds of the tree. The phytochemical screening revealed various secondary metabolites with diverse biological effects, including flavonoids, alkaloids, saponins, tannins, steroids, triterpenoids, polyphenols, anthraquinones, coumarins, glycosides and polysaccharides. Seven compounds *viz.* ethyl 3,3-dimethyl-13-hydroxytridecanoate (1), stigmasta-5,22-dien-3β-ol (2), tert.butyltridecanoate (3), 6-α-hydroxy stigmast-20(21)-en-3-one (4) of dichloromethane extract and 18-(22 , 32 -dihydroxyphenyl)nonadec-17-en-2-ol (5), 1-(N-propyl amino)-2-henecosanone (6), and stigmast–5(6), 20(21)-diene-3-one (7) were isolated from the barks of *A. pavonina*.

Of these compounds, 1, 4, 5, 6, and 7 appear new. The structures of these compounds were elucidated by spectroscopic techniques, mainly by NMR.Five new and two known compounds have been isolated and characterized from the bark of *A. pavonina*. The isolated compounds could be a potential template for the synthesis and development of new lead compounds with interesting pharmacological properties.The oil was found to be rich in neutral lipids and low in polar lipids. The neutral lipids consisted mainly of triaglycerols. Unsaturated fatty acids were found

Figure 26.1: (A and B) Pods and inflorescence; (C) Trunk of the tree; (D) Pods and seeds; (E and F) Seeds of the plant (*Adenanthera pavonina*).

as high as 71 per cent, while the percentage of saturated fatty acids was only 29 per cent. GC and GC/MS analyses revealed linoleic, oleic and lignocerotic acid to be predominant among all fatty acids in the *A. pavonina* oil, whereas stigmasterol was the major steroid identified.

The seeds of *A.pavonina* contained appreciable amounts of proteins, crude fat and minerals, comparable to commonly consumed staples. Total sugar was low while starch constitutes the major carbohydrates.The oil was found to be rich in neutral lipids and low in polar lipids. The neutral lipids consisted mainly of triaglycerols. Unsaturated fatty acids were found as high as 71 per cent, while the

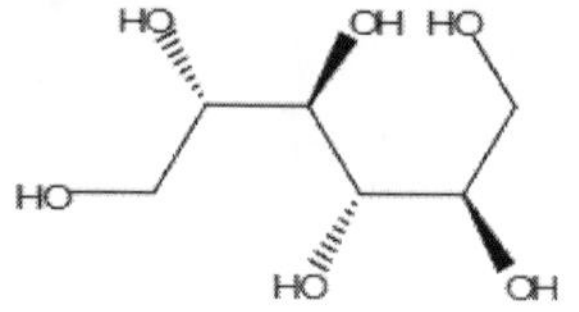

Echinocystic acid{ 5,10-dihydroxy-2,2,6a,6b,9,9,12a-heptamethyl-1,2,3,4,4a,5,6,6a,6b,7,8,8a,9,10,11,12,12a,12b,13,14b-icosahydropicene-4a-carboxylic acid }

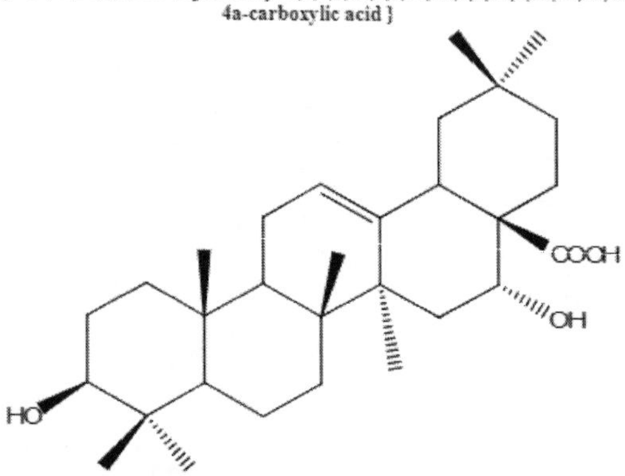

Eicosenoic acid{ (E)-icos-10-enoic acid }

Chemical constituent of *Adenanthera pavonina*

Chalcone

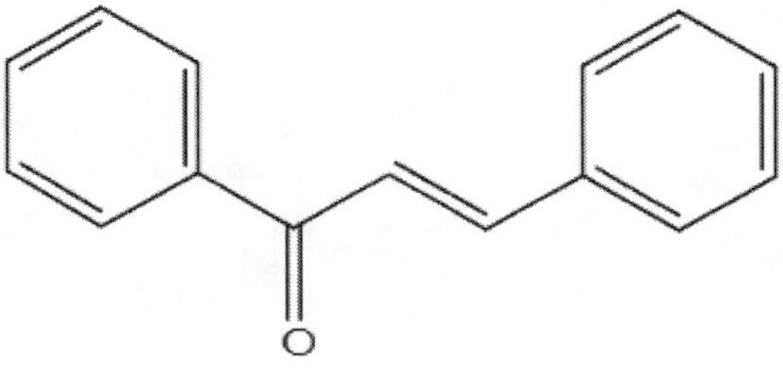

Arginine

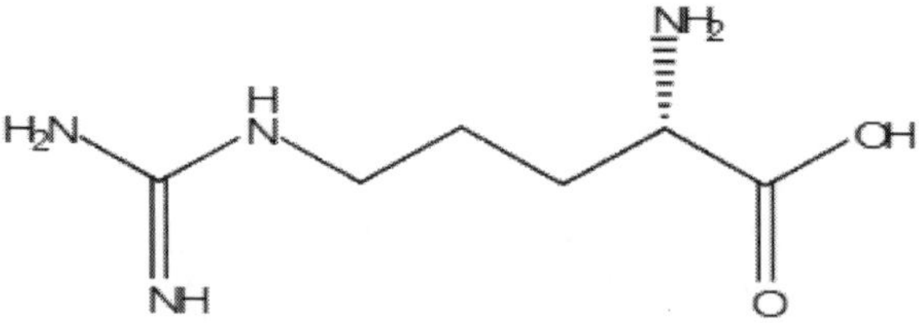

percentage of saturated fatty acids was only 29 per cent. GC and GC/MS analyses revealed linoleic, oleic and lignocerotic acid to be predominant among all fatty acids in the *A. pavonina* oil, whereas stigmasterol was the major steroid identified.

Several amino acids *viz.* arginine, cystine, histidine, isoleucine, leucine, lysine, methionine, phenylalanine, tyrosine and valine ¯methylene glutamic acid, ¯-methylene glutamine and traces of ¯-ethylendine glutamic acid have

Oleanolic acid {10-hydroxy-2,2,6a,6b,9,9,12a-heptamethyl-1,2,3,4,4a,5,6,6a,6b,7,8,8a,9,10,11,12,12a,12b,13,14b-icosahydropicene-4a-carboxylic acid }

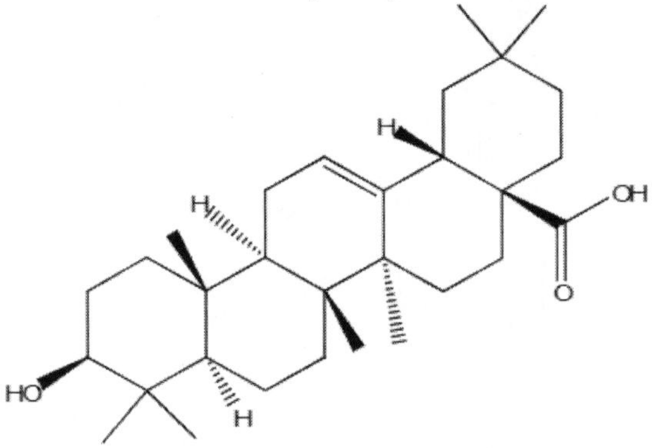

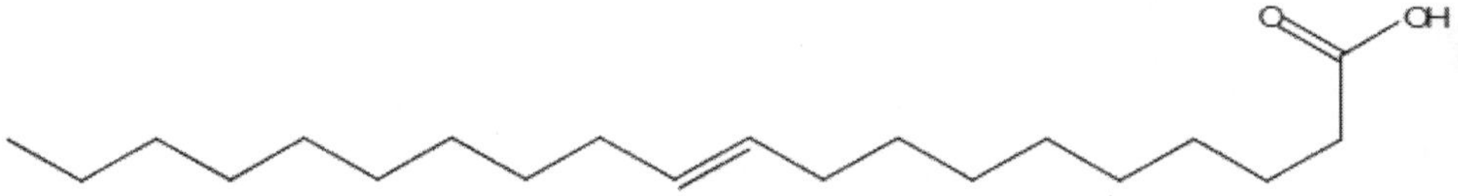

Glutamic acid{2-aminopentanedioic acid }

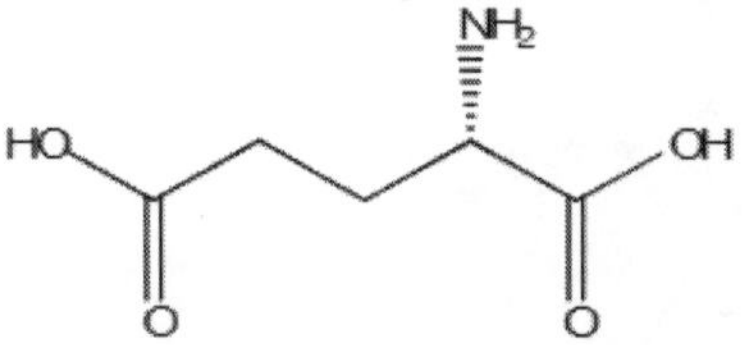

Histidine {2-amino-3-(1H-imidazol-4-yl)propanoic acid}

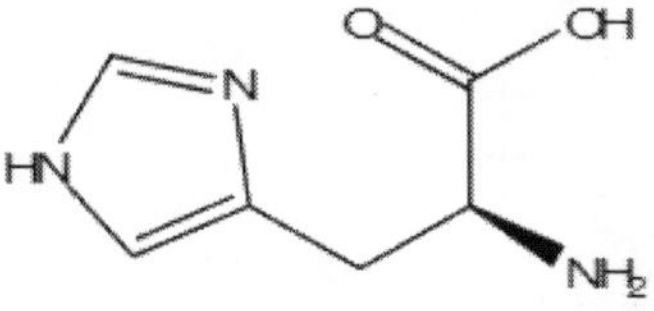

Isoleucine {2-amino-3-methylpentanoic acid}

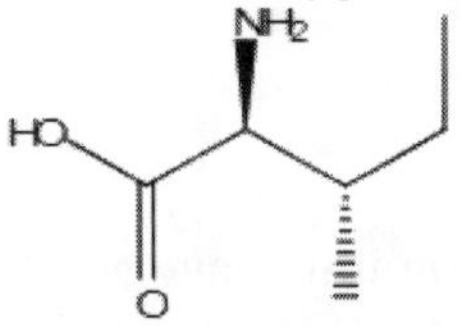

been investigated in seed of this tree. The kernels contain pale yellow fat. The fatty acid presents are palmitic, stearic, arachidonic, lignoceric, eicosenoic. The kernels also contain stigmasterol and its glycoside, dulcitol and a polysaccharide. Oleanolic and echinocystic acid.Octacosanol, dulcitol, glucosides of β-sitosterol and stigmasterol. From the chloroform extract of dried powdered leaves of A. pavonina the hydrocarbon nonacosane and hentriacontane, the triterpenoid squalene, and

Butein {1-(2,4-dihydroxyphenyl)-3-(3,4-dihydroxyphenyl)prop-2-en-1-one}

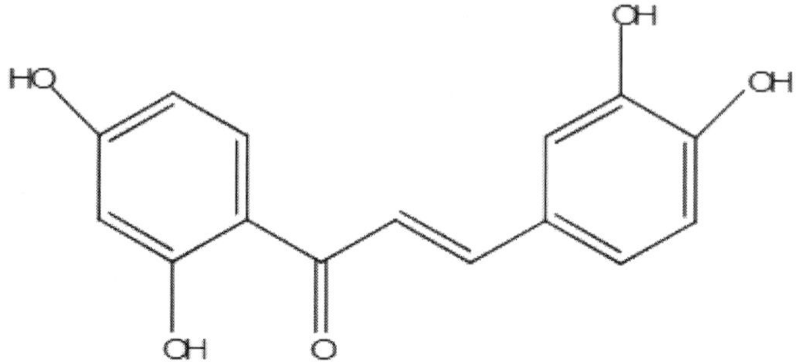

Cysteine

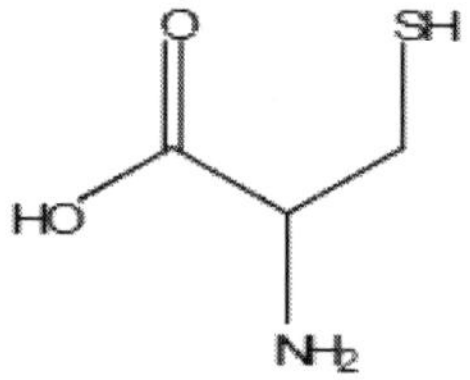

Arachidonic acid

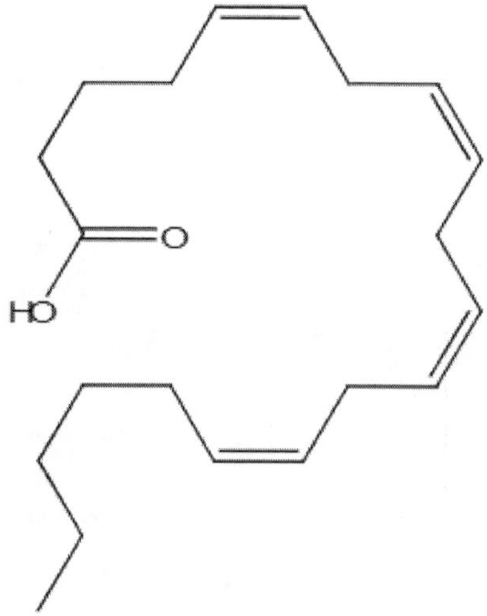

the long chain fattyacid ester palmitate have been isolated. The methanolic extract yielded β-sitosterol, β-sitosterol-3β -D- glucoside.

26.3 Pharmacological Attributes

Experimental studies of this tree have shown reported to demonstrate numerous pharmacological activities such as antidiarrheal, anti-inflammatory, antinociceptive, antioxidant, anticancer, anti-inflammatory, analgesic activities, antihypertensive effect, antifungal, hepatoprotective, renal protective, CNS depressant, anticonvulsant, hyperlipidemic, antibacterial and BP-reducing activities of the bark, leaf, and seed extracts and their isolated compounds have been investigated. Seeds are anticephalgic and also used for the treatment of paralysis and blood pressure. *A. pavonina* constituted a number of phytochemicals, which disclose its uses for various therapeutic purposes. Pavononin and various flavonoids can be used for the treatment of various health illnesses. However attempts are required to evaluate the mechanism of actions with therapeutic activity for *A. pavonina*. In general, the literature reports its empirical use for diabetes, lipid disorders, diarrhea, ulcers, stomach bleeding, hematuria, rheumatism, asthma, hypertension, pulmonary infections and chronic ophthalmia. These reports of traditional use were most common in Asian countries widespread use of specific parts of the tree, bark and/or leaves derivate are used as anthelmintic, in colonorrhea, ulcers, pharyngoplasty, gout, rheumatism and different painful conditions, heartwood is beneficial in dysentery and haemorrhages. Seed extract is used for the treatment of boils, inflammations, blood disorders, arthritis, rheumatism, cholera, paralysis, epilepsy, convulsion, spasm, indigestion, gout, burning sensation, hyperdipsia, vomiting, fever and giddiness. Seed powder is used as a poultice to induce abscess suppuration.

26.4 Traditional Uses and Benefits of Red Bead Tree

Different parts of this tree have been used in traditional medicine for the treatment of various diseases. The bark and leaves are used as a remedy for diarrhea, gout, hematuria, hematemesis, and chronic rheumatism. A red powder made from the wood is also used as an antiseptic paste. In Ancient Indian medicine, the ground seeds are used to treat boils and inflammations. A decoction of the leaves is used to treat gout and rheumatism. The bark was used to wash hair. Their medicinal properties are due to the presence of flavonoids, glycosides, saponins and steroids. Decoction of the leaves is used in the treatment of rheumatism and gout. The bark is used to treat leprosy. Decoction of the bark and leaves is used to treat dysentery, diarrhea and tonsillitis. Pulverized wood, mixed with water, is taken orally for treating migraines and headaches. Red bead tree leaf extract has antibacterial activity against the intestinal pathogen *Campylobacter jejuni*.In India, Malaysia and Indonesia, this species is used in traditional medicine against rheumatism, migraines, headaches and dysentery. Seeds of this tree have been found to be effective in treating cardiovascular diseases during pregnancy. Red powder made from the wood is also used as an antiseptic paste. In Ancient Indian medicine, seeds are made into a paste with water and applied externally to cure boils and swellings. The ground seeds are used to treat inflammation. Decoction of leaves used for chronic rheumatism and gout; also useful for bowel hemorrhage and hematuria. Powdered seeds are used externally to hasten suppuration of boils. Powdered seeds are also used for headaches. Seeds are used for cholera and also used for general paralysis. Bark is used in rheumatism, hematuria, ulcers, and gonorrhea. Powdered seeds are also used for treatment of abscesses.

26.5 Ayurvedic Health Advantages of Red Bead Tree

Antiseptic

Sprinkle Red Sandalwood powder on the damaged Skin.; Boils: Grind Seeds of Red Sandalwood to make a powder. Sprinkle it on the Boils twice a day.; Sprinkle powdered seeds of Red Sandalwood on the inflamed areas.; Prepare a decoction of Red Sandalwood leaves. Drink once a day.; Boil leaves of Red Sandalwood in a cup of water. Drink once a day.; Make an infusion of Red Sandalwood bark. Wash hair

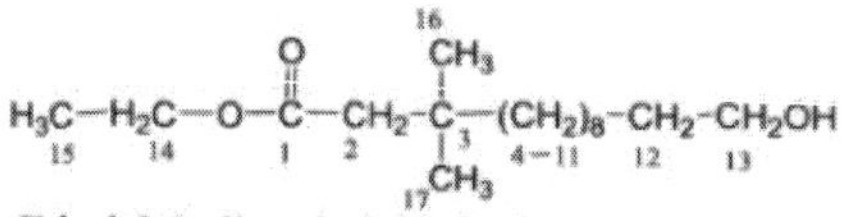

Ethyl 3,3-dimethyl-13-hydroxy tridecanoate (**1**)

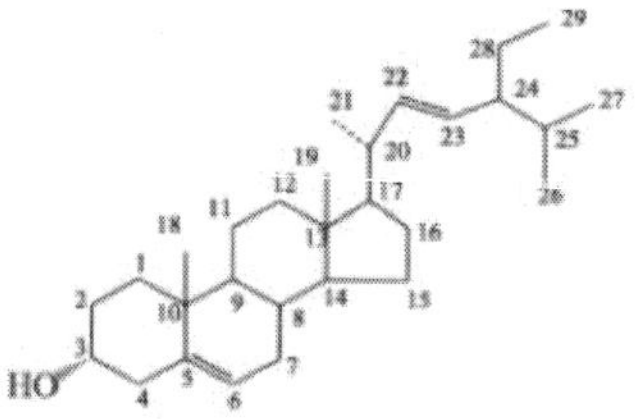

Stigmasta-5,22-dien-3 -ol (**2**)

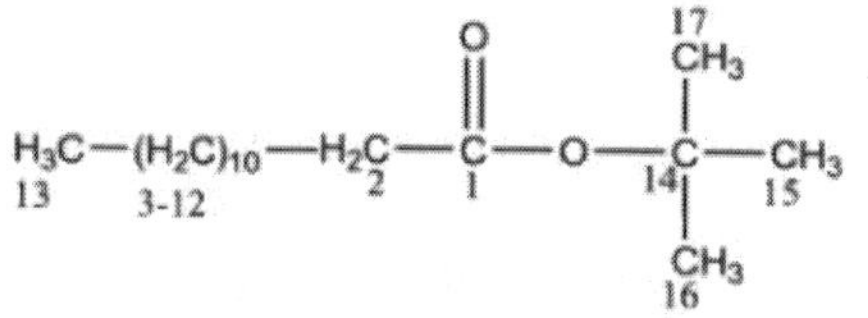

Tert butyl tridecanoate (**3**)

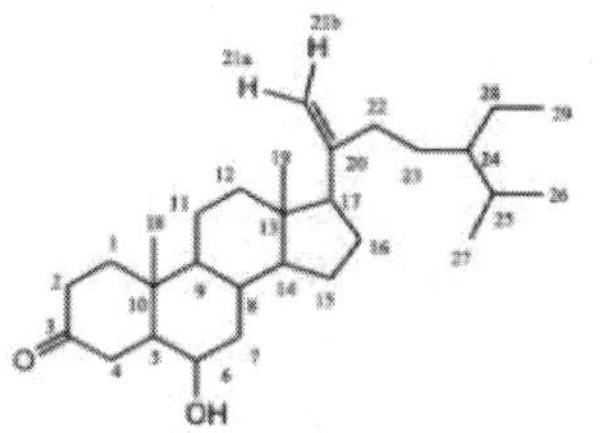

6- α -hydroxy stigmast-20(21)-en-3-one (**4**)

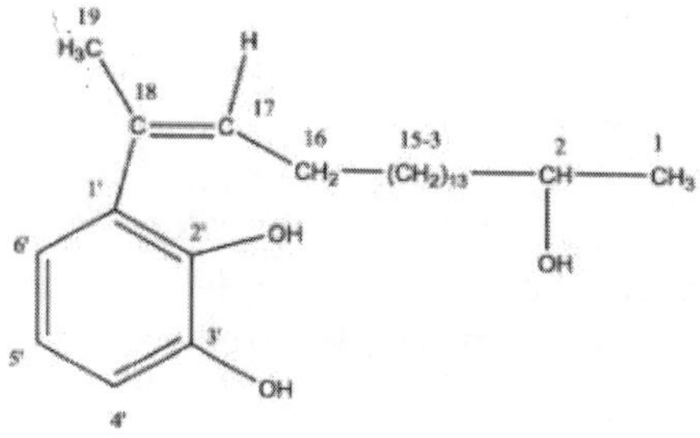

18-(2' ,3' -Dihydroxyphenyl)nonadec-17-en-2-ol (**5**)

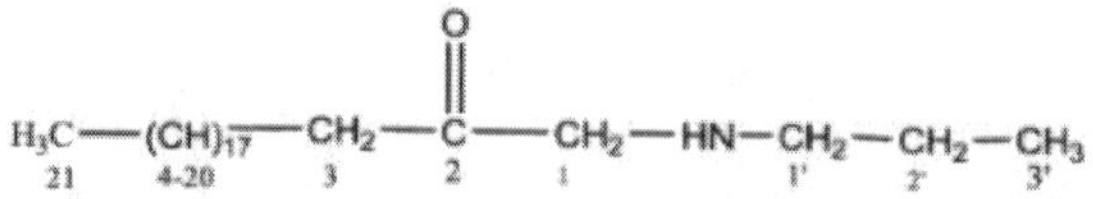

1-(N-propyl amino)-2-henecosanone (**6**)

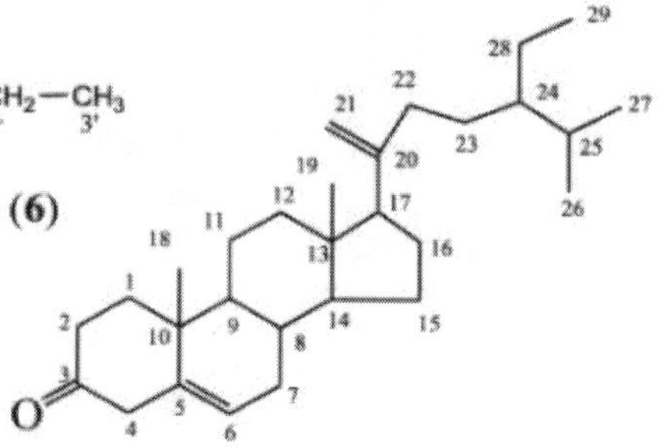

Sigmast-5(6), 20(21)-diene-3-one (**7**)

with it.; Prepare a mixture of Red Sandalwood, Glycerine and Rose water. Apply it on your face for 20 minutes. Wash after it.; Take equal amount of powdered Indian madder, Red sandal wood, Chonemorpha Fragrans. Add adequate amount of Cow's ghee. Boil for 15-20 Mins. Apply 3-4 times a day; (i)Take equal quantity of Saffron, Turmeric, Neem, Red Sandalwood, Basil and Fuller's Earth. Mix them in Butter Milk or Milk. Apply on the affected part twice a day.; (ii) Take 1 teaspoon Neem, 1 teaspoon Red Sandalwood and 1 teaspoon Yogurt. Mix all ingredients. Apply on the face for 20 minutes. Rinse with cold water.; (iii) Take 1 teaspoon Powdered Indian Madder, 1 teaspoon Red Sandalwood, 1/4 teaspoon Turmeric and 1 teaspoon Snow Lotus. Mix all ingredients. Add water to make a paste. Apply on affected parts for 20 minutes. Wash with cold water; Take equal quantity of Indian Madder, Red Sandalwood and Snow Lotus. Grind them together. Add 1/4 teaspoon of Turmeric. Make a paste by adding water. Apply on the affected part until it dries. Wash with normal water.; Take one teaspoon powder of each Indian Madder, Red Sandalwood and Snow Lotus. Add half teaspoon of Turmeric. Make a paste by adding water. Apply for 20 minutes and wash.

26.6 Culinary Uses

Young leaves are cooked, eaten and considered as a Famine food. Seed can be roasted, shelled and then eaten with rice. Bright red seeds taste like soya beans and contain 25 per cent oil plus 39 per cent protein. Flowers are also eaten. Roasted or boiled seeds eaten with rice in Java.

26.7 Conclusion

There has been a growing interest in products derived from higher trees around the world. Besides being used in modern therapies, these tree derivatives are relevant for the synthesis of more complex molecules. *A. pavonina* is a tree that presents a complex and chemically varied group of compounds. Preliminary phytochemical studies on this tree revealed the presence of various secondary metabolites including mainly alkaloids, flavonoids, glycosides, carbohydrates, saponins, steroids, tannins, terpenoids and proteins. Saponin having glucose as the sugar moiety and oleanic acid and echinocystic acids as sapogenin. Pavonin: A new five-membered lactone named pavonin with an exo-cyclic double bond has been isolated from the methanol soluble part of *Adenanthera pavonina*.

REFERENCES

1. Olajide OA., Echianu CA., Adedapo ADA., Makinde JM. (2004). Anti-inflammatory studies on *Adenantherapavonina* seed extract. *Inflammopharmacol.,* 12(2): 196–202.

2. Ara A., Arifuzzaman M., Ghosh CK., Hashem MA., Ahmad MU., Bachar SC., Nahar L., Sarker SD. (2010a). Anti-inflammatory activity of *Adenantherapavonina* L., Fabaceae, in experimental animals. *Rev. Bras.deFarmacog.,*20(6): 929–932.

3. Ara A., Saleh-e-In MM., Ahmed NU., Ahmed MU, Hashem MA., Bachar SC. (2010b.) Phytochemical screening, analgesic, antimicrobial and antioxidant activities of bark extracts of *Adenantherapavonina* L. (*Fabaceae*). *Adv. Na.t Appl. Sci.,* 4(3): 352–360.

4. Ara A., Saleh-e-In MM., Ahmed NU., Hashem MA., Bachar SC. (2013). Anti-diarrheal, acute toxicity activity of methanolic bark extract and elemental composition of*Adenantherapavonina* L. (*Fabaceae*). *Turk. J. Pharma. Sci.,*10(2): 263–272.

5. Yadev N., Misra G., Nigam SK. (1976). Triterpenoids of *Adenantherapavonina* bark. *Treea. Med.,* 29(2): 176–178.

6. YeohHH., Wee YC., Watson L. (1984). Systematic variation of leaf amino acid compositions of leguminous trees. *Phytochem.,* 23(10): 2227–2229.

7. Chandra SM., Saxena VH. (1982). Triterpenoids of*Adenantherapavonina* root. *Int. J. Crude Drug Res.,* 20: 165–167.

8. Mesbah UA., Rahman MA., Tabassum R., Nahar K. (2002). Chemical constituents of the leaves of *Adenantherapavonina* L. *J. Bang. Chem. Soc.,*15(2): 194–199.

9. Shaiq AM., Ahmed F., Azhar I., Pervez MK. (2005).Pavinin: a new five membered lactone from *Adenantherapavonina* L. (Mimoaceae). *Nat. Prod. Res.,* 19(1): 37–40.

10. Enuo Y.; Shishan PY. (2007). Studies on chemical constituents from stems and leaves of Adenantherapavonina L. *ZhongguoZhongyaoZazhi,* 32(20): 2135–2138.

11. Adeyemi OA., Adedapo AD., AdepapoAA., Moody JO. (2015). Evaluation of the antimicrobial activity of crude extracts and chromatographic fractions of Adenantherapavonina Linn (Leguminosae) seeds.*Afri. J. Biotechnol.,*14(12): 1067-1073.

12. Afolabi IS., NwachukwuIC.,Ezeoke CS., Woke RC., Adegbite OA., Olawole TD., Martins OC. (2018). Production of a new Tree-Based Milk from Adenantherapavonina seed and evaluation of its nutritional and health Benefits. *Fro. Nut.,* 5: 1-13.

13. Ali MS., Ahmed F., Azhar I., Pervez MK. (2005). Pavonin: a new fivemembered lactone from Adenantherapavonina Linn. (Mimoaceae).*Nat. Prod. Res.,*9: 37-40.

14. Aquino LRC.,Macêdo AAM., Graça MPF., Valente MA., Silva C. (2017). Preparation and characterization of cement-based hydroxyapatite and galactomannan extracted from Adenantherapavonina L. seeds. Revista*Latino. de Metal. Mat.,* 37(1): 102-110.

15. Ara A., Msaleh-E-In MM., Ahmed NU., Ahmed M., AHashem MA, Bachar SC. (2010a). Phytochemical Screening, Analgesic, Antimicrobial and Antioxidant Activities of Bark Extracts of Adenantherapavonina L. (Fabaceae).*Adv. Appl. Sci. Res.,* 4: 352-360.

16. Ara A., Arifuzzaman M., Ghosh CK., Hashem MA., Ahmad MU., Bachar SC., Nahar L., Sarker SD. (2010b). Anti-inflammatory activity of Adenantherapavonina L., Fabaceae, in experimental animals. *Braz. J. Pharmacog.,* 20: 929-932.

17. Chourasia OP., Rao JT. (2005). Chemical and antimicrobial examination of the fixed oil from the seeds of *Adenanthera pavonina* Linn.*Asia. J. Chem.,* 17(1): 289-292.

18. Dash S., Das C., Sahoo DC.(2010). Phytochemical and anthelmintic screening of crude bark extract of Adenantherapavonina Linn.PharmacieGlobale. *Int. J. Compreh. Pharma.,* 1: 1-4.

19. Das S., Dash S., Sahoo AC., Giri RK., Sahoo DC., Guru P. (2011). Antihyperlipidemic activity of Adenantherapavonina Linn.ethanolic bark extract fractions. *Nat. Pharmaceut. Technol.,*1: 1-4.

20. Geronço MS., Melo RC., BarrosHLM., Aquino SR., Fátima de Cássia de Oliveira E., Islam MT., do Ó Pessoa C., dos Santos Rizzo M., da Costa PM.(2020).Advances in the research of Adenantherapavonina: From traditional use to intellectual property. *J. Med trees Res.,* V14(1): 24-53, doi: 10.5897/JMPR2019.6819

21. Mujahid M., Ansari VA., Sirbaiya AK., Kumar R., Usmani A. (2016). An insight of pharmacognostic and phytopharmacology study of Adenantherapavonina. *J. Chem. Pharmaceut. Res.,* 8(2): 586-596.

22. Zarnowskia R., Jarominb A., Certikc M., Fontaine TCJ., Jakubike T., Iqbal MCM., GrandmouginAF, Arkadiusz., Stanislaw JP. (2004). The Oil of Adenantherapavonina L. Seeds and its Emulsions. *Zeitschrif.Für. Naturforschung.,*59: 321-325.

Chapter 27

Versatile Uses and Therapeutic Properties of a Miraculous Tree: Drumstick (*Moringa oleifera* Lam.)

Popularly known as the "drumstick tree" in India, moringa was named the 'plant of the year' in 2008 by the "National Institute of Health". Researchers found that every part of this deciduous tree starting from its leaves, fruits, seeds, pod, roots, flowers to bark have benefited man since ages in some way or the other which led the ancient Romans, Greeks and Egyptians highly value it mainly for its medicinal benefits. Moringa oleifera is the most well known species of moringa genus that belongs to the Moringaceae family. It is native to India and is widely cultivated in the subtropical and tropical areas of Asia, South America and Africa. It has whitish-gray thick bark with fragile, droopy branches and large green oval leaflets. It can grow up to 10-12 cm in height with a diameter of around 45cm. Due to its high nutritious value, this plant helps in the general human well being and hence is aptly called "The Miracle Tree" in many Asian, African and Caribbean countries.

Keywords: *Therapeutic uses, Drumstick, Moringa oleifera, Nutritious.*

27.1 Introduction

Undoubtedly, the health benefits of moringa or drumstick make it the new superfood, which you can vouch on. It is good for our heart, liver, lung, kidney, eye, stomach and can also fight several diseases like cancer, diabetes, anemia *etc.*

Hence, it is mandatory to include it in our diet chart. Often seen floating in our bowl of *sambhar,* Drumstick is scientifically known as *Moringa oleifera Lam*. It belongs to the family *Moringaceae*. It is a small, fast-growing, evergreen tree that grows in tropical regions. It is found in the Himalayan areas of India, Bangladesh, Afghanistan and Pakistan. Drumstick is also known as horseradish tree or drumstick tree (in English), Subhanjana (in Sanskrit), Haritashaaka or Akshiva (in Ayurveda) and Sainjna or Saguna (in Hindi). Every part of the drumstick might contain essential properties for human health; hence, it may be one of the crucial nutritional herbs. For many years, the drumstick has been used as traditional medicine. According to the Ayurvedic system of medicine, drumsticks may have potential use and be helpful for various diseases due to their high nutritional value, water retention and capacity for purification.

Figure 27.1. A Tree of Drumstick Showing Leaves and Fruits.

All the parts of the moringa plant are good for humans in some way or the other and hence you can incorporate it in your food regime in several ways (Anwar *et al.,* 2007; Berglund 2018). You can consume it in the form of tea, or use it in a shake or smoothie or can use it in various recipes (Mishra *et al.,* 2012; Mane 2024). But, despite its good effects, you must never ignore the precautions. As you know nothing in excess is good, so try to eat them in moderate amounts and if you are pregnant, please consume only on doctor's advice. High levels of blood sugar is very dangerous and may cause various health complications. It may affect your heart and the other important organs of the body. Therefore, maintaining blood sugar levels is mandatory and with moringa, you can put your blood sugar levels easily at check. The chlorogenic acid content in moringa has been proved to regulate the blood sugar levels Until then, stay healthy, stay safe.

27.2 Phytochemicals in *Moringa*

Drumstick provides large and rare varieties of vital minerals and nutrients. Pods, leaves, flowers, barks, roots and seeds of drumsticks also contain bioactive compounds.

Table 27.1. Nutritional Value of Raw Drumstick Pods per 100 grams

Nutritional Components	*Value/100 g*	*Nutritional Components*	*Value/100 g*
Energy	37 kcal	Protein	2.1 g
Fat	0.2 g	Carbohydrate	8.53g
Fibre	3.2 g	Calcium	30 mg
Iron	0.36 mg	Magnesium	45 mg
Phosphorus	50 mg	Potassium	461 mg
Sodium	42 mg	Zinc	0.45 mg
Copper	0.084 mg	Manganese	0.259 mg
Selenium	0.7 µg	Vitamin C	141 mg
Thiamine	0.053 mg	Riboflavin	0.074 mg
Vitamin B6	0.12 mg	Folate	44 µg
Vitamin A	4 µg		

Source: US Department of Agriculture. Drumstick pods, raw; 2019 Jan 4. https://fdc.nal.usda.gov/fdc-app.html#/food-details/170483/nutrients.

27.3 Potential Uses of Drumstick

Drumstick can be used in the following ways (Moyo *et al.*, 2011): The entire plant might have potential use for muscle spasms or muscle cramps. It might also act as a cardiac and circulatory tonic and may acts as an anti-septic and a diuretic. Juice of leaf of drumstick might help with hiccoughs (contractions of the diaphragm). Leaves of drumstick might be helpful for influenza (viral infection of the respiratory system). Juice of leaf of drumstick might help with hiccoughs (contractions of the diaphragm). Leaves contain antioxidants (Chumark 2008). The dried root bark of drumstick might help with piles, goitre (swelling of the thyroid gland) and internal abscess (formation of pus) (Mishra *et al.*, 2011). The bark of the stem and flowers of the drumstick might be helpful to lower glucose levels in the blood. Charak Samhita mentioned use of *Moringa bark* when poured in ear releaves from otorrhoea or ear problems (Saxena 2023 p.303). Gum obtained from moringa also helps in remedy of earache. Flowers of drumstick might stimulate gallbladder contraction and increase the flow of bile. Pods of drumstick might be helpful for high fever. The seeds of drumsticks are acidic and contain many stimulants that might be helpful for the central nervous system (CNS).

27.3.1 Use in Diabetes

The leaf extract of drumstick might show potential anti-diabetic properties that might help lower blood glucose levels. An animal study (Gupta *et al.,* 2012) showed that the extract of leaves of drumsticks might help reduce the progression of diabetes and may induce a reduction in serum glucose levels by increasing the production of protein and insulin hormone (Paikra *et al.,* 2017). Another animal study (Ndong *et al.,* 2007) showed that the extract of drumstick leaves might help control glucose levels, urinary sugar and protein levels, haemoglobin and total protein in the blood (Mishra *et al.,* 2011). High levels of blood sugar is very dangerous and may cause various health complications. The chlorogenic acid content in moringa has been proved to regulate the blood sugar levels. Another study on 30 women also proved that by consuming moringa leaf powder for 3 months, there was significant reduction in blood sugar levels by 13.5 per cent. Therefore, moringa became a natural cure for diabetes also, which occurs due to imbalance in blood sugar levels, and there were several studies to back the fact. *"International Journal of Food Science Technology"* also confirmed that diabetic patients when consumed moringa as a part of high carbohydrate diet showed positive results on insulin and blood glucose levels. In fact, not just blood sugar, the immunoglobulin and glycosylated hemoglobin, the other two markers of diabetes also lessen with moringa intake.

27.3.2 Uses in Kidney Diseases

Barks, leaves, seeds, flowers and roots of drumsticks might have diuretic activity that may be helpful for urine formation in patients with kidney dysfunction. It might also help to lower the deposition of oxalate salts (stones forming units) in the kidneys. An animal study (Kamal 2008) found that the extract of drumstick root lowers the kidney retention of salts and excretion in the urine. Moreover, these extracts also reduce elevated serum uric acid and creatinine levels (Mishra *et al.,* 2011). This information is insufficient as these studies were done on animals. However, more studies on humans are required to back this claim. Kidney diseases are serious and must be diagnosed and treated by a doctor, so kindly consult a doctor. We recommend do not use drumsticks to self-medicate yourselves before taking medical advice. Kidney failure or kidney problems can lead to many health problems, so the early you protect your kidney, the better and moringa will help you in it. Studies show that you can prevent nephrotoxicity, a kidney problem that mainly occurs due to the toxin or drug exposure with proper intake of moringa. Besides, the presence of antioxidants in moringa extracts gives protection against the renal injuries and also intensifies the recovery process in body. It is an ideal bio-absorbent that helps your body to wash off the harmful toxins and heavy metals that are bad for your kidneys.

Charak Samhita mentioned use of *Moringa* flowers when mixed with mishri in form of sweet drink helps in better release of urine (Saxena 2023 p.469).

27.3.3 Uses in Digestive System

Moringa boosts your digestive health and since ancient times, it is considered the perfect remedy for treating stomach disorders as well as other digestive issues. Due to the presence of isothiocyanates in it, it has the ability to treat several abdominal issues like the problem of gastritis, constipation, diarrhea and ulcerative colitis. The anti-inflammatory properties of moringa enable it to fight various germs and make it the perfect detoxifying agent that flushes out the harmful metal toxins from your body. It inhibits the growth of disease causing pathogens and cures various bacterial, yeast or fungal infections. The Vitamin B present in its leaves further aids in the digestive process and enables smooth conversion of food into energy instead of storing them as fats.

27.3.4 Uses in Protection of Liver

Moringa extract can protect the health of liver. Various studies have proved that moringa is highly beneficial and often recommended in case of the liver damages that are caused by the anti-tubular drugs. Phytochemicals like catechin, ferulic acid, epicatechin and vitamin C present in moringa helps your body speed up the recovery process. They facilitate the process of restoration of glutathione levels in your body as well as prevent hepatic lipid peroxidation induced by radiation.

27.3.5 Uses in Heart Related Problems

Heart is the most important organ in your body and with moringa, you can protect it and prevent the occurrence of several cardiovascular diseases. The various antioxidants present in it helps to check the risk of heart attacks and strokes that has taken innumerable lives in the recent past. Moringa helps to inhibit the lipid peroxidation to increase in the myocardial tissues and hence keeps our heart healthy. With adequate moringa intake, you can control high blood pressure, high cholesterol and hypertension, those that put your heart at risk largely.

- ☆ **Reduces cholesterol levels**– In order to keep your heart healthy, the cholesterol levels in body must be regulated. Moringa helps to lower the high levels of cholesterol in body and hence reduce the risk of strokes and heart attacks.
- ☆ **Controls your high blood pressure levels**– Having high blood pressure also puts your heart at risk but with moringa you may keep the pressure levels under control. The powerful antioxidant quercetin present in moringa reduces the levels of blood pressure in human body.
- ☆ **Reduces hypertension**– Excessive tension can affect your heart severely but moringa helps you to control it. The presence of bioactive components like niaziminin and isothiocyanate prevents the arteries from thickening and inhibits the development of pulmonary hypertension.

27.3.6 Uses in Strengthening Autoimmunity

It has number of health benefits with strengthening your immune system being one of them. The ethanolic content present in the leaves of moringa enhances the immune response of the cells. It also facilitates the functioning of other immune system values like antibodies, leucocyte count and hypersensitivity reaction. Moringa being a perfect antimicrobial agent prevents the growth of the disease causing microbes. They also give protection against water-borne microorganisms and foodborne microorganisms as well. All these together boost the immune system and help to lead a healthy, diseases free life.

27.3.7 Uses in Prevention of Cancer

Studies show that the powder from the Moringa leaves register a higher ORAC (Oxygen Radical Absorbance Capacity) score than many any other foods. With an ORAC score of 157,000, *M. oleifera* tops other antioxidant superfoods, including Acaci berries, green tea, blueberries, dark chocolate, garlic, goji berries, pomegranates, *etc.* Herbal medicines have a vital role in the prevention and treatment of cancer and medicinal herbs are commonly available and comparatively economical. The *M. oleifera* extract shows high antioxidant activity, potent cancer cell anti-proliferation, and induction of quinonereductase (Chumark 2008).

This deadly disease is currently topping the list of the life threatening diseases that has been the cause of several deaths all over the world. However, researchers have found out that the moringa extracts are beneficial to prevent and cure cancer, which is one of the most important health benefits of moringa. The phenolic compounds like kaempferol and quercetin present in moringa possess chemo-preventive properties and hence moringa is widely used in tumor therapy. Niazimicin, the bioactive compound present in it has also been proved to restrain the growth and the development of cancer cells. In addition, the chemo-modulatory effects of moringa make it a perfect cure for various types of cancers like ovarian cancer, skin papillomagenesis and hepatic carcinogenesis. *Moringa* inhibits the rapid growth of malignant cancer cells and hence is safe for the cancer patients. It is also a cure for breast and colorectal cancer as well.

27.3.8 Uses in Eye Cure

Moringa oleifera is good for the eyes. Several studies proved that moringa has retinoprotective ability and can aid your body to cure any retinal damage. Moringa prevents the retinal vessels to dilate, thereby facilitates the retinal vasculature to remain intact. Besides, moringa prevents the capillary membranes to thicken which in turn lessens the risk of retinal dysfunction. You can treat your problem of cataract with moringa as well.

27.3.9 Uses in Brain Functioning

Adequate intake of moringa is beneficial for better cognitive function. It is good for better retention of memory and helps in alteration of brain monoamines such as norepinephrine, dopamine and serotonin. Besides, it provides protection against Alzheimer's disease and its monoaminergic deficiencies. In addition, moringa being rich in protein and amino acid tryptopohan produces serotonin hormone that in turn helps to control your mood swings, maintain energy levels, and help fight depression, fatigue and insomnia.

27.3.10 Uses in Treatment of Asthma

People who suffer from asthma or has asthma tendencies may take help of moringa to ease their problem. Studies have shown that it cures any kind of inflammation in the airways and also relieves you of bronchial asthma. Many treatments of asthma involve moringa as it reduces the frequency of asthmatic attacks and its other symptoms like wheezing, dyspne, cough, and chest contraction that accompany it. Various studies have proved moringa's effectiveness in treating several allergic diseases like anaphylaxis and allergic rhinitis. It also gives you protection from bronchial constrictions as well as aids in the proper functioning of your lung and respiratory system.

27.3.11 Certain other Uses

The presence of calcium and phosphorous present in it contribute to bone and teeth health and prevents them from breaking fast. Also, due to the anti inflammatory property of moringa, you can treat the problem of arthritis with it as well. Besides, various bone related problems like jawbone fracture or mandibular fracture can be treated with moringa extracts. Last but not the least, the magnesium content also keep the teeth strong and prevent its falling early. Monera *et al.* (2017) found that leaf extract of *M. oleifera* inhibits 6Î^2-hydroxylation of testosterone by CYP3A4. Mahajan and Mehta (2009) found anti-arthritic activity of hydroalcoholic extract of flowers of *M. oleifera* in Wistar rats.

Moringa extracts helps to prevent the formation of these stones. Various studies have shown that the moringa roots possess certain anti-urolithiatic property that can reduce the urinary oxalate levels in body. Moringa extracts also inhibits the development of stone deposits and regulates the synthesis of endogenous oxalate. Researches have shown that moringa promotes the absorption of the iron by your body, which in turn helps to increase the levels of red blood cells and maintain the overall blood parameters in your system. Moringa is good for all the pregnant mothers as well. The nutrients present in moringa have the potential to increase the stamina and give a lot of energy to the would-be mothers. Moringa also provide them with the perfect prenatal nutrition mainly the supply of iron and calcium that is highly beneficial during pregnancy. Moreover, it helps the lactation or breast milk production in mothers.

Moringa spp. have excellent wound healing properties. it has the ability to reduce the clotting time. In other words, it lessens the time taken for the wounds to stop bleeding. Besides, it fades the marks of the wounds and the scars quickly leaving behind clear glowing skin. Also, it increases the skin- breaking strength and reduces the scar area. Moringa extracts can treat the problem of edema in which there is swelling due to the fluids getting trapped in the tissues. It is believed to cure it and according to studies, it is the anti-inflammatory property of the moringa roots, which inhibit the development of edema.

27.4 Uses of Drumstick

Moringa oleifera can be used in various ways to treat several health issues. Undoubtedly, the health benefits of moringa or drumstick make it the new superfood,

- Consume moringa in the form of tea. Steep the dried moringa leaves in hot water. Strain it and then drink it. Try to avoid boiling it in order to retain the maximum nutrients in it. The moringa tea is generally light and has an earthy taste.
- You may also add moringa powder to your smoothie and then drink it. This is one of the healthiest smoothies that taste good as well as is highly beneficial for health.
- You may also pop moringa seeds like popcorn with some butter. You can cook the seeds like peas and consume them along with other meals.
- The leaves can be eaten as salads or may be added to pasta, rice or some other dish. Moringa leaf powder may be sprinkled on soups or added to any other foods while cooking. However, try to add them at the end of cooking or just before you eat them in order to get the maximum nutrients.
- The moringa pods may be boiled, fried or steamed. You may consume them in any way just like green beans.
- Lastly, there are various moringa or drumstick recipes. You may include them in your diet chart in the way you want.

Figure 27.2. Use of the Seeds in Water Treatment.

27.5 Use of the Seeds in Water Treatment

Researchers from all around the world agree that *Moringa* seeds can give a more efficient purification process than conventional synthetic materials in use today. In many countries it is still difficult to obtain clean water. But not only Third World countries can benefit from Moring seeds as a water purifier. Even developed countries can benefit from a process that treats waste water without addition of further synthetic chemicals. Results from studies show that the clusters of material (flocs) that are produced with the protein are much more tightly packed than those formed with conventional flocculating agents. This is better for water purification as such flocs are more easily separated. Crushed Moringa seeds clarify and purify water to suit domestic use and lower the bacterial concentration in the water making it safe for drinking. Water purification: The study on the *Moringa oleifera* as a natural gift point out that, A billion people across world like; Latin America Africa, and Asia are assessed to depend on untreated water sources for their daily needs. Kumar and Sheeja (2013) used seeds of this plant for removal of heavy metals from waste water

Numerous research reports shown that, *Moringa* seed powdered can be used cleaning dirty water in simple and quick method simple. The *Moringa* seed powder joins with the solids in the dirty water and sinks to the bottom. This action can remove bacteria contained in water up to 90-99 per cent. Rather than using $Al_2(SO_4)_3$, which are dangerous to people and the environment, using *Moringa* seed powder to purify water is cheap. Water can be purified by adding 2 g of *Moringa* seed powder to 20 l into a bottle and shake for 5 min. Dirty water that is to be treated can be filtered through a clean cloth into the container. Until the water becomes clear and the impurities have sunk to the bottom leave the bucket undisturbed for one hour then filter the water through a clean cloth boil the water before drinking.

27.6 Conclusion

Drumsticks might have a wide range of uses with high nutritional value. Various parts of this plant act as stimulants that might possess various activities employed to help with different diseases in the indigenous systems of medicine, particularly in South Asia. Some of the potential uses of drumstick are described as follows, which you can vouch on. It is good for your heart, liver, lung, kidney, eye, stomach and can also fight several diseases like cancer, diabetes, anemia *etc.* All the parts of the moringa plant are good for humans in some way or the other and hence you can incorporate it in your food regime in several ways. You can consume it in the form of tea, or use it in a shake or smoothie or can use it in various recipes. But, despite its good effects, you must never ignore the precautions. As you know nothing in excess is good, so try to eat them in moderate amounts and if you are pregnant, please consume only on doctor's advice. But make sure you do not overdo it as moringa if eaten in large amounts can have laxative effects and may lead to stomach upset, gaseous distension, heartburn and diarrhea. As far as seeds are concerned, they has a certain level of toxicity in it, so it is best to eat only a handful of seeds that is around 2-3 a day. Lastly, although researches have proved its efficiency in case of pregnant mothers. This is because the chemicals in the bark, roots or flowers of moringa may lead to uterus contraction that in turn may create pregnancy complications. Therefore, it is always better to be safe from before.

REFERENCES

Anwar F, Latif S, Ashraf M, Gilani AH. (2023) *Moringa oleifera*: A food plant with multiple medicinal uses. Phytother Res. 2007 Jan 1; 21(1): 17-25. Available from: https: //onlinelibrary.wiley.com/doi/epdf/10.1002/ptr.2023

Anonymous source: fdc.nal.usda.gov

Berglund LJ. (2018) Anaphylaxis to *Moringa oleifera*: First description. Australas Med J. 2018 Apr 1; 11(3): 176-7. https: //amj.net.au/index.php/AMJ/article/viewFile/3344/1655

Chumark, P. Khunawat P. Sanvarinda Y. Phornchirasilp S. Morales N. P. Phivthongngam L. Ratanchamnong P. Srisawat S. Pongrapeeporn. K. U. (2008) The *in vitro* and *ex vivo* antioxidant properties, hypolipidaemic and antiatherosclerotic activities of water extract of *Moringa oleifera* Lam. Leaves. J. Ethnopharmacol., 116 (2008), pp. 439-446

Gupta R, Mathur M, Bajaj VK, Katariya P, Yadav S, Kamal R, (2012) Evaluation of antidiabetic and antioxidant activity of *Moringa oleifera* in experimental diabetes. J Diabetes. 2012; 4(2): 164-71

Kamal M. (2008). *Moringa oleifera* Lam - The miracle tree. http: //www.pharmainfo.net/reviews/moringa-oleifera-lam-miracle-tree

Kumar R.K. Sheeja A. K. (2013) Heavy metal removal from water using *Moringa oleifera* seed coagulant and double filtration.Int. J. Sci. Eng. Res., 4 (2013), pp. 10-13.

Mahajan G. S., Mehta A. A. (2009) Anti-arthritic activity of hydroalcoholic extract of flowers of *Moringa oleifera* lam. In Wistar rats. J. Herbs Spices Med. Plants, 15 : 149-163.

Mane G. (2024) Drumstick (Sahjan): Uses, Benefits, Side Effects and More!Last updated: Apr 4, 2024Array

Mishra G, Singh P, Verma R, Kumar S, Srivastav S, Jha KK, (2011) Traditional uses, phytochemistry and pharmacological properties of Moringa oleifera plant: An overview. Pharm Lett. 2011 Jan 1; 3(2): 14164.

https: //www.researchgate.net/publication/281577396_Traditio nal_uses_phytochemistry_and_pharmacological_properties_of_Moringa_oleif era_plant_An_overview

Mishra S. P.Singh P. Singh S. (2012) Processing of *Moringa oleifera* leaves for human consumption. Bull. Environ. Pharmacol. Life Sci., 2 (2012), pp. 28-31.

Monera TG, Wolfe AR, Maponga CC, Benet LZ, Guglielmo J. (2017) *Moringa oleifera* leaf extracts inhibit 6Î²-hydroxylation of testosterone by CYP3A4. J Infect Dev Ctries. 2017 Apr 10; 2(5): 379 83. https: //www.ncbi.nlm.nih.gov/pmc/articles/PMC5385924/pdf/nihms8545 27.pdf

Moyo B. Masika P. Hugo A. Muchenje V. (2011) Nutritional characterization of Moringa (*Moringa oleifera* Lam.) leaves. African J. Biotechnol., 10 (2011), pp. 12925-12933.

Ndong M, Uehara M, Katsumata S, Suzuki K. (2007) Effects of oral administration of *Moringa oleifera* Lam on glucose tolerance in gotokakizaki and wistar rats. J of Clin Biochem and Nutri. 2007; 40: 229-33.

Paikra BK, Dhongade HKJ, Gidwani B. (2017) Phytochemistry and Pharmacology of *Moringa oleifera Lam*. J Pharmacopunc. 2017 Sep 24; 20(3): 194-200. from: https: //www.ncbi.nlm.nih.gov/pmc/articles/PMC5633671/pdf/2093-6966-v20-n03-194.pdf

Saxena O.P. (Nidar) (2023) Charak Samhita in Hindi by DPB Pub. Delhi pp.303 and 469

US Department of Agriculture [Internet]. Drumstick pods, raw; (2019) Jan 4 [cited 2022 Jun 6]. Available from: https: //fdc.nal.usda.gov/fdc-app.html#/food-details/170483/nutrients [59].

Chapter 28

A Religious Plant with Medicinal Importance: Kadamba [*Neolamarckia cadamba* (Roxb.) Bosser]

Ayurvedic science is deeply rooted in India and its neighboring countries. It was developed even before the medieval period, when people had little knowledge of science. There are a number of ancient therapeutic measures based on medicinal plants that have been developed in India. The cure for several diseases and ailments such as diabetes, cardiovascular disorders, cancer, and liver damage are mentioned. A variety of plants are used for medicinal treatments; either whole or in specific parts (bark, root, leaves, fruit, flowers, seeds), in the dried state. The Neolamarckia cadamba is an important plant having tremendous medicinal properties. Most surprisingly, despite being an astounding plant, very few studies have been done on Kadamba. There are very few derived products from the N. cadamba, which have been commercialized or been recommended in daily life for people. There is an urgent requirement for intensive studies on this plant to exploit it for the treatment of various incurable diseases prevalent across the world. This review has show cased various pharmacological activities of the Cadamba. It has been reported to be used for its hepatoprotective activity. The hepatoprotective activity is due to the presence of chlorogenic acid (CGA), the leaves and bark have great significance. Antitumor activity of defatted methanol extract of Cadamba (MEC) on Ehrlich ascites carcinoma (EAC) has been evaluated. The ethanolic extracts of bark powder extract at a dosage of 0.5 g/kg produced a fall of 23.8 per cent in the blood glucose levels of diabetic rats. Bark decoction is used for fevers. Powdered bark boiled with fruits and inhaled through the mouth for toothache and the sap is used by various Indian tribes for treatment of jaundice.

The Cadamba is one of the important medicinal plants belonging to the Rubiaceae family. It is crucially significant as it has the largest number of phytochemicals and secondary metabolites (viz., cadambagenic acid, cadamine, quinovic acid, â-sitosterol, cadambine, etc.) having pharmacological and biological properties. It can be used as an alternative to various synthetic chemical compounds in the prevention as well as the treatment of several incurable diseases. More than 100 years of research has been done to discover various phytochemicals and their implications. Very few of them, i.e. d"2% have been commercialized due to the lack of a suitable model system as well as various associated controversial issues. The solubility of phytochemicals is another major concern: Further response that will be generated due to the solvent used is also unpredictable. Moreover, the Cadamba is one of the ornamental plants with religious significance. Here we have made an effort to summarize all the phytochemicals and their significance to render the interest that would help in their commercialization.

Keywords: *Antioxidant, Antihelminthic, Anti-inflammatory, Hyperimmune antivenom, Penta cyclic terpenes, Kadamba.*

28.1 Introduction

The *Cadamba* is commonly known as "Kadamba" in Sanskrit and Hindi and as "Kodom" in Bengali. The other names of the plant are *Neolamarckia cadamba, Nauclea cadamba* (Roxb.),

Anthocephalus cadamba (Roxb.) Miq., *Samama cadamba* (Roxb.) Kuntze, *Anthocephalus morindifolius* Korth., *Nauclea megaphylla* S. Moore, *Neonauclea megaphylla* S. Moore, *etc.* The genus name is in honour of French naturalist Jean-Baptiste Lamarck. The common names are: Bur-flower tree, Laran (English); Kadamba, Kaduavalatige (Kannada); Kadam, Kadamb (Hindi); Kadambamu (Telugu); Kadambai, Vellaikadambu (Tamil); Attutek (Malayalam). The species have been widely but incorrectly called *Anthocephalus chinensis* as it has scented orange flowers present in dense globe-shaped clusters, which are used in the preparation of perfumes. It is an ornamental plant that is also used for timber- and paper-making.

This species is globally distributed from Indo-Malesia to Australia and Pacific islands. Within India, it is found all over, common in the damp places along large streams. It is also frequently cultivated for ornament, and as a shade-tree in plantations throughout the country. In India, it is found in the temperate Himalayas (which extend from Kashmir to Bhutan), Garhwal, Himachal Pradesh, Sikkim, Assam, and Manipur (Dwevedi *et al.,* 2015). In Garhwal, it is particularly distributed in the temperate zones of Pauri, Tehri, Chamoli, and Uttarkashi districts, while in Himachal Pradesh it is prevalent in the districts of Chamba, Kangra, Manipur, Bilaspur, Kullu, Sirmour, and Simla at the elevation of ~ 2 km. Besides India, the Kadamba is found in Nepal, Myanmar, and western China (Sandhya *et al.,* 2016).

It has crucial significance in Indian mythology and religion. Various religions in India have strongly believed that God lives inside a *Cadamba* tree based on its enormous significance to humankind. It has been said in the Sanskrit shloka,

"Ayi Jagadamba Mad-Amba Kadamba Vana-Priyavaasini Haasa-Rate,"

that is, Goddess Durga likes to live in the forest of *Cadamba* trees. The *Cadamba* has been known to cure a number of diseases; particularly, the extract prepared from the bark and leaves is crucial. Various phytochemicals as well as secondary metabolites *i.e.* saponins, indole and quinoline alkaloids, secoiridoids, and triterpenes with pharmacological significance from the *Cadamba* are described.

Figure 28.1. (A) Part of Kadam tree with leaves and flowers; (B) Bark of tree; (C) Section of ball like inflorescence of plant; (D) Seeds of *Cadamba.*

28.2 Botanical Features

Kadamba tree is an evergreen, tropical tree about 45 m with a broad crown and straight cylindrical bole. Bark is dark gray in color, rough and frequently longitudinally fissured, exfoliating in thin scales. Leaves are glossy green, opposite, simple more or less sessile to petiolate, ovate to elliptical, 30 cm long 10-15 cm broad with prominent veins (Bhandary *et al.,* 1995). Flowering usually begins when the tree is 4–5 years old. Kadamda flowers have sweet fragrance, occurring in dense, globular heads of ~5.5 cm diameter. The fruits of the *Cadamba* are small, containing fleshy capsules packed closely together to form a yellow-orange infructescence (Dwevedi *et al.,* 2015). Fruits are round like small balls, hard containing approximately 8000 seeds and are green when young turn yellow when ripe. Seeds are triangular or irregularly shaped. The Kadamba tree, which is considered the 'sthalavruksham' (Tree of the place) of the city that is otherwise known as 'Kadambavanam' (Kadamba forest) and is present in Meenakshi Temple and a few trees at the Meenakshi Government College for Women at Madurai. A withered relic of the Kadamba tree is also preserved in the precincts of the Madurai Meenakshi temple.

28.3 Religious Significance of Kadamba

Kadam features in Indian mythology and is mentioned in Bhagavata Purana. In North India, it is associated with Lord Krishna while in the South Goddess Parvathi. It is believed that Sri Krishna performed many of his divine acts (leelas) in Vrindavan under the canopy of Kadamba trees. He used to play with his friends around these trees in Vrindavan, holding the unique globular flowers of the tree. Hence the tree is also known as Haripriya, God's favourite. He is depicted as playing his flute under its canopy. Radha and Krishna were supposed to have conducted their love play (romance) in Kadambavana. The episode representing one of the naughtiest acts of Lord Krishna: the stolen clothes of Gopis by Krishna were hidden on branches of a tree related to this tree!

मानुस हौं तो वही रसखान, बसौं मिलि गोकुल गाँव के ग्वारन।

जो पसु हौं तो कहा बस मेरो, चरौं नित नंद की धेनु मँझारन।।

पाहन हौं तो वही गिरि को, जो धर्यो कर छत्र पुरंदर धारन।

जो खग हौं तो बसेरो करौं मिलि कालिंदीकूल कदम्ब की डारन।।

Subhadra Kumari Chauhan wrote about the tree in one of her poems:

यह कदंब का पेड़ अगर माँ होता यमुना तीरे।

मैं भी उस पर बैठ कन्हैया बनता धीरे-धीरे।।

ले देतीं यदि मुझे बांसुरी तुम दो पैसे वाली।

किसी तरह नीची हो जाती यह कदंब की डाली।।

Many indigenous trees are designated as sacred trees as they are associated with mythology, religion and folklore, and is mentioned in Bhagavata Purana. A few are symbolic of specific Gods. 'Kadamba' is one such tree associated with Lord Krishna. The flowers are offered in temples and used in tribal festivals. The tree is enlisted in 'Nakshathra Vana' representing 'Satabhisha' nakshatra roughly corresponding to Aquarii (a binary star system). It lends its name to the 'Kadamba Dynasty' with Banavasi as capital, the first ruling Kingdom of Karnataka. It was considered a holy tree by the dynasty. The spring festival, 'Kadambotsava' is celebrated in honour of Kadamba kingdom by the Government of Karnataka at Banavasi every year. It is said Varuna, the sea-god, had forbidden nude bathing in rivers, ponds and other public places, but 'gopis' often resorted to it. One day to teach them a lesson Krishna reached the bank of the pond where they were taking bath and before any one of them noticed him he took away their garments and spread them on the branches of nearby 'Kadamba' tree. He himself climbed the tree and hid there behind a branch.

Another interesting fact is that 'Kadamba' flower was the emblem of Athmallik State, an erstwhile princely state of India, now part of Odisha. The tree has immense medicinal values besides being a beautiful ornamental tree with scented golden ball like floral heads. A postal stamp has also been issued by the Department of Posts to commemorate its importance. According to another belief since the wood is soft in *Neolamarckia cadamba* or *Anthocephalus* the other genus *Mitragyna* (family: Rubiaceae) may be the actual plant growing in Vrindavan, these consists of trees growing exclusively in humid conditions. Several of these plants have been used in local folklore medicine for a wide variety of diseases such as fever, colic, muscular pains and for the expulsion of worms (Shellard and Phillipson 1964) and commercially in the timber and paper industry (Chatterjee *et al.,* 1982).

It is believed that Sri Krishna performed many of his divine acts (leelas) in Vrindavan under the canopy of Kadamba trees (Pandey 2020). He used to play with his friends around these trees in Vrindavan, holding the unique globular flowers of the tree. Hence the tree is also known as Haripriya, God's favorite. He is depicted as playing his flute under its canopy. Radha and Krishna were supposed to have conducted their love play (romance) in Kadambavana. The episode representing one of the naughtiest acts of Lord Krishna: the stolen clothes of Gopis by Krishna were hidden on branches of a tree related to this tree! Devi is the radiant beauty who dwells in the Kadamba forest: Kadamba-vana-vasini or Kadamba-vana-nilaye, whose presence is sensed if the koel (Cukkoo) sings in the Kadamb-van (forest). In the Southern Tamil Nadu, Kadamba tree is considered the 'sthalavruksham' (Tree of the place) in Madurai and a withered relic of the tree is preserved at Meenakshi Temple. The tree is also associated with a local deity called Kadambariyamman and the place was once a Kadambavanam (Kadamba forest). During Sangam period, Lord Murugan of Tirupparankundram Hill Temple (a centre of nature worship) near Madurai is believed to be in the avatar of a spear under a Kadamba tree. Many festivities are associated with the tree and 'Karam-Kadamba' also known as 'Posatt' or the 'Koral parba' is a popular harvest festival, celebrated on the eleventh lunar day of Bhadra. A twig of the tree is worshipped in the courtyard of the house. Later in the day, young ears of grain are distributed among friends and relatives – a custom adopted by Tulu people. In West Bengal and Odisha, agricultural communities celebrate Kadam festival by planting Kadam saplings ceremoniously during the month of Bhadra. In Theravada Buddhism, the Kadam tree was where Sumedha Buddha achieved enlightenment. The impact of the 'koral parba' on the regional populace is so deep that it is also celebrated by local Christians, converted from Hinduism; Onam (Kerala) and Huttari (Kodagu) are regional variants of this festival. *Kadambotsava* ("The festival of Kadamba") is also the festival that is celebrated every year by the Government of Karnataka in honor of the Kadamba kingdom, the first ruling Kingdom of Karnataka, at Banavas], as it was here that the Kadamba kings organized the spring festival every year.

Here is an extract of well known Rabindra Sangit, which is sung during the rainy season–

Badala-diner pratham kadam phul

Amay karecho dan

Ami to diechi bhara sravaner

Meghamallar gan.... (Tagore 1961).

English version-

You made a gift to me of the first blossoms of kadamba of the rainy days. I too have given you a song of numbling cloud when Sravan (rainy season) was in full spate.

Kadamba is associated with Lord Krishna. It is a temple tree held sacred by the Hindus and the Buddhist monks of India. When the rains set in and the poet's mind was transported by joy and his heart danced like a peacock he could think of no better place than the woodlands of the kadamba flowers.

Rain clouds wet my eyes with blue collyrium

I spread out my joy on the ṣhaded

New woodland grass

My soul and Kadamba trees blossom together. (Tagore 1987).

In Jayadeva's Gitagovindam, Song of Govinda, (a poetic work on Lord Krishna composed in 1200 AD by Jayadeva stanza 1, says "He who is mixed up or mingled in the darkness at a peaceful Kadamba tree, pre-set by me, – deserve supreme love and affection of the Supreme and hence I reminisce about him." Variousr eligions in India have strongly believed that God lives inside a *Cadamba* tree based on its enormous significance to humankind.

Kadamb trees have great importance in kunj lanes of Vrij. In Shravan month ladies use to swing on branches of these trees. It is favorite tree of Kamdeva, goddess Parvati and Kali. Banabhatta's Kadambari is one of the best romantic fictions. It is a detailed, exquisite novel in the Sanskrit language. It is one of the world's earliest novels. The title of this fiction is named after its heroine. The presentation is made in seven chapters. It is an imaginative romantic story of love in the katha form. The plot of Kadambari is very impressive. It deals with the three lives of the characters. It presents a beautiful love story. Kadambari, the heroine of the novel, was the daughter of a king. Mahasweta was one of her companions.

28.4 Phytochemicals Present in Kadamba

Root bark contains ursolic acid, stigmasterol, prunetinoside, glucogenkwanin

Seed constitutes naringenin-5-O-α-L-rhamnopyranoside, 4′-O-methylliquiritigenin-7-O-α-L-rhamnopyranoside, naringenin 4′-methylether 7-xyloside, β-sitosterol-3-O-D-galactopyranosid. Branches constitute substitute of hydrocyanic acid, amygdalin. Stem bark constitutes padmakastein and its derivatives, β-sitosterol behenate, tectochrysin, genistein, leucocyanidin, 4′-glucoside of genkwanin, chrysophenol, emodin, 8 β-D glucosides, orientalone, physcion, β-sitosterol glucoside, amygdalin, prunasetin, sakuranetin, puddumetin, flavanone, sakuranetin (5, 4′-dihydroxy-7-methoxy flavone) and its 5-glucoside, neosakuranin (2, 4′-dihydroxy-4-methoxy-6-glucosidoxychalcone), leucocyanidin, puddumin B, naringenin-4′-methylether-7-O- β-D-galactoside), taxifolin. Leaves of kadamb contains Quercetin-3-rhamnoglucoside, kaempferol. Heartwood constitutes dihydrotectochrysin, dihydrowogonin, pinocembrin, chrysin, naringenin, kaempferol, aromadendrin, quercetin, taxifolin, 7-hydroxy-5, 2′, 4′-trimethoxyflavanone, 2′-hydroxy 2, 4, 4′, 6′- tetramethoxychalcone, 2′, 4′ dihydroxy-2, 4, 6′- trimethoxychalcone. stem bark of *Cadamba*; (a) Padmakastein (b) Tectochrysin (c) Leucocyanidin (d) Chrysophenol (e) Emodin (f) Physcion (g) Sakuranetin (h) Puddumin B (i) Taxifolin. II. Phytochemicals from the stem of *Cadamba*; (a) Apigenin (b) Sakuranetin (Brown and Chapple 1976; Banerji 1977). Studies by Mondal *et al.* (2009) proved that bark extract is diuretic and has laxative activity.

Table 28.1. Phytochemicals Present in different Plant Parts of Kadamba

Sl.No.	*Parts*	*Phytochemicals*
1.	Flower	Linalool, geraniol, geranylacetate, linalyl acetate, Selinone, 2-nonanol,
2.	Leaves	Cadambine, Cadamine, Isocaamine, 3 dihydrocadambine,3 isodihydrocadambine,3 dihydrocadambine,3 isodihydro-cadambine, Cinchotannic acid
3.	Stem bark	Sapogenins, Cadambogenic acid, quinonic acid, Saponin A,B,C,D, Tannins
4.	Wood	Cellulose, Lignin, Cadambine, dihydrocadambine, isodihydro-cadambine

28.5 Pharmacological Research Findings

It has been found that methanolic extract of the root bark of the *Cadamba* can be used as an antidote against snakebite. It is used in neutralizing *Vipera russellii* and *Naja kaouthia* venom, which can induce hemorrhage, cardiotoxicity, neurotoxicity, defibrinogenation, and inflammation. The pentacyclic triterpenes (free or as glycosides) have a crucial significance in providing ~20 per cent protection against snake venom (Halliwell and Gutteridge 1990; Rios *et al.,* 2009).

Studies on antioxidants are crucial, particularly in the food industry and in therapeutic research. The reactive oxygen species (ROS) and free radicals are byproducts of biological metabolism, which is responsible for cell membrane breakdown, membrane protein damage, and DNA mutation. These can further initiate the development of many diseases such as cancer, liver injury, cardiovascular diseases, cellular damage, and the aging process. The indigenous enzymatic systems (superoxide dismutase, glutathione peroxidase, catalase), chemical scavengers and dietary antioxidants (α-tocopherol, β-carotene, ascorbic acid, glutathione, uric acid), and hormones (estrogen, angiotensin) are able to remove free radicals formed in cells and thus protect against oxidative damage. Antioxidant properties in the ethanolic extract of the *Cadamba* leaves were assayed by estimating liver and kidney tissue enzymes using the 2'- diphenyl-1-picrylhydazyl (DPPH) assay, the superoxide anion radical scavenging assay, and DNA damage. It was found that the *Cadamba* possesses potent antioxidant properties. Further, UPLC-ESI-QTOF/MS has also confirmed the presence of various bioactive compounds from the *Cadamba* leaves having antioxidant properties (Ito *et al.,* 1993; Halliwell 2007; Gupta *et al.,* 2013).

Recently, the antihelminthic activity of the *A. cadamba* has been elucidated. It was examined on adult Indian earthworms, *Pheritima posthuma*, due to its anatomical and physiological resemblances with the intestinal roundworm parasites of the human beings. Each group was treated with aqueous and ethanolic extracts of the mature bark of the *A. cadamba* with varied concentrations ranging from 10 mg/ml to 25 mg/ml along with vehicle (piperazine citrate, 15 mg/ml, prepared in 1 per cent tween-80). It was observed that paralysis and subsequently death of an individual worm have taken almost 4 h. Here paralysis was said to occur when the normal worm did not revive in saline, while death was concluded to have occurred when the worms lost their motility followed by the fading of their body color (Acharyya *et al.,* 2011; Chandel *et al.,* 2012).

The alcoholic and aqueous extracts of *cadamba* fruits have shown significantly higher antibacterial activity against microorganisms (*Staphylococcus aureus, Escherichia coli, Pseudomonas aeruginosa, Micrococcus luteus, Bacillus subtilis, Staphylococcus aureus, Pseudomonas aeruginosa, Salmonella typhi, Klebsiella pneumonia, Proteus mirabilis, Candida albicans, Trichophytonrubrum, Asperagillus niger, A. flavus,* and *A. nidulans*). An experiment demonstrated by Mishra and Siddique (2011) concluded that the antibacterial properties of *cadamba* have zones of inhibition as 22.0 cm and 24.0 cm against *E. coli* and *P. aeruginosa*, respectively, at minimum inhibitory concentration (MIC) of 1.00 mg/ml. Some researchers have demonstrated that the extract of the bark and leaf of the *A. cadamba* showed antifungal activity against *Aspergillus fumigatus* and *Candida albicans.* They have also found that the leaf extract shows higher antifungal activity than the bark extract (Bhardwaj and Laura 2007; Mishra and Siddique 2011). Mosquito-borne diseases like malaria, dengue, chikungunya, filariasis, and Japanese encephalitis cause thousands of deaths per year in India as well as in other developing countries. Therefore, mosquito control is a serious concern and necessary to enhance the health and quality of life of the country's residents and visitors. The management of vector-borne diseases has failed due to their increased resistance and revitalization against synthetic chemicals. There is a number of reports on the use of plant extracts for killing mosquito larvae. Recently, it was reported that *cadamba* leaf extract has excellent larvicidal and pupicidal activities against the filarial vector, *Culex quinque fasciatus*, even at low concentrations. It was found that the methanolic extract of *Cadamba* leaves was most effective against stage I, *i.e.* the larval stage, where 46 per cent mortality was observed at 5 ppm and increased to 100 per cent at 80 ppm. The observed LC_{50} and LC_{90} values were 12.15 ppm and 56.62 ppm, 15.15 ppm and 64.72 ppm, 21.82 ppm and 79.52 ppm, and 31.29 ppm and 102.13 ppm against stages I, II, III, and IV of larval pupae, respectively. Moreover, the addition of gold nanoparticles to the extract has proved to be more lethal, leading to 100 per cent mortality at the larval stage at a very low concentration with LC_{50} at 0.61 ppm. There is another study, which found that the dimethyl sulfoxide extract of the *A. cadamba* showed antimalarial activity with LC_{50} of 3.7 µg/ml against *Plasmodium falciparum*, K1 strain (Kumar *et al.*, 2013).

Researchers demonstrated an experiment to determine the antidiabetic study of *A. cadamba* in which they have used the stem bark which contains flavonoids (7.83 mg) and phenolic acids (12.26 mg) per 100 g of dry weight of the bark powder. The effect of the different doses of ethanolic extract of bark on the fasting blood glucose levels of both normal and diabetic mice was studied. It was found that the fasting blood glucose levels of diabetic untreated mice were significantly higher than those of normal untreated rats. The ethanolic extracts of *A. cadamba* bark powder at a dosage of 0.5 g/kg produced a fall of 23.8 per cent in the blood glucose levels of diabetic rats after 5 h of treatment (Devgan *et al.,* 2012). Plant-based products have been frequently examined as potential anticancer agents. The screening of various medicinal plants has found several bioactive compounds which are effective chemopreventive as well as chemotherapeutic agents. The phytochemical screening of Kadamba has revealed the presence of lupeol and betulinic acid-type triterpene which have antineoplastic activity. Antitumor activity of defatted methanol extract of *A. cadamba* (MEC) on Ehrlich ascites carcinoma (EAC) has been evaluated. There was extensive *in vitro* cytotoxicity as found by trypan blue and *in vivo* antitumor activity was evaluated by inoculating mice with EAC (Dolai *et al.,* 2012).

Anti-inflammatory activities of Kadamba are studied (Bachhav *et al.,* 2009) using active enzyme expressions of cyclooxygenase and lipoxygenase. Further, intact lysosomal membrane is important as the release of lysosomal constituent of activated neutrophil such as bacterial enzymes and proteases occurs during tissue inflammation. It has been reported that ethanolic extract of *A. cadamba* leaves exhibited significant membrane stability as found from heat induced hemolytic effect on erythrocyte membrane. The dry hydroethanolic extract of the flowering tops of the Kadamba has exhibited a dose-dependent decrease in the frequency of fecal droppings in castor oil-induced diarrhea in mice. The extract also produced a dose-dependent reduction in intestinal fluid accumulation. It has been found from experimental studies that alloxan has the capacity of reducing lipid levels by 30 per cent as observed in diabetic mice. In comparison to this drug, the oral administration of root extract of the *Cadamba* for 30 days in dyslipidemic animals resulted in a significant decrease by 80 per cent in total cholesterol, phospholipids, triglycerides, and lipid peroxides, with a reduction in lipid levels in diabetic mice (Bachhav [et al.,] 2009).

The hepatoprotective activity is due to the presence of chlorogenic acid (CGA) isolated from the *Cadamba* plant. It was also found that the intraperitoneal administration of CGA to mice at a dose of 100 mg/kg for 8 days exhibited better liver protective action than silymarin (SM) in CCl_4. The study concluded that, surprisingly, the antioxidative activity of CGA prepared in CCl_4 is responsible for its hepatoprotective nature in mice model with liver injury. Some researchers studied the extracts of bark of the *Cadamba* in various concentrations in different solvents and demonstrated its diuretic and laxative activities. They have reported that the methanolic extract of *Cadamba* bark showed significant increase in urinary output as compared to aqueous, chloroform, and petroleum ether extracts. Moreover, there was higher laxative activity in case of the chloroform extract than with respect to the methanol, petroleum, and aqueous extracts (Ambujakshi *et al.,* 2009).

28.6 Other Clinical Uses of Kadamba

The paste of the leaves of *Cadamba* are tied over the wound or area affected with localized pain and swelling to reduce the complaints. The decoction of the bark of *Cadamba* is used to wash the infected wounds. The decoction of the bark of the plant is used for gargling to treat mouth ulcers and inflammation of the gums. The decoction of *Cadamba* is taken in a dose of 30-40 ml to treat diarrhea and irritable bowel syndrome. The powder of the bark of the tree is given with sugar candy in a dose of 5-6 g to treat nausea and vomiting. The juice of the fruit is given in a dose of 40-50 ml to treat excessive sweating, thirst and burning sensation of the body. The decoction of the root is taken in dose of 30-40 ml to treat urinary tract infection and renal calculi. The decoction of the bark is consumed in a dose of 30-40 ml to treat fever. The paste of the bark is applied over black spots and pimples. The fresh juice of the leaf is consumed in a dose of 10-15 ml to treat leucorrhoea and increased menstrual flow. The fresh juice of the fruit is useful to increase the breast milk in lactating women. The paste prepared from the bark and leaves is useful to treat pain, redness and itching due to insect bite (Razafimandimbison and Sylvain 2002).

28.7 Conclusion

The *Cadamba* is one of the important medicinal plant, which is crucially significant as it has the largest number of phytochemicals and secondary metabolites (*viz.*, cadambagenic acid, cadamine, quinovic acid, β-sitosterol, cadambine, *etc.*) having pharmacological properties. It can be used as an alternative to various synthetic chemical compounds in the prevention as well as the treatment of several incurable diseases. So many researchers have been done to discover various phytochemicals and their implications. Very few of them, *i.e.* ≤2 per cent have been commercialized due to the lack of a suitable model system as well as various associated controversial issues.

REFERENCES

Acharyya S, Rathore DS, Kumar HK, Panda N. (2011) Screening of *Anthocephalus cadamba* (roxb.) miq. root for antimicrobial and anthelmintic activities. Int J Res Pharm Biomed Sci. 2: 297–300.

Ambujakshi HR, Antony ST, Kanchana Y, Patel R, Thakkar H.(2009) Analgesic activity of *Anthocephalus cadamba* leaf extract. J Pharm Res. 2: 1279–80.

Anonymous (2020) The Hindu : Metro Plus Madurai : Nature's unsung heroes

Bachhav RS, Buchake VV, Saudagar RB. (2009) Analgesic and anti-inflammatory activities of *Anthocephalus cadamba* Roxb. leaves in wistar rats. Res J Pharm Technol. 2: 164–7.

Banerji N. (1977) New saponins from stem bark of *Anthocephalus cadamba* MIQ. Indian J Chem B. 15: 654–5.

Bhakuni DS, Dhar ML, Dhar MM, Dhawan BN, Mehrotra BN. (1969) Screening of Indian plants for biological activity. II. Indian J Exp Biol. 7: 250–62.

Bhandary MJ, Chandrashekar KR, Kaveriappa KM. (1995) Medical ethnobotany of the siddis of Uttara Kannada district, Karnataka, India. J Ethnopharmacol. 47: 149–58.

Bhardwaj SK, Laura JS. (2007) Antibacterial properties of some plants-extracts against plant pathogenic bacteria *Rathyibacter tritici*. Int J Biosci Biotechnol Res Asia. 4: 693–8.

Brown RT, Chapple CL. (1976) *Anthocephalus* alkaloids: Cadamine and isocadamine. Tetrahedron Lett. 19: 629–30.

Chandel M, Sharma U, Kumar N, Singh B, Kaur S. (2012) Antioxidant activity and identification of bioactive compounds from leaves of *Anthocephalus cadamba* by ultra-performance liquid chromatography/electrospray ionization quadrupole time of flight mass spectrometry. Asian Pac J Trop Med. 5: 977–85.

Devgan M, Bhatia L, Kumar H. (2012) *Anthocephalus cadamba*: A comprehensive review. Res J Pharm Technol. 5: 1478–83.

Chatterjee, A., Dhara, K.P., Banerji, J. (1982) Alkaloids of *Mitragyna parvifolia* (Roxb) Korth. and their transformations. J. Indian Chem. Soc. LIX, 1360–1363.

Dogra SC. (197) Antimicrobial agents used in ancient India. Indian J Hist Sci. 22: 164–9.

Dolai N, Karmakar I, Suresh Kumar RB, Kar B, Bala A, Haldar PK. (2012) Evaluation of antitumor activity and *in vivo* antioxidant status of *Anthocephalus cadamba* on Ehrlich ascites carcinoma treated mice. J Ethnopharmacol. 142: 865–70.

Dwevedi A, Sharma K, SharmaY K. (2015). *Cadamba*: A miraculous tree having enormous pharmacological implicationsPharmacogn Rev. 29(18): 107–113. doi: 10.4103/0973-7847.162110

Gupta A, Anand M, Yadav S, Gautam J. (2013) Phytochemical studies and antioxidant activity of different leaves extracts of *A. cadamba*. Int J Futur Sci Engg Technol. 1: 21–5.

Halliwell B, Gutteridge JM. (1990) The antioxidants of human extracellular fluids. Arch Biochem Biophys. 280: 1–8.

Ito N, Fukushima S, Hagiwara A, Shibata M, Ogiso T. (1983) Carcinogenicity of butylated hydroxyanisole in F344 rats. J Natl Cancer Inst. 70: 343–52.

Kumar AN, Jeyalalitha T, Murugan K, Madhiyazhagan P. (2013) Bioefficacy of plant-mediated gold nanoparticles and *Anthocepholus cadamba* on filarial vector, *Culex quinquefasciatus* (Insecta: Diptera: Culicidae) Parasitol Res. 112: 1053–63.

Mishra RP, Siddique L. (2011) Antibacterial properties of *Anthocephalus cadamba* fruits. Asian J Plant Sci Res. 1: 1–7.

Mondal S, Dash GK, Acharyya A, Acharyya S, Sharma HP. (2009) Studies on diuretic and laxative activity of bark extracts of *Neolamarckia cadamba* (Roxb.) Bosser. Drug Invent Today. 1: 78–80.

Pandey R. S. (2020) Uttar Pradesh Ke Vraksha avam Vrakshropan (In Hindi). Pub. By Bharat Book centre, Lucknow. 230 pp.

Razafimandimbison, Sylvain G. (2002) A Systematic Revision of Breonia (Rubiaceae-Naucleeae). Annals of the Missouri Botanical Garden. Pharmacog Rev. 89 (1): 1–37. doi: 10.2307/3298655.

Rios AO, Antunes LM, Bianchi ML. (2009) Bixin and lycopene modulation of free radical generation induced by cisplatin-DNA interaction. Food Chem. 113: 1113–8.

Sandhya T, Lathika KM, Pandey BN, Mishra KP. (2006) Potential of traditional ayurvedic formulation, Triphala, as a novel anticancer drug. Cancer Lett. 231: 206–14.

Shellard, E.J., Phillipson, J.D. (1964) The Mitragyna species of Asia. Part I. The alkaloids of the leaves of *Mitragyna rotundifolia* (Roxb.) O. Kuntze. Planta Med. 12, 27–32.

Tagore R. (1961) *Deoa-neova* (Sanai 1940), Rabindra Rachanabali vol. 3 English rendering by P.K. Ghosh (Birth Centenary Publication, West Bengal Govt.)1961.742pp.

Tagore R. (1987) *Naba Barsha* (*Kshanika*), Selected poems translated by Radice William (Penguin Books, London).

Zaidan MR, Noor Rain A, Badrul AR, Adlin A, Norazah A, Zakiah I. (2005) *In vitro* screening of five local medicinal plants for antibacterial activity using disc diffusion method. Trop Biomed. 22: 165–70.

Chapter 29

Phytochemistry and pharmacological Actions of Gudhal (*Hibiscus rosa-sinensis* L.)

Hibiscus rosa-sinensis (shoe flower) is a perennial shrub belonging to the family Malvaceae and genus Hibiscus native to tropical and south Asia and are widely distributed in different regions of world. Many Hibiscus species are mostly cultivated as ornamental plants. They produce vibrantly colored showy flowers, throughout the year.

This plant has various important medicinal uses for treating wounds, inflamation, fever and coughs, diabetes, infections caused by bacteria and fungi, hair loss, and gastric ulcers in several tropical countries. Phytochemical analysis documented that the main bioactive compounds responsible for its medicinal effects are namely flavonoids, tannins, terpenoids, saponins, and alkaloids. Experiment from recent studies showed that various types of extracts from all H. rosa sinensis parts exhibited a wide range of beneficial effects such as hypotensive, anti-pyritic, anti-inflammatory, anti-cancer, antioxidant, anti-bacterial, anti-diabetic, wound healing, and abortifacient activities. Hibiscus is a rich source of hydroxy acids. This review has presented a comprehensive summary of recent studies on the phytochemistry, pharmacological actions and traditional uses of H. rosa sinensis precisely.

Keywords: *Anthocynidins, Terpenoids, Gentisic acid, Malvalic acids, Pharmacology, Phytoconstituents*

29.1 Introduction

Hibiscus is a flowering plant that belongs to the genus *Mallow* in the *Malvaceae* family. In general, hibiscus is not a name for just one plant, but a collection of several hundred species, which is why the botanical name of all the plants in it begins with the name *Hibiscus*. Native to the tropical, subtropical, and temperate regions of the world, this species is famous for its exotic, large, and colorful flowers. It is also known as Tropical *Hibiscus*, Rose of Sharon, and Hardy *Hibiscus*. All perennial herbaceous plants of this genus grow also as annual plants, and this species includes short shrubs, wood shrubs, small trees, and trees. Plants such as *Hibiscus rosa-sinensis* and *H. syriacus* are often widely cultivated as ornamental plants (Fig. 1). Gudhal often a popular name for the gorgeous flowering plant *H. rosa sinensis*, as it is mainly found in south-east China and some islands in the Pacific and Indian Ocean. Current scientific literature suggests that more than 50 per cent of today's clinical medications were of natural product origin. Many of them have played a significant role in pharmacological industry and in developing better therapies for various diseases. This plant is economically very essential owing to the herbal products and medicinal uses. Because of insufficient current pharmacological information, there is not much scientific research or clinical trials conducted on the chemical extracts of Gudhal that could be crucial in exploring its fast potential medicinal applications (Mak *et al.*, 2013; Bhakta and Das 2017). Each part of *H. rosa sinensis* contains a wide range of compounds. It was reported that phlobatannins, glycosides, saponins, flavonoids, terpenoids including other compounds such as thiamine, riboflavin and niacin are present in leaves, flowers, stem and roots of this plant (Kumar and Singh 2012; Salem *et al.*, 2014; Kumari *et al.*, 2015).

Figure 29.1. *Hibiscus rosa sinensis* Plant Possessing Red Coloured Flowers.

Traditionally, *Hibiscus* flowers have been reported to possess antitumor properties, as well as have been used as analgesic, antipyretic, anti-asthmatic, and anti-inflammatory (Begam *et al.*, 2015) agents. Several studies have proved the presence of anti-oxidant, anti-fungal, and antimicrobial properties in flowers of *H. rosa-sinensis* (Figure 29.2).

Figure 29.2. Red (most common), Pink, Orange, Yellow, White and Multi Shaded *Hibiscus rosa-sinensis* Flowers.

This review summarizes the phytochemistry, pharmocology, and medicinal uses of this flower with the purpose of finding gaps demanding for future research and investigating its therapeutic potential through clinical trials.

29.2 Chemical Constituents in *Hibiscus rosa-sinensis*

Generally, the edible flowers contain moisture, nitrogen, fat, crude ca, P and Fe. The yellow flowers contain several flavones such as cyanidin-3, 5-diglucoside, cyaniding-3-sophoroside3-5- glucoside quercetin-3,5-diglucoside, and quercetin-3,7diglucoside. Including the mentioned compounds, kaempferol-3-xylosylglucoside isolate can be found in white flowers. In addition to fatty acids, fatty alcohols, hydrocarbons, the leaves also contain β-carotene, as well as gentisic acid (El Sayed *et*

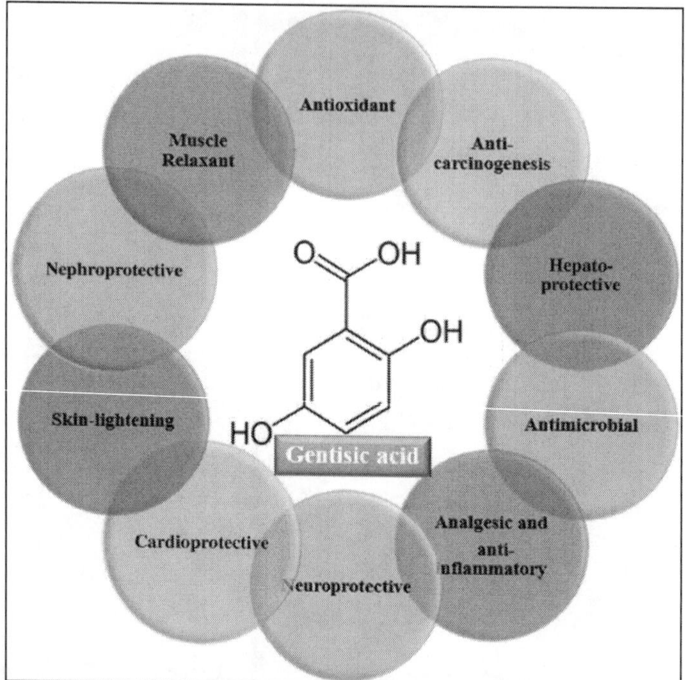

Figure 29.3. Chemical Structure of Gentisic Acid.

Figure 29.4. Chemical Structure of Malvalic Acids.

al., 2012), mucilage, and catalase. Cyclopropenoids can be found in roots. Although flowers, stems, and leaves contain minor amounts of cyanin and cyanidin chorides, quercetin can be found in all parts of Hibiscus rosa sinensis. However, β-sitosterol, teraxeryl acetate, and malvalic acids (Begam *et al.,* 2015) can be found only in stems and leaves. Using mass spectrum GC-MS interpretations, it was found that methanol extracts of Hibiscus flowers contained components such as Ethanimidic acid, ethyl ester, Propanal, 2,3dihydroxy, Propanamide, Nethyl-, Ethylenediamine, O-Methylisourea hydrogen sulfate, Ethene, ethoxy-, Hexadecanoic acid, methyl ester, 7-Formylbicycloheptanes, 2-Butanamine, 1,3,5-Triazine-2,4,6-triamine, N-Formylβ-alanine, 9-Pentadecadien-1-ol, Butanedial, 1-Propanol, 2- methyl-, and Methane carbothiolic acid (Hammad 2009). These components were shown to have anti-cancer, antioxidant, pesticide, hypocholesterolemic, dermatitigenic, and anemiagenic properties. Moreover, Propanal, 2,3-dihydroxy, Ethylenediamine, o-Methylisourea hydrogen sulfate, 2-Butanamine and 9-Pentadecadien-1-ol all have antioxidant and antimicrobial activities. Recently, another study on antioxidant effects of *H. rosa sinensis* showed that there was a strong correlation between antioxidant activity and flavonoids, phenolics, as well as anthocyanins found in the extracts. This confirmed that these components were responsible for the observed antioxidant activity. Although most of these phytochemicals have future potential medicinal uses, which will be more discussed in the next section, anthocyanins have special natural dyeing properties. It was demonstrated that using metal mordant such as Cu, Sn and Al in conjunction with Gudhal's anthocyanin extract, the dye uptake process of cotton and silk fabrics was enhanced as compared with the controlled sample in terms of fastness. This is could be mainly due to its antioxidant activity mentioned earlier. Another molecule, known as cyanidin-3-sophoroside, was also extracted from Hibiscus (Udo *et al.,* 2016; Noman *et al.,* 2017).

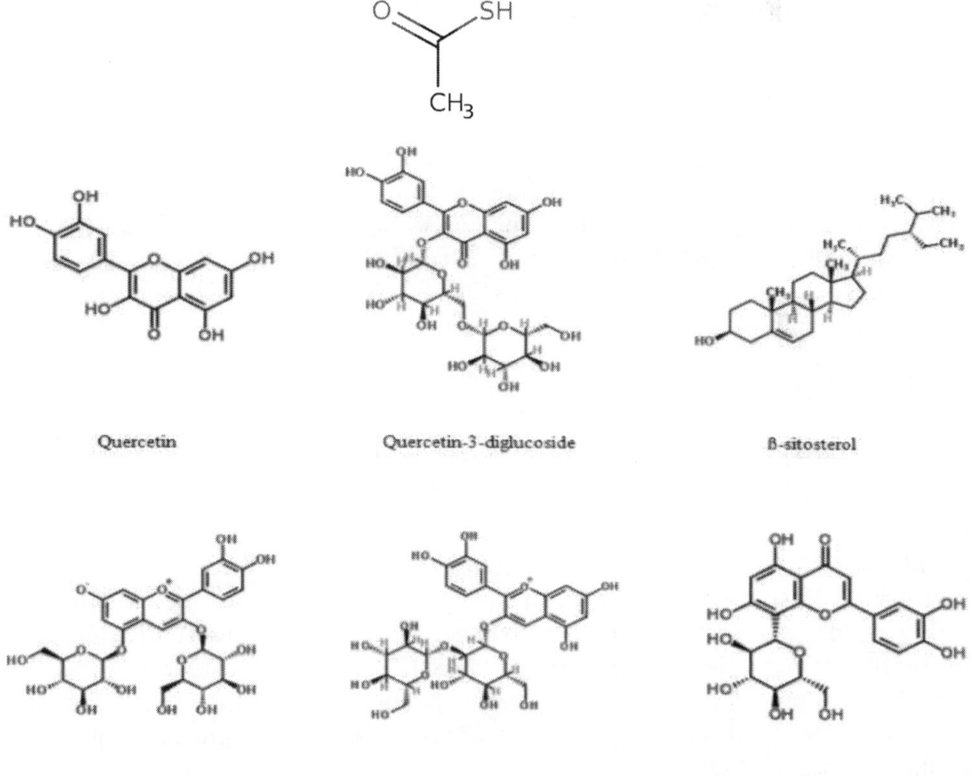

Figure 29.5. Chemical Structures of Various Bioactive Compounds Occurring in *H. rosa sinensis*; Quercetin, Quercetin-3-diglucoside, β-sitosterol, Cyanidin-3,5-diglucoside, Cyanadin 3-sophoside, Luteolin-8-C-glucoside.

29.3 Culinary Uses

Hibiscus Tea (Figure 29.7) is highly preferred by people all over the world for its medicinal properties. This beverage is red in color and has a sour taste and is high in Vitamin C. The tangy juice extracted from the flowers and stems of the hibiscus plant is used in various traditional medicine around the world as a natural abortion and contraceptive.

Figure 29.6. Hibiscus Tea from Dried Flowers.

Hibiscus helps greatly in promoting proper digestion. The flowers help to break down the complex molecules of food and thereby provide energy. This procedure can easily treat digestive disorders such as indigestion and abdominal pain. In addition, appetite is stimulated by regular. In addition, the juice extracted from the leaves and flowers has been used since a long time ago as natural remedy for some diseases and painful symptoms, as well as in herbal cosmetics as wilted.

29.4 Health Benefits of *Hibiscus* Plant

The leaves, stems, and flowers of the *Gudhal* plant are often used to treat a variety of ailments. In particular, various research results have demonstrated that tea made by dried hibiscus flowers offers numerous health benefits. Today, more than 100 million women worldwide utilized Hormonal Contraceptives, which is the artificial birth control method. Although it has been proven that this artificial method of preventing pregnancy can cause various side effects for most people, this method is the most sought after. However, studies have shown that *Gudhal* has anti-fertility properties, making it effective as a natural contraceptive with no side effects when prescribed properly. Flowers contain anthocyanin and flavonoids. These are used for treatment of menstrual disorder, menorrhigia and other gynic problems (Daniel 2008).

It is noteworthy that ancient Ayurvedic and Unani references contained extensive texts on the contraceptive and abortion properties of hibiscus flowers. These herbal plants contain plant-based compounds called phytoestrogens which are responsible for various potential functions of the body. In a randomized study, some men were given the juice of hibiscus flowers for 45 consecutive days, eventually, it proved to have an antispermatogenetic effect on them. Gudhal is very effective in lowering blood pressure and regulating blood flow. For people with heart problems and high BP, the daily use of hibiscus extract or its dried flowers can make a difference. In addition, the intake of stem extract prevents the accumulation of fat in the blood vessels and reduces the amount of cholesterol in the blood. Due to this process, the blood flow is better improved and therefore the risks of heart attack are effectively reduced. Hibiscus extract has both anti-cancer and anti-tumor properties. Studies have shown that hibiscus juice helps greatly in preventing and suppressing cancerous and life-threatening tumors (Jadhav *et al.,* 2009).

Gudhal flowers or leaf powder containing mild diuretic can easily cleanse your body naturally. The diuretic property of hibiscus flowers is useful in relieving various conditions such as kidney disease, heart disease, high blood pressure, liver cirrhosis (Gupta *et al.,* 2012). In addition, these plants carry out processes that reduce the level of uric acid in the blood. Kidney stones are caused by the accumulation of uric acid in the kidneys, and studies found that *Hibiscus* extract inhibits the growth of kidney stones by effectively reducing the level of uric acid. *Hibiscus* is used traditionally as an excellent medicine to reduce liver inflammation, protect the liver, and enhance liver health (Khan *et al.,* 2017).

Hibiscus root extract was given to people with Hepatic Steatosis (accumulation of triacylglycerols and cholesterol in the liver). At the end of this study, the persistence and development of cholesterol-dependent liver steatosis were found to be significantly lower. The common plant pigment Anthocynidins is abundant in the petals of hibiscus flowers, providing the protective effects needed by the liver (Vankar and Shukla 2011).

Various parts of the hibiscus plant have the ability to effectively lower lipid and blood sugar levels. In particular, the flowers of this plant are famous for treating diabetes. Diabetics have been shown to have potent results in effectively lowering their blood sugar levels by drinking tea made from hibiscus flowers daily. Besides, another animal experiment proved that the extract obtained from the leaves of the *Hibiscus* plant has high antidiabetic properties. Research shows that hydro citric acid is found in the flowers of the hibiscus and this acid is commonly found in medicines prescribed for weight loss processes. So overweight and obese people can significantly reduce their body weight by consuming tea made by dried hibiscus flowers daily. The main reason for the increment of body fat is the increase in the amount of starch and sugar (Missoum 2018).

The enzyme inhibitors in hibiscus flower extract greatly stimulate the release of amylase, which breaks down sugar and starch and prevents fat accumulation in the body. In a study of overweight mice, the flowers of this plant were shown to significantly reduce weight, and they also pave the way for reducing the absorption of starch and sugar. *Hibiscus* flowers act as a natural laxative and reduce constipation problems and have diuretic properties that expel excess fluid from the body. High levels of antioxidants in these plants help slow down the aging process by neutralizing free radicals. In addition, hibiscus plants contribute to the improvement of cognitive and brain functions. Hibiscus flowers contribute greatly for the treatment of various respiratory problems such as asthma, bronchitis, and difficulty breathing. These greatly help in effectively eradicating respiratory tract infections and improving the health of the respiratory tract.

Hibiscus, which has antimicrobial properties, is very helpful in killing the germs in the airways and reducing their damage (Meena *et al.,* 2014). It is noteworthy that the hibiscus has a unique place in the fight against respiratory infections from ancient times to the present day. The extract of hibiscus flowers is an excellent source of vitamin C and is a beneficial remedy for coughs, colds, flu, sore throats, and mouth ulcers. *Hibiscus* is used as a traditional medicine in the treatment of bronchial cataracts. Vitamin C plays an important role in promoting enzymatic production of certain neurotransmitters, repairing tissue, and promoting tissue regeneration. The flower extract strengthens the body's natural defenses as it is an excellent source of high concentrations of vitamin C. It gives strength to the body to fight effectively against infections and physical ailments. The strong antimicrobial properties of these plants help fight viruses and bacteria that weaken the body's immune system. In-vitro studies upon humans demonstrated that the extract of hibiscus flowers has antibacterial activity (Rassem *et al.,* 2017; Sobhy *et al.,* 2017; Ananad and Sarkar 2017).

The flavonoids found in the flowers of *Hibiscus* react appropriately with monoamines to exhibit antidepressant properties. Hibiscus is important in relieving stress, insomnia and relaxing the body, thereby stimulating deep sleep. Those who have difficulty in sleeping and want to experience deep sleep, take the dried flowers or leaf powder of this plant orally. The combination of various chemical compounds in hibiscus leaf extracts accelerates the wound healing process. Furthermore, they help greatly in recovering damaged tissue and healing injured muscle areas. Hibiscus stimulates collagen growth because increasing collagen growth in the injured areas accelerates the healing process (Pekamwar *et al.,* 2013).

Gudhal is very beneficial for women. That is, these plant flowers help in controlling excessive bleeding during menstruation and reducing menstrual cramps. *Hibiscus* helps to soothe hot flashes that suddenly occur on the chest, face, and neck during menstruation, so women can take it in any form not only during menstruation but also on normal days as it has beneficial effects on the body. A decoction of hibiscus flowers is used in many parts of the world, including India and Bangladesh, to regulate the menstrual cycle and to control the problems that occur during it. In China, hot juice made by flowers and bark of *Hibiscus* is used to treat irregular menstruation.

29.5 Skincare Products

The extract from the flower of the hibiscus plant has been used extensively since ancient times to treat skin problems. Hibiscus extract is still used today as an important ingredient in various. The leaves of the hibiscus plant treat various types of hair and scalp problems such as dandruff, itching, hair loss, dull hair, dry hair, and scalp patches. The paste of the leaves moisturizes, brightens, nourishes, strengthens the hair, and controls hair loss. Also, the extract of the leaves gives a darker color to the hair. The tonic made from the petals of these flowers acts as a natural dye and prevents graying of the hair. Research has shown that the petroleum ether extract of the leaves is more powerful than the extract of hibiscus plant flowers, and is more active in promoting hair growth.

Various traditional medicines of the world including Ayurveda highly recommend *Hibiscus* plant leaves to treat hair and scalp problems. However, both the flower and the leaves of this plant promote strong hair growth (Daniel 2008). The tonic of hibiscus flowers contains hyaluronic acid and linolenic acid which can improve the strength of hair and stimulate its growth. Applying hibiscus oil (Figure 29.7) on the scalp stimulates strong hair growth. *Hibiscus* extract, which is a powerful antioxidant, is a rich source of vitamin C. Dark flowers' extract is used to make eyeliners, and in shoe-blacking.

Figure 29.7. Hibiscus Oil.

Hibiscus flowers are a great supplement in the process of preventing acne caused by bacteria and excessive secretion of sebum. High in anti-inflammatory properties, these flowers reduce the appearance of wrinkles and blackheads on the skin. Besides, these enhance the elasticity of the skin thus helping to keep it soft and firm. *Hibiscus* flowers also carry out the process of removing lines, marks, wrinkles, *etc.*, thus retaining the youthful appearance. It acts as a natural moisturizer and treats various skin conditions such as eczema, psoriasis, red patches, white patches, rashes, and scabies. *Hibiscus* keeps the skin glowing, moisturized, supple, and fresh by removing toxins, dead cells, and impurities from the skin.

In the studies, hibiscus has been successfully used as an indicator (Figure 29.8) for a wide range of acid base titrations and even used lower volumes of acids and bases compared to the conventional indicators, such as methyl orange and phenolphthalein (Senathirajah *et al.,* 2017). Moreover, natural pigments, such as cyanidin-3-Sophoroside, are biodegradable, eco-friendly, and non-carcinogenic compared to synthetic pigments.

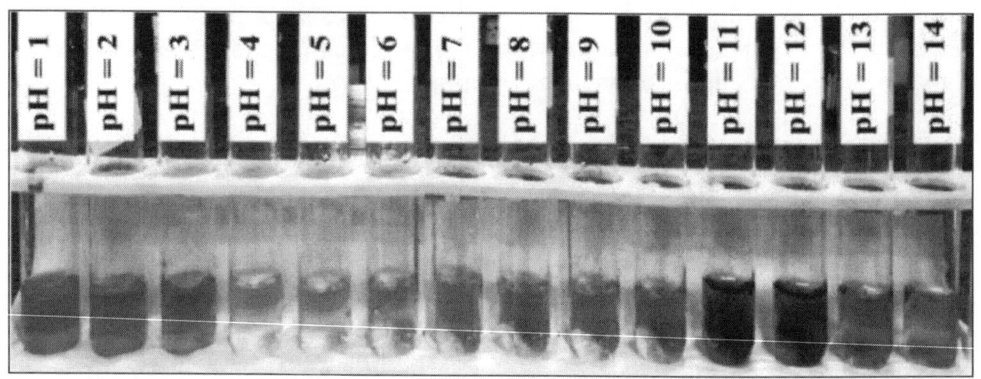

Figure 29.8. Hibiscus Extract as an Indicator at Varying pH.

29.6 Conclusion

Hibiscus, has been widely used as a traditional remedial plant in China and several tropical countries. All of its parts have been used in the treatment of fever, inflammation, bacterial infections, and even as contraceptive agent. Flavonoids, tannins, terpenoids, saponins, and alkaloids are the main phytochemicals present in different extracts, and are more likely responsible for their biological activities. Lower toxicity of this plant can be an advantage to qualify it to be used as new therapeutic agent. *Hibiscus* flowers has been reported to possess antitumor properties, as well as have been used as analgesic, antipyretic, anti-asthmatic, and anti-inflammatory agents. Several studies have proved the presence of anti-oxidant, anti-fungal, and antimicrobial properties in the flowers of Gudhal.

REFERENCES

Anand A., Sarkar B. (2017). Phytochemical screening and antioxidant property of anthocyanins extracts from *Hibiscus rosa-sinensis.* Applicat. Biotechnol. Sustain. Develop., 139–147.

Begam Z., Younus I., Ali SM. (2015). Anti-inflammatory, analgesic and antipyretic activities of *Hibiscus rosa sinensis* Linn and phytochemicals. World J. Pharm. Pharmaceut. Sci., 4(12): 116–123.

Bhakta S., Das SK. (2017). A review on *Hibiscus rosa sinensis* : a rural traditional medicine for contraception. Int. J. Engg. Inf. Syst., 1(9): 77–79.

Daniel M.(2008) Healing herbs for eves. Dept. of Botany. M.S. University of Baroda, Vadodara pp. 111.

El Sayed ZI., Ateya A-MM., Fekry M. (2012). Macro- and micromorphological study of the leaf, stem, flower and root of *Hibiscus rosa-sinensis* l. J. Appl. Sci. Res., 8(1): 34–56.

Gupta P., Jain P., Jain, PK. (2012). Isolation of natural acid base indicator from the flower sap of *Hibiscus rosa sinensis*. J. Chem. Pharm. Res., 4: p. 4957-4960.

Hammad I. (2009). Genetic Variation among *Hibiscus rosa-sinensis* (Malvaceae) of different flower colors using Issr and Isozymes. Aust. J. Basic Appl. Sci., 3(1): 113–125.

Jadhav VM., Thorat RM., Kadam VJ., Sathe NS. (2009). *Hibiscus rosa sinensis* Linn – "Rudrapuspa": A Review. J. Pharma. Res., 2(7): 1168–1173.

Khan IM., Rahman R., Mushtaq A., Rezgui M. (2017). *Hibiscus rosa-sinensis* L. (Malvaceae): Distribution, Chemistry and Uses Int. J. Chem. Biochem. Sci., 12: 147-151

Kumar A., Singh A. (2012). Review on *Hibiscus rosa sinensis*. Int. J. Res. Pharmaceut. Biomed. Sci., 3(2): 534–538.

Kumari OS., Rao NB., Reddy VK. (2015). Phyto-chemical analysis and antimicrobial activity of *Hibiscus rosa-sinensis.* World J. Pharma. Pharmaceut. Sci., 4(5): 766–771.

Mak YW., Chuah LO., Ahmad R., Bhat R. (2013). Antioxidant and antibacterial activities of Hibiscus (*Hibiscus rosa-sinensis* L.) and Cassia (*Senna bicapsularis* L.) flower extracts. J. King Saud. Uni. Sci. 25(4): 275–282.

Meena AK., Jain A., Pandey K., Singh RK. (2014). Acute toxicity and genotoxic activity of *Hibiscus rosa sinensis* flower extract. Am. J. Phytomed. Clin.Therapeut., 2(4): 524–529.

Missoum A. (2018). An update review on *Hibiscus rosa sinensis* phytochemistry and medicinal uses. *J.* Ayur. Herb. Med., 4(3): 135-146

Noman A., Aqeel M., Javed MT., Zafar S., Ali Q., Islam W., Irshad MK., M. Buriro M., Kanwal H., Khalid N., Khan S. (2017). Histological changes in *Hibiscus rosa-sinensis* endorse acclimation and phytoremediation of industrially polluted sites. The J. of An. Plant Sci., 27(5): 1637–1648.

Pekamwar SS., Kalyankar TM., Jadhav AC. (2013). *Hibiscus rosa-sinensis* : a review on ornamental plant. World J. Pharma. Pharmaceut. Sci., 2(6): 4719–4727.

Rassem HHA., Nour AH. Yunus RM. (2017). GC-MS analysis of bioactive constituents of Hibiscus flower. Aust. J. Basic Appl. Sci. (3): 91–97.

Salem MZM., Olivares-Pérez J., Salem AZM. (2014). Studies on biological activities and phytochemicals composition of Hibiscus species- A review. Life Sci. J., 11(5): 1–8.

Senathirajah T., Rasalingam S., Ganeshalingam S. (2017). Extraction of the Cyanidin-3-Sophoroside from *Hibiscus rosa-sinensis*: An efficient natural indicator over a wide range of Acid-Base Titrations Nat. Prod. Plant Resour., 7(3): 1-7

Sobhy EA., Abd Elaleem KG., Abd Elaleem HG. (2017). Potential antibacterial activity of *Hibiscus rosa sinensis* Linn flowers extracts. Int. J. Curr. Microbiol. Appl. Sci., 6(4): 1066–1072.

Udo IJ., Ben MG., Etuk CU., Tiomthy AI. (2016). Phytochemical, proximate and antibacterial properties of *Hibiscus rosa-sinensis* L. Leaf. J. Med. Plants Stud., 4(5): 193–195.

Vankar PS., Shukla D. (2011). Natural dyeing with anthocyanins from *Hibiscus rosa sinensis* flowers. J. Appl. Poly. Sci., 122(5): 3361–3368.

Chapter 30

Beneficial Aspects of Supari: Betel Nut (*Areca catechu* L.)

From religious ceremonies to mouth-fresheners, 'paan supari' has been an inseparable combination. We know a lot about the 'paan' leaves, but not enough attention has been given to 'supari' or betel nuts. Betel nuts are the seeds of the fruits of palm trees which are mostly cultivated in eastern countries like Bangladesh, India, Malaya, Ceylon, Philippines and Japan. Scientifically called Areca catechu, is botanically a nut chewed by large numbers of people throughout the world. Traditionally, betel nut was regarded as useful to oral hygiene, appetite as well as saliva production. Scientists now realize that chewing betel nut boosts the chance of cancers of the mouth and also the esophagus, and that it can adversely affect the health of kids within the womb. However, there are many possibly beneficial aspects of betel nut, many experts agree that the risks result in the substance unhealthy. In spite of the name "betel nut," the substance that individuals chew in lots of areas of the world is generally a combination of two plant items as well as lime minerals. The nut portion is definitely the areca nut, which will come through the palm tree Areca catechu. The nut and lime is used within a wrapping of leaves from the betel vine (Piper betel). In other preparations all of the components are combined together as powder. Chemically it has isoguvocine, choline and alkaloids arecaine, arecoline, arecaidine and Guvacine. Nut contains tannin, oil, alkaloids, polyphenols such as flavonoids. Other alkaloids found in it are caidine, arecolidine, guvacoline, guvacine, norarecoline and isoguvacine norarecaidine. Some research suggests that betel nut utilize may relieve a few of the symptoms of schizophrenia, although not enough evidences, are available. Betel nut extract is a great idea for stroke recovery, in accordance with Inteli-Health. Improvements have been mentioned in speech, bladder control as well as muscle strength in patients taking betel nut extract, even though Inteli-Health explains that studies so far have already been small and flawed. In vitro researches show that water extract of betel nuts exhibit anti-microbial activity towards particular strains of oral bacteria. They're good for making up for severe iron deficiency and low blood glucose levels. Frequent, however moderate, usage of betel nut might help combat an iron deficiency.

***Keywords**: Beneficial aspects, Supari, Betel nut, Areca catechu, Chinese medicine.*

30.1 Introduction

Betel nut is the dried seed of the fruit from the Areca palm, which is cultivated particularly in South and Southeast Asian countries. According to the statistics, India, Malaysia, and Bangladesh have the world's largest annual production in the world. Meanwhile, betel nut is an important traditional Chinese medicine (TCM), which has been used since antiquity (Peng 2017). It is estimated that about more than 600 million people chew and eat betel nut (Volgin *et al.,* 2019). In addition to its medicinal use, there is a widespread abuse of addiction of betel nut, which is considered the fourth most used stimulant.

Other areas of medicine where the substance might offer small benefits consist of recovery from stroke and also the intestinal situation ulcerative colitis, however, not enough proof exists to demonstrate this. The nut is considered to have somewhat beneficial effects on the protection against tooth cavities. Although betel nut might have obvious desired biological effects on your body, the serious chances of the nut should also be taken into consideration. Chewing betel nut may cause cancers of the mouth, pharynx and esophagus. Just before cancer develops, the person utilizing betel nut can have problems with serious mouth problems to the extent where the mouth cannot open. Additionally, pregnant women who use betel nut can slow the development of their baby therefore it might be born underweight. The roots and leaves of the tree have got excellent therapeutic uses (Chusri *et al.,* 2013). The nut of the tree also offers therapeutic uses. The nut shouldn't be taken whenever young. Young fresh nuts can cause gidding. Once the nuts get ripe, they may be dried and utilized for numerous reasons. Listed here are some health advantages of betel nuts:

It's been discovered that those who have suffered a brain injury have demonstrated some reaction to betel nuts. Usage of betel nuts helps to improve speech as well as muscle control in such people. However, it may also have unwanted effects when utilized on other sorts of medicine. Consuming betel nuts assist in treatment of stomach worms just like the tapeworms and also the roundworms. It will help avoid the problems associated with the stomach worms. Betel nut is among the best vermifuge that can help expel the stomach worms. Arecoline is among the bio-chemicals present in betel nut. This particular bio-chemical has been analyzed to have a considerable control on diabetes. Tests on lab animals showed control of blood glucose levels for the considerable time period. Other research has linked chewing betel nut with an increase of chance of diabetes (Amudhan *et al.,* 2012).

Betel nuts include antioxidants which help to avoid cellular degeneration within the older ones. Studies have shown that the tannins contained in betel nut are helpful in managing high blood pressure activity by inhibiting the reaction to both angiotensin I and II. Gnawing betel nuts does make a feeling of well-being and euphoria. This may be one more reason behind its increased use. The elated feeling can make users happy. Users also have reported experiencing a warm feeling in the body after consuming this particular psychoactive substance. Chewing betel nuts helps prevent nausea. Betel nuts could be chewed just before traveling to avoid vomiting sensation. Stomach problems just like indigestion, diarrhea and stomach aches is often curable. Consuming betel nuts provides rest from dysentery and in addition helps prevent it when consumed regularly. Chewing the betel nuts help reduce excessive heat through the body of a human. Consuming betel nuts with betel leaves can be quite beneficial in the summers. In accordance with Ayurveda, betel nut is considered to encourage the nervous system (Shen *et al.,* 2017). Research into betel nut shows that it does have got anti-depressant qualities. Tests on lab rats recommended that betel nut extracts decreased stress brought on by swimming and tail-suspension.

Taking betel nut is assigned to rise in the immunomodulatory response. The extracts of betel nut demonstrated modulation of arachidonic acid metabolism which influences the rise of platelet production. The reason behind its increasing popularity is its capability to boost energy. Thanks to its stimulating effects, users experience elevated levels of energy quickly upon chewing these types of nuts. Many state that it wakes them up and offers energy to continue with their day-to-day routine. So chewing it might work to keep you wide awake for a long time, that is especially beneficial whenever driving over night. The heightened alertness as well as increased ability to handle task comes from its active ingredient are coline, an alkaloid along with a powerful stimulant (Inokuchi 1987), the betel nut is *A. catechu*. In India, it is commonly known as gue and kwai in Meghalaya, Tamul in Assam and Nagaland, kua and kuhva in Mizoram and Manipur, supari in Maharashtra. Chali, parcha, neetadaka, Iylon, Nuli, Kalimak, Bura tamul and tamol are the common varieties of betel nuts. In this article, we will learn all that you might want to know about betel nuts, including its health benefits, side effects and precautions, to name a few Amudhan *et al.* (2012), Peng (2017) and Volgin *et al.* (2019).

Therefore, it is stated that its consumption might safeguard teeth from cavities. Chewing on betel nuts might prevent dental plaque. Betel nut might have anti-bacterial effects, also it formerly was included like a toothpaste component to prevent cavities, in accordance with InteliHealth (Asthana *et al.,* 1996). Due to its toxic effects, however, betel nut is probably less beneficial compared to other restorative agents for dental reasons. Chewing pan along with betel nut has become a common habit. It really is considered to avoid tooth decay. However this must be taken from time to time. Regular intake leads to inflammation resulting in cancer (Nouri *et al.,* 2014).

Preliminary research signifies enhancements in symptoms for schizophrenia patients who take betel nut, in accordance with the NIH. Standard schizophrenia medications create unpleasant negative effects, fueling further efforts to find out new treatments, however the NIH notes that betel nut also created side effects of tremors as well as stiffness within the schizophrenia research (Maurizio and Rafaella 2011; Graziano *et al.,* 2017). Individuals who chew betel nut have a tendency to create considerable amounts of saliva, in accordance with Inteli Health. This might help individuals with dry mouth brought on by health problems like diabetes and Sjogren's syndrome. Inteli Health explains that safer medicines are for sale to reduce this symptom. Betel nuts have been in use for several years like a medication to deal with anemia in pregnant women.

30.2 Chemical Constituents

Betel nuts are also rich in phytochemicals which have a high medicinal value. The phytochemicals present in betel nuts include alkaloids like arecoline, guvacoline, guvacine, arecaidine and polyphenols which mostly include flavonoids like leucocyanidins, catechins, epicatechin and tannins. Betel nuts have importance in Hindu religious ceremonies and are also called the fruit of divine origin. Apart from this, there are numerous benefits of betel nuts, some of which are described below. Nutritional contents and phytochemicals are presented in Tables 30.1 and 30.2 respectively.

Table 30.1. Nutritional Value of Betel Nuts

Nutritional Components	*Percentage*
Carbohydrates	20 per cent
Fats	15 per cent
Phosphorus	0.13 per cent
Calcium	0.05 per cent
Iron	0.01 per cent

Table 30.2. Phytochemical Constituents of Betel Nuts

Phytochemical Components	*Percentage*
Leucocyanidins	12 per cent
Catechins	10 per cent
Arecoline	0.75 per cent
Guvacoline	0.75 per cent
Epicatechin	0.25 per cent
Guvacine	0.29 per cent
Arecadine	0.1 per cent

30.3 Areca Nut Based Cropping System

Areca catechu is an economically important plantation crop predominantly grown in West coast and North East regions of India. The economic part of the palm is called as 'betel nut' and is mainly used for masticatory purpose in many parts of Asia. It has several alternate uses and all parts of the palm are useful. It is cultivated in 0.42 million ha in India with a total production of 0.52 million tonnes and productivity of 1249 kg/ha. Compatibility of different crops is an important factor in the success of any cropping system. The criteria for selection of component crops are: 1) ability of crop to grow under shade, 2) ability to withstand heavy dripping during monsoon, 3) availability of local market and 4) local weather condition. A large number of annuals, biennials and perennial crops like banana, ginger and chilli (Bhat *et al.,* 2014).

Figure 30.1. Nuts of *Areca* (Showing hard endosperm).

Evaporation, air and soil temperatures are lower in mixed cropping than in sole arecanut and open conditions. The least diurnal variation in microclimate within the mixed cropping is an important feature besides lower evaporative demand, reduced wind velocity and higher soil moisture content. This is very important as the synergistic effects of wind and solar radiation can cause mechanical injury in shade plants like cocoa. The arecanut palms provide excellent protection to the intercrops against wind and vice versa. Crops like cocoa and banana limit the damage of sun scorching effect in initial years. The high-density multi-species cropping system involving arecanut, pepper, cocoa and banana indicates that recycling of organic matter could meet 50 per cent each of N and P requirement. Similarly, a six-year study to quantify the feasibility of economising the fertilizer use in the cropping model involving arecanut-banana-cocoapepper reveals that fertilizer requirement could be scaled down to 66 per cent of the recommended dose. Generally, mixed cropping/multiple cropping does not increase the water requirement. The transpiration losses may increase in mixed cropping but evaporation and run-off losses are likely to be reduced because of crop cover and presence of residues with increased soil moisture storage and water-use efficiency. No serious pest and disease problems are observed due to inter/mixed cropping in areca plantations (Bhat *et al.*, 2014).

Based on the comparative yield trials of indigenous and exotic accessions, promising types have been identified and released as varieties for commercial cultivation. They are Mangala, Sumangala, Sreemangala, Mohitnagar, Swarnamangala, Kahikuchi, Nalbari and Madhuramangala. Hybridization and exploitation of dwarfing genes resulted in breeding dwarf and high-yielding varieties. Dwarf hybrids with high yield potential directly benefit the growers by way of enhanced returns and reduced cost of various cultural operations like harvesting and spraying and reduced damages to palms due to sun-scorching and heavy wind. Selection of typical hybrid seedlings in the nursery is must. For the purpose of hybridization, parental blocks of Hirehalli dwarf and released high yielding tall varieties, *viz.* Sumangala and Mohitnagar have been established at CPCRI Regional station, Vittal and Research centre, Kidu (Ananda and Chowdappa 2014).

30.4 Potential Uses in High Blood Pressure and Glucose

By exercising regularly and following a healthy diet consisting of fruits, vegetables, and nuts may have a positive impact on blood pressure. A study conducted by Chen *et al.* (2021) stated that betel nuts might improve blood circulation and helps in reducing high blood pressure. The exact mechanism behind this effect is not known.

The findings of Musdja *et al.* (2020) showed that betel nut consumption in diabetic rats helped in reducing blood glucose. This effect is attributed due to the presence of alkaloids like arecoline and guvacoline. This indicates that betel nuts have the potential to reduce blood glucose levels (Yuan *et al.*, 2012).

30.5 Uses of Betel Nuts on Lipid Profile

A lipid profile is a complete cholesterol test which measures total cholesterol, triglycerides, low-density lipoprotein (LDL), high-density lipoprotein (HDL), *etc.* Peng *et al.* (2015) conducted a review, and the summary estimates of this review state that betel nuts may help in lowering elevated lipid levels. However, we need more studies to claim these results with greater reliability. Also, it is advised not to consider the consumption of betel nuts as an alternative to modern medicine; you should always consult your doctor for proper management of an abnormal lipid profile (Sen *et al.*, 1989; Bhat *et al.*, 2017).

30.6 Uses of Betel Nuts in Infections

An infection occurs when germs like bacteria, viruses, fungi invade the body. Ayurveda or the Indian system of medicine, supports the use of household remedies for managing infections. According to Peng *et al.* (2015) stated that betel nuts might help inhibit the growth of bacteria, viruses, fungi and parasites like tapeworms and ringworms. This effect is due to the presence of polyphenols and alkaloids in betel nuts. This indicates that these nuts may have the potential to manage infections.

30.7 Uses in Allergies and Skin Irritation

Betel nut helps in reducing allergies. Peng *et al.* (2015), summarised estimates of which state that betel nuts may have the potential to manage skin irritation and provide relief from allergies. The exact mechanism behind these effects is unknown; however, it can be attributed to their property to reduce inflammation.

30.8 Betel Nuts Useful in Digestion

Chen *et al.* (2021) stated that arecolinem present in betel nuts, increases the secretion of human saliva and increases the contractions of the digestive system, which improves digestion. Traditionally, they have been used for managing constipation. Therefore, betel nuts may positively impact the digestive system (Sericuly *et al.,* 2021).

30.9 Conclusion

It may help reduce cough and control other diseases. Betel nuts may provide relief from toothache. Betel nuts may provide relief from skin irritation.It may help manage anxiety and depression. It may help in removing excess salt and water from the body and help in managing oedema (Kumar *et al.,* 1976; Peng *et al.,* 2015). Antioxidant and cytotoxic phenolic compounds are present in areca nut (Zhang *et al.,* 2010).

REFERENCES

Amudhan M.S, Begum, V.H., Hebbar,K.B. (2012) "A review on phytochemical and pharmacological potential of *Areca catechu* L. Seed," International Journal of Pharmaceutical Sciences and Research, 12, (8) 2012.

Ananda K.S and Chowdappa P.(2014)Improved varieties to make India self-sufficient in arecanut Indian horticulture 59: 63-65

Arjungi KN. (1976) Areca nut: a review. Arzneimittel-forschung. 1976 Jan 1; 26(5): 951-6.

Asthana S, Greig, NH H. W. Holloway H.W. *et al.* (1996) Clinical pharmacokinetics of arecoline in subjects with Alzheimer's disease, Clinical Pharmacology and therapeutics, 60(3): 276–282.

Bhat S.K., M. Sarpangala M., and D. Ashwin D,(2017) "Antilipidemic activity of arecanut, *Areca catechu* L.: a valuable herbal medicine," International Journal of Herbal Medicine, 5(1): 35–38. https: //pubmed.ncbi.nlm.nih.gov/786304/

Bhat R., Sujatha S. and Chowdappa P. (2014) Arecanut-based cropping system fetches more return. Indian horticulture 59: 67-69

Chen X, He Y, Deng Y. Chemical composition, pharmacological and toxicological effects of betel nut. Evidence-Based Complementary and Alternative Medicine. 2021 Aug 18; 2021. Available at: https: //www.ncbi.nlm.nih.gov/pmc/articles/ PMC8387188/

Chusri S, S. Settharaksa,S. J. Chokpaisarn,J. S. Limsuwan, S and S. P. Voravuthikunchai S.P. (2013), "+ai herbal formulas used for wound treatment: a study of their antibacterial potency, anti-in- flammatory, antioxidant, and cytotoxicity effects, Journal of Alternative and Complementary Medicine, 19(7)671-676

Graziano S, Orsolini L, Rotolo MC, TittarelliR, Schifano F, Pichini S. (2017) Herbal Highs: review on psychoactive effects and neuropharmacology," Current Neuropharmacology, 15: 750–761.

Inokuchi,E (1987) Antihypertensive substance in seeds of Areca catechu L," *Journal of Pharmacobio- Dynamics*, 10: 62,

Kumar A, Oswal K, Singh R, Kharodia N, Pradhan A, Sethuraman L, Venkataramanan R, Purџshotham A. Assessment of areca nut use, practice and dependency among people in Guwahati, Assam: a cross-sectional study. Ecancermedicalscience. 2021; 15. Available at: https: //www.ncbi.nlm.nih.gov/pmc/articles/ PMC8043683/

Maurizio C. and Raffaella M. (2011) Potential action of betel alkaloids on positive and negative symptoms of schizophrenia: a review," *Nordic Journal of Psychiatry*, vol. 66, 2011.

Musdja MY, Nurdin A, Musir A. (2020) Antidiabetic effect and glucose tolerance of areca nut (*Areca catechu*) seed ethanol extract on alloxan-induced diabetic male rats. In IOP Conference Series: Earth and Environmental Science 462, (1): 012036). IOP Publishing. Available at: https: //iopscience.iop.org/ article/10.1088/1755-1315/462/1/012036/pdf

Nouri L., A. Mohammadi N. A, and Karim A.A. (2014) "Phytochemical, antioxidant, antibacterial, and α-amylase inhibitory properties of different extracts from betel leaves," Industrial Crops and Products, 62: 47–52.

Peng, W.(2017) "Using the "target constituent removal combined with bioactivity assay" strategy to investigate the optimum arecoline content in charred areca nut," *Scientific Reports*, vol. 7, Article ID 40278, 2017.

Peng W, Liu YJ, Wu N, Sun T, He XY, Gao YX, Wu CJ. (2015) *Areca catechu* L. (Arecaceae): A review of its traditional uses, botany, photochemistry, pharmacology and toxicology. Journal of ethno pharmacology. 164: 340-56.

Sen S., Talukder G, and Sharma A. (1989) "Betel cytotoxicity," Journal of Ethnopharmacology, 26 (3): 217–247.

Serikuly N., E. T Alpyshov E.T., D Wang D. (2011) Effects of acute and chronic arecoline in adult zebrafish: anxiolytic-like activity, elevated brain monoamines and the potential role of microglia, Progress in neuro-psychopharmacology and biological psychiatry, 104

Shen X., W. Chen W., Y. Zheng Y. (2017) Chemical composition, antibacterial and antioxidant activities of hydrosols from different parts of *Areca catechu* L. and *Cocos nucifera* L, Industrial Crops and Products, 96: 110–119

Volgin, A.D., Bashirzade,A. Amstislavskaya T.G. (2019) "DARK classics in chemical neuroscience: arecoline," ACS Chemical Neuroscience, 10, (5): 2176–2185.

Yuan, Y., Yang D, Liang Y. (2012) Alkaloids from areca (betel) nuts and their effects on human sperm motility in vitro. Journal of Food Science, 77 (4): T70–T78.

Zhang X., J. Wu J., Z. Han Z., W. L. Mei W.L., and H. F. Dai H.F. (2010) Antioxidant and cytotoxic phenolic compounds of areca nut (*Areca catechu*)," Chemical Research in Chinese Universities, 26: 161–164.

Chapter 31

Extensively Consumed Beverage: Boisterous Coffee Beans (*Coffea arabica* L.)

Globally coffee (Coffea arabica L. C. robusta) are considered amongst the commonly used beverages due to the presence of chemical components like caffeine, cafestol, kahweol and many micronutrients. It shows beneficial effects when consumed adequately. Coffee beans are roasted and extolled for its flavor and smell. Besides caffeine it contains numerous components such as, alkaloids, nitrogenous compounds, carbohydrates, minerals, vitamins, phenolic and lipids. Coffee consumption is supposed to inhibit the absorption of calcium, iron and zinc. Many epidemiological studies demonstrate the health risk of large consumption of caffeine due to its strong relation with inadequate behavior including sedentary lifestyle and smoking. Most of the researches in humans on health effects are observational. However, results of recent studies exhibited that drinking coffee in recommended amounts may decrease the risk of chronic ailments. Results of many studies indicated that consumption of coffee provide the prevention against diabetes, Parkinson's diseases and colorectal cancer. However, coffee intake is also contributing the negative impact on body and increased the risk of developing heart diseases, stroke, cardiac arrhythmia's and hypertension due to the presence of lipid content in it which enhances the blood lipid profile of human body. The substances accountable for the weight-controlling results of coffee are usually anti-oxidants known as the chlorogenic acids. In the present review useful medicinal aspects of an important non alcoholic beverages are highlighted.

Keywords: *Diabetes, Heart disease, Caffeine, Chlorogenic acids, Diterpenes, Stimulant, Coffea arabica, Coffea robusta.*

31.1 Introduction

Coffea is a genus of flowering plants whose seeds, are popularly called as coffee beans. It belongs to Rubiaceae family. They are shrubs or small trees, which usually grow to a height of 3-3.5 m, and are native to tropical and southern Africa and tropical Asia. Coffee is one of the most commonly consumed beverages prepared from *C. arabica* beans worldwide. As such, even small individual health effects could be important on a population scale. There have been mixed conclusions as to whether coffee consumption is beneficial or harmful to health. Roasted coffee is a complex mixture of over 1000 bioactive compounds, some with potentially therapeutic antioxidant, anti-inflammatory, antifibrotic, or anticancer effects that provide biological plausibility for recent epidemiological associations (Bae *et al.,* 2014).

Fundamental active compounds include caffeine, chlorogenic acids, and the diterpenes, cafestol and kahweol. Existing research has explored the associations between coffee as an exposure and a range of outcomes including all cause mortality, cancer, and diseases of the cardiovascular, metabolic, neurological, musculoskeletal, gastrointestinal, and liver systems. Coffee polyphenols, such as chlorogenic acids contribute to reducing mortality and morbidity risks. Coffee also has anti-inflammatory properties, and is protective against oxidative damage. Chlorogenic acid is really a main element of coffee which could supply much more of a reason for coffee's impact on chance for type 2 diabetes (Dam and Feskens 2002). Coffee has been for decades the most commercialized food product and most widely consumed beverage in the world. Intake of moderate quantity (150-200ml) of coffee per day seems to be beneficial and is related with favorable outcomes for human body. Many clinical trials indicated that six to seven cups of coffee enhanced the severity of cardiac arrhythmias. Caffeine consumption is negatively associated with Ca levels (Heckman *et al.,* 2010). Excessive intake of coffee also related to mineral deficiencies including Fe and Zn. Polyphenols and phytates contents present in coffee beans inhibit the iron and zinc absorption/bio-availability from intestine. Coffee is combination of complex chemical compounds which play significant roles in human body. It is considered as a primary source of caffeine intake among teenagers and adults (Esquivel and Jiménez 2012; Brian *et al.,* 2017).

Results of many studies indicated that consumption of coffee provide the prevention against diabetes, Parkinson's diseases and colorectal cancer, while The purpose of this review article is to understand the theory of recent researches on the health benefits and risks in human beings by consuming coffee. The increased consumption of coffee have association of developing heart diseases, stroke, cardiac arrhythmias and hypertension, still birth and miscarriages. Coffee provides the highest proportion of antioxidants in the diet of the populations of some European countries, including Italy, Finland, France and Spain, and is one of the beverages that has been most widely studied in epidemiological studies with respect to health benefits (Matos *et al.,* 2016; Poole *et al.,* 2017).

Figure 31.1. Coffee Plant, Leaves, Flowers, Beans (Seeds).

31.2 Chemical Constituents

The complex beverage, coffee contains thousands of compounds including flavor, aroma, bioactive components such as methylxanthines (*e.g.,* caffeine, theobromine, and theophylline) diterpenes (*e.g.,* cafestol, kahweol), chlorogenic acids (CGAs) (*e.g.,* caffeoylquinic acids, feruloylquinic acids, p-coumaroylquinic acids), flavonoids (*e.g.,* catechins, anthocyanins), hydroxycinnamic acids (*e.g.,* ferulic acid, caffeic acid (CA), p-coumaric acid), tocopherols, and melanoidin (Ludwig *et al.,* 2014). However, the amount of bioactive components mainly the caffeine depends on species, type, processing, storage as well as roasting process and temperature. In a home-prepared coffee generally, it contains 30 to 175 mg of caffeine (150 ml). The most common polyphenols in coffee are phenolic acids, mainly CA, a type of trans-cinnamic acid, and its derivative, CGA. A single serving of coffee provides between 20 and 675 mg of CGAs depending on the type of roast, and the volume

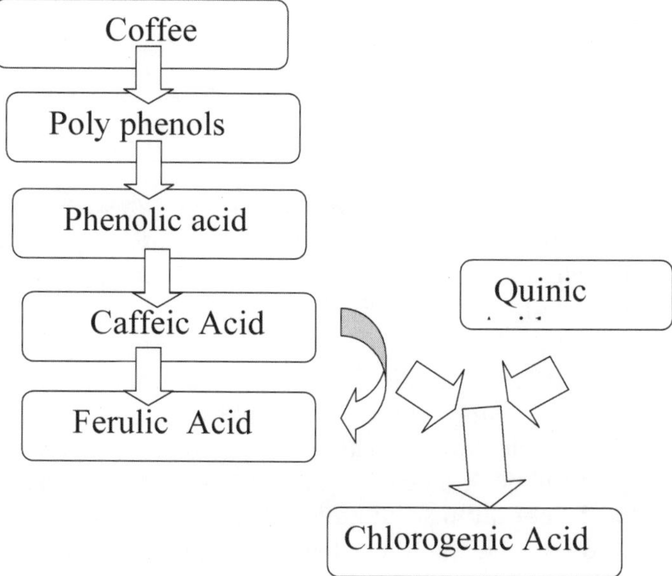

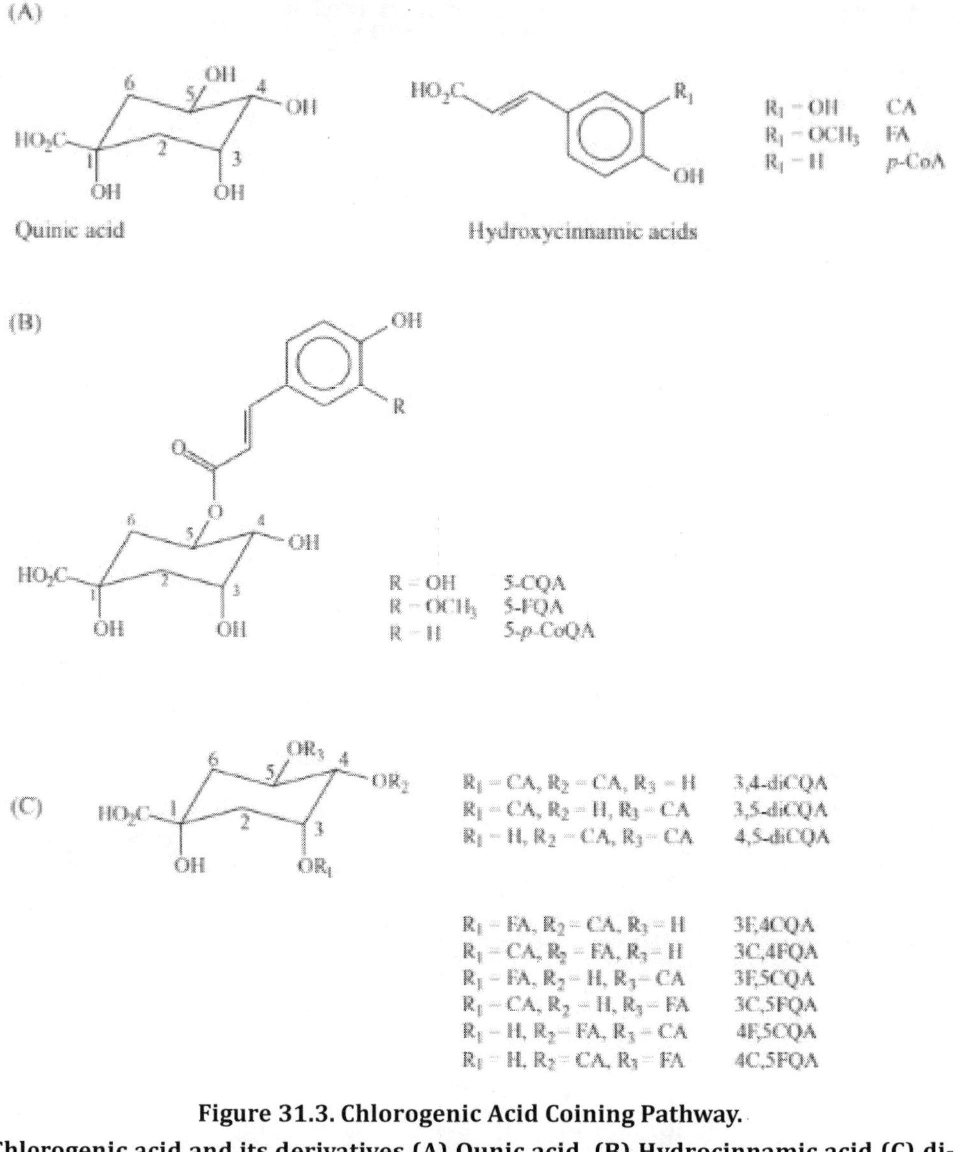

(A)

Quinic acid

Hydroxycinnamic acids

R_1 = OH CA
R_1 = OCH_3 FA
R_1 = H p-CoA

(B)

R = OH 5-CQA
R = OCH_3 5-FQA
R = H 5-p-CoQA

(C)

R_1 = CA, R_2 = CA, R_3 = H 3,4-diCQA
R_1 = CA, R_2 = H, R_3 = CA 3,5-diCQA
R_1 = H, R_2 = CA, R_3 = CA 4,5-diCQA

R_1 = FA, R_2 = CA, R_3 = H 3F,4CQA
R_1 = CA, R_2 = FA, R_3 = H 3C,4FQA
R_1 = FA, R_2 = H, R_3 = CA 3F,5CQA
R_1 = CA, R_2 = H, R_3 = FA 3C,5FQA
R_1 = H, R_2 = FA, R_3 = CA 4F,5CQA
R_1 = H, R_2 = CA, R_3 = FA 4C,5FQA

Figure 31.3. Chlorogenic Acid Coining Pathway.

Chlorogenic acid and its derivatives (A) Qunic acid, (B) Hydrocinnamic acid (C) di-esters of quinic acid with caffeic acid, and mixed esters; diCQA (dicaffeoylquinic acid); FQA (Feruloylquinic acid); p-coumaroylquinic acid).

consumed. CGAs constitute a family of esters made up of transcinnamic acids, mainly CA andferulic acid, and quinic acid. The resulting forms of CGAs include caffeoylquinic acids, feruloylquinic acids that may be found in several isomeric forms depending on the position of the ester link. Additionally, there can be one or two phenolic acids per quinic acid moiety. 5-O-caffeoylquinic acid is the most common form of CGA, which is often called 'chlorogenic acid.' The metabolism of

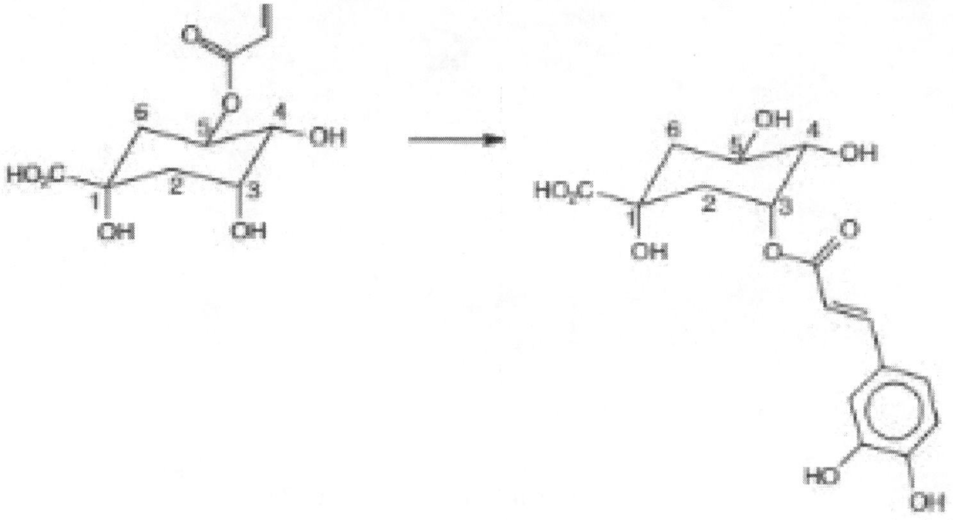

Epimerization

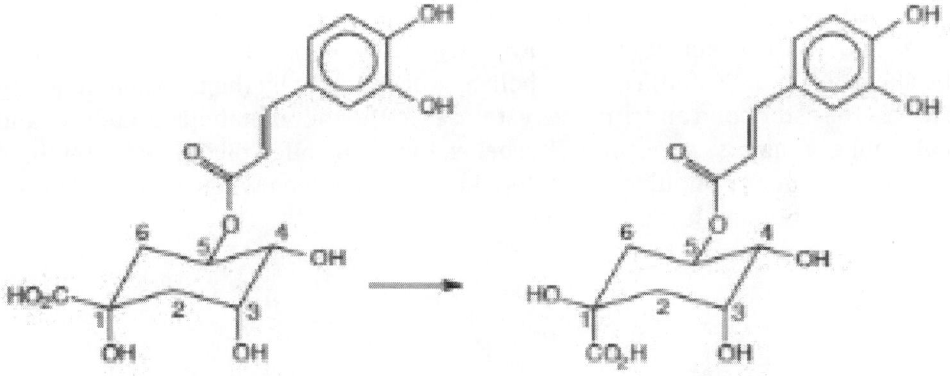

Lactonization

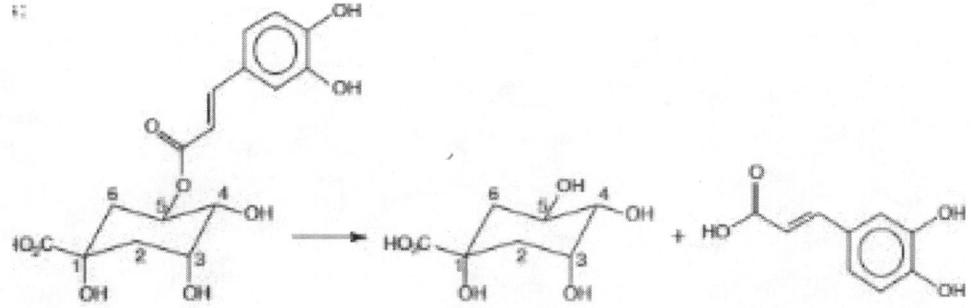

Hydrolysis

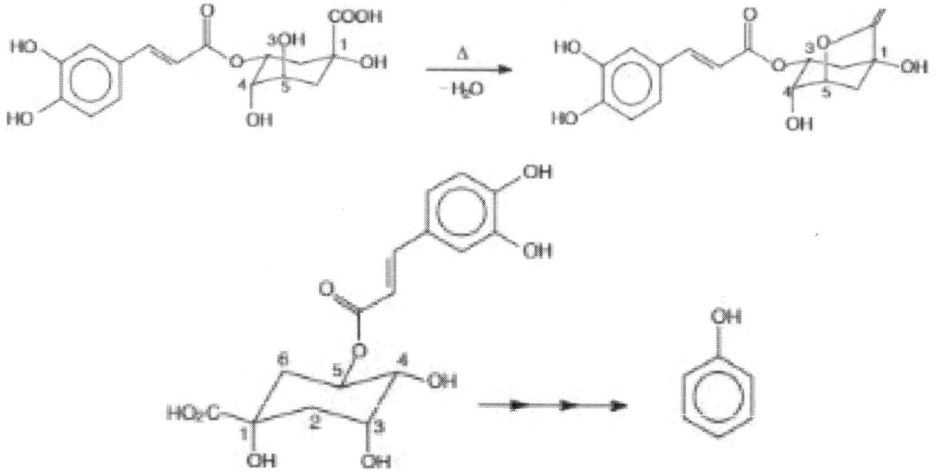

Degradation in smaller compounds

Figure 31.4. Changes in Chlorogenic Acid during Roasting of Coffee Beans.

CGAs is still unclear but thought to be in the small intestine and colon. The first step is carried out by the active esterase enzymes, which generate the original phenolic acids in both the small and large bowel. Around two third of CGAs are thought to be absorbed in the colon. The metabolism is carried out by the microbiota, which cleaves the ester bond and provides esterases for further metabolism. Cafestol and kahweol are the two important diterpenes found in coffee oil. They are retained in part by paper or popular sock filters but are preserved when coffee is directly prepared by boiling the ground beans.

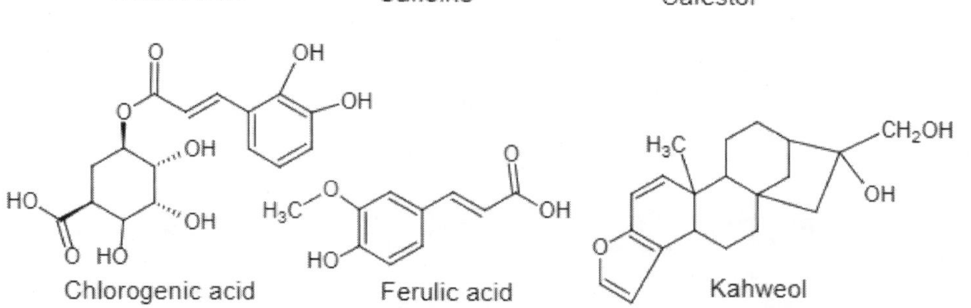

Caffeic acid Caffeine Cafestol

Chlorogenic acid Ferulic acid Kahweol

Main Chemical Constituents Occurring in Coffee Beans.

Main Coffee Diterpenes (A) Cafestol (B) Kathweol.

(A) $H_2N-(CH_2)_4-NH_2$

(B) $H_2N-(CH_2)_4-NH-(CH_2)_3-NH_2$

(C) $H_2N-(CH_2)_3-NH-(CH_2)_4-NH-(CH_2)_3-NH_2$

Chemical Structures of Selected Bigenic Amines in Coffee (A) Putrescine (B) spermidine, and (C) Spermine.

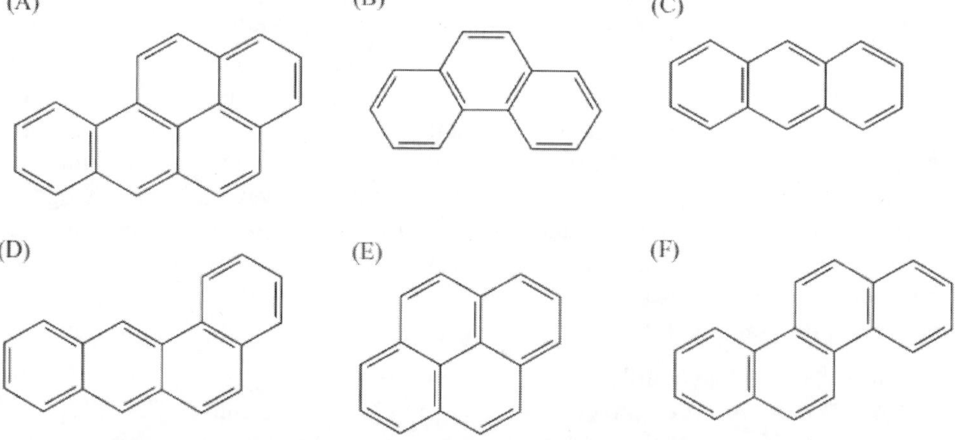

Figure 31.5. Identified Polycyclic Aromatic Hydrocarbons in Coffee (A) Benzopyrene, (B) Phenanthrene, (C) Anthracene, (D) Benzoanthracene, (E) Pyrene, and (F) Chrysene.

Coffee polyphenols, such as chlorogenic acids contribute to reducing mortality and morbidity risks (Fukushima *et al.,* 2014). Coffee also has anti-inflammatory properties, and is protective against oxidative damage. Chlorogenic acid is really a main element of coffee which could supply much more of a reason for coffee's impact on chance for type 2 diabetes. Chlorogenic acid has numerous suggested health benefits on glucose metabolism. In vitro research as well as animal research has shown that chlorogenic acid derivative can easily reduce hepatic glucose output through inhibition of glucose-6-phospatase. Chlorogenic acid furthermore plays a role in anti-oxidant results.

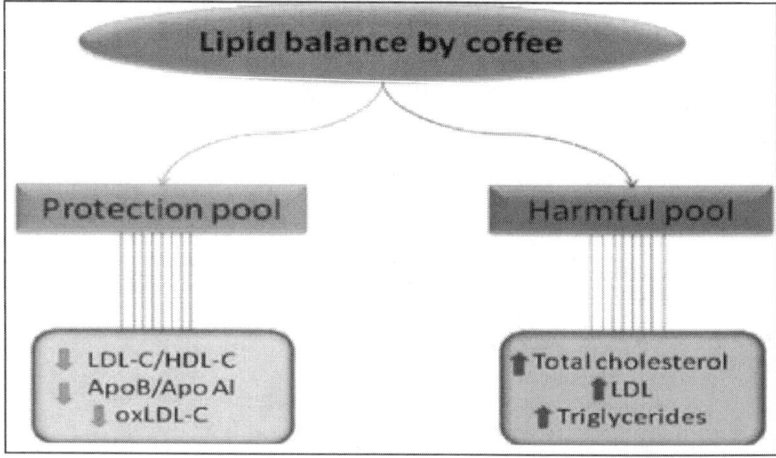

Figure 31.6. Net Lipid Balance Results of Coffee.

Most of these outcomes of chlorogenic acid assist the body to process sugar as well as the cells to reply far better to insulin, that really help deal with problems brought on by type 2 diabetes. A couple of additional elements of coffee that may decrease chance of type 2 diabetes are usually lignans and magnesium. There are actually huge amounts of various lignans within coffee which could have an effect on glucose metabolism by means of anti-oxidant as well as (anti) estrogenic qualities. Coffee is definitely considered to be a protective element from moderate depressive disorders, Parkinson's disease, as well as colon as well as rectal cancers. Right now it seems that substances in coffee furthermore assist to control blood glucose levels, decrease fat manufacturing, and let regular weight reduction.

The substances accountable for the weight-controlling results of coffee are usually anti-oxidants recognized jointly as the chlorogenic acids. These types of acids seem to slúggish the creation of glucose within the body right after a meal, simply by adjusting the action of particular enzymes within the liver. In addition, the chlorogenic acids result in aa lot more sluggish as well as continual discharge of glucose in the body after consuming, therefore decreasing the manufacture of new fat cells. Espresso, produced by steam expressing carefully ground coffee, is abundant with taste as well as smell along with chlorogenic acids, although not very focused at all in caffeine. Consuming an espresso after consuming leads to a reduction of glucose manufacturing as well as release, together with resulting in the body to create a lot more gastric juices; this usually helps digestion of food. In addition, coffee also seems to behave as a precautionary element in type 2 diabetes (Reunanen *et al.*, 2003). Consuming coffee decreases the probability of liver cancer. Cirrhosis is believed to be the main reason for liver cancer. Even though consuming coffee decreases the chance of alcoholic cirrhosis, an easy method to reduce chance of the disease is always to prevent excessive drinking.

Table 31.1. Chemical Composition of *Coffea arabica*

Chemical Constituents	*Raw Coffee g/100g*	*Brewed Quantity g/100g*
Carbohydrates/Fiber		
Sucrose	6-6.9	4.2
Reducing sugars	0.1	0.3
Polysachharides (arabinogalactan, mannon, glucan)	34-44	31-33
Lignin	3	3.0
Pectins	2	2.0
Nitrogenous compounds		
Protein	10-11	7.5-10
FAAs (Free Amino Acids)	0.5	ND
Cafeine	0.9-13	1.1-1.3
Trigonelline	0.6-2	1.2-2.0
Nicotinic acid		0.016-0.026
Lipids		
Coffee oil [triglycerides with USM (unsaponifiable matter)]	15-17	17.0
Diterpene esters	0.5-12	0.9
Minerals	3-4.2	4.5
Acid and Esters		
Chlorogenic acid	4.1-6.9	1.9-2.5
Aliphatic acids	1.0	1.6
Quinic acid	0.4	0.8
Melanoidins		25

31.3 Decaffieination

The process by which caffeine is removed from coffee beans and tea leaves. Coffee and teas made from *Camellia sinensis*, such as black and green teas, naturally contain caffeine, which is a chemical used by the plants to deter herbivores. A number of coffee and tea drinkers relish the taste of their beverage of choice but cannot tolerate the jolt from caffeine. The main methods of coffee decaffeination are based on chemical solvents, carbon filtering, carbon dioxide extraction, or triglycerides. In all cases, to make "decaf," the caffeine is removed in the green bean stage, before the coffee is roasted. Tea typically is decaffeinated by using chemical solvents or carbon dioxide extraction. Regardless of the method of decaffeination, some adulteration of the coffee beans or tea leaves occurs along the way, and in no case is 100 per cent of the caffeine removed. In the United States the Food and Drug Administration (FDA) requires that at least 97 per cent of the caffeine be removed for a tea or coffee to be sold as "decaffeinated." Herbal and fruit teas labeled "caffeine-free" are not decaffeinated but rather are made of plants that do not contain caffeine.

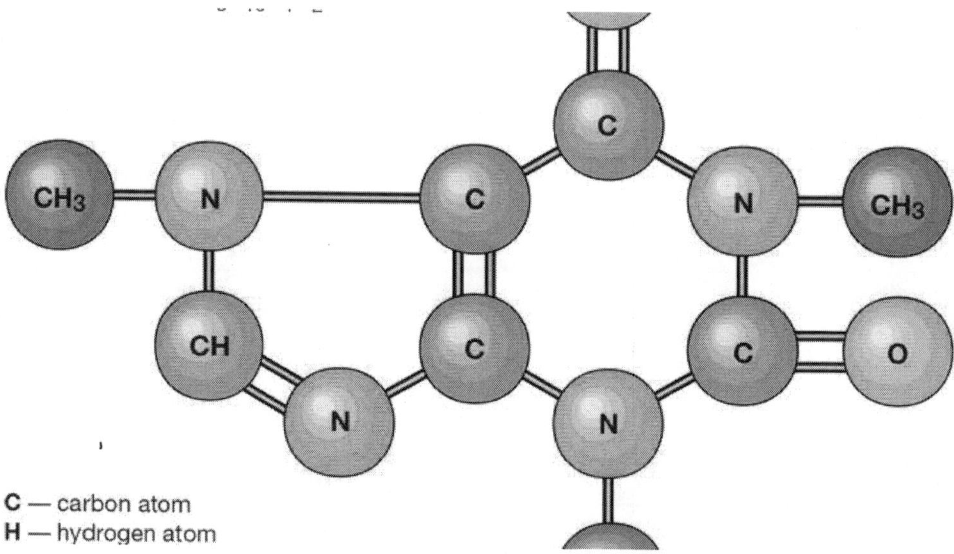

Figure 31.7. Caffeine Molecule.

The process of decaffeination in the early 20th century was initially solvent-based–using benzene, which is now recognized as a carcinogen. Modern solvent-based decaffeination is done mostly with ethyl acetate or, to a lesser extent, methylene chloride. In the direct method, the coffee beans or tea leaves are steamed and then rinsed by the chemical agent. In the indirect method, the chemical agent never touches the beans or leaves but rather treats the water-based coffee or tea solution in which the beans or leaves are first soaked. After the caffeine-laden solvent is removed from the solution, the solution is reapplied to the beans or leaves to restore some of the flavours and oils that were initially removed. Tea makers commonly decaffeinate bagged tea with ethyl acetate, and either of the two solvents is used by some high-volume coffee companies to decaffeinate their products, though other decaffeination processes have gained in popularity. The process is regulated to ensure that all but trace amounts of the chemicals are removed before the coffee is roasted or the tea is packaged. Some countries do not allow the importation of tea or coffee that has been decaffeinated with methylene chloride, as it is considered a possible carcinogen; the FDA has deemed its use to be safe for decaffeination (Tavani 2004).

31.4 Coffee Roasting

The aromatic and gustatory qualities of coffee are developed by the high temperatures to which they are subjected during roasting or broiling. Temperatures are raised progressively from about 180 to 250 °C and heated for anywhere from 7 to 20 minutes, depending on the type of light or dark roast desired. The most important effect of roasting is the appearance of the characteristic aroma of coffee, which arises from very complex chemical transformations within the bean. Roasting too long can destroy volatile flavour and aroma compounds, and *Robusta* beans are often deliberately over roasted (as in the dark French and Italian roasts) to rid the coffee of its natural harshness.

31.5 Traditional Uses of Coffee

The coffee plant was introduced into Cuba in 1748, but coffee farming did not develop into a proper industry until the 1790s, when the island received an influx of French citizens fleeing the Haitian Revolution. Mostly cultivated on plantations in Cuba's mountainous regions, Cuba's coffee boom saw the country become the top global coffee exporter by the 1940s and produce a record 60,000 tons in 1960–61, but the Cuban Revolution of 1959 resulted in a nationalized economy that rapidly decimated the island's coffee industry. Despite coffee's decline as a lucrative crop, *café cubano* remains an integral part of Cuban culture.

Traditionally, different parts of the coffee plants are used for influenza, anemia, edema, asthenia and rage, hepatitis and liver troubles, externally for nervous shock, as a stimulant for sleepiness and drunkenness, as an antitussive in flu and lung ailment, as a cardiotonic and a neurotonic and for asthma. The fruit is taken orally for hepatitis and liver troubles. The soaked fruit is used externally for nervous shock. Hot water extract of the dried fruit is used as a stimulant for sleepiness and drunkenness[9] and as antitussive in flu and lung ailment. The hot water extract of the dried seed is taken orally as a cardiotonic and a neurotonic. The hot water extract of the seed is taken orally for asthma. The root juice is taken orally for scorpion sting. Infusion of green seeds contains caffeine and is drunk in French Guiana for remedying migraine headaches. Coffee is considered to be analgesic, an aphrodisiac, anorexic, antidotal, cardio tonic, stimulant, counterirritant, diuretic, hypnotic, lactagogue and nervine. It is commonly used in folk medicine as a remedy for headache, asthma (Schwartz and Weiss 1992), flu, tropine poisoning, jaundice, migraine, narcosis, nephrosis, malaria, sores, opium toxicity, snake-bite and vertigo. Leaf poultices are used to treat sores in and root sap or root infusions are drunk to relieve scorpion stings in Trinidad. Infusion of green beans is drunk for migraine headaches in French Guiana. Infusion contains tannins which is useful for gout and as a febrifuge. Coffee drinkers have a lower risk of Alzheimer's disease (Prediger 2010). Drinking a cup of caffeinated coffee significantly improves blood flow. Two cups of coffee a day were found to reduce the risk of alcohol-related cirrhosis. Coffee has been found to reverse liver damage caused by alcohol. Drinking 2 to 4 cups of coffee daily has been found to drop the risk of suicide. Coffee improves the reaction time of older adults.

31.6 Instant Coffee

In the manufacture of instant coffee (called soluble coffee in the industry), a liquid concentration of coffee prepared with hot water is dehydrated. This can be done by spray-drying (by drying with a hot gas) or by freeze-drying (a dehydration process known as lyophilization). The operations are complex, and methods vary among manufacturers. The resulting soluble powder, on the addition of hot water, forms reconstituted coffee. The average yield is 25 to 30 per cent by weight of the ground coffee, thereby lowering shipping costs. Instant coffee is obviously quicker to make than brewed coffee, and it enjoys a longer shelf life than coffee beans, but it picks up moisture readily and must be kept dry. The taste of instant coffee is also widely considered inferior to brewed coffee.

31.7 Disadvantage of taking Coffee

Caffeine is really a medicine which is naturally manufactured in the leaves as well as seeds of numerous plants. Additionally it is made synthetically as well as included with particular foods. Caffeine is described as a medicine mainly because it energizes the central nervous system, leading to elevated overall health. Caffeine provides a lot of people a short-term energy boost as well as improves feelings (McCusker 2006). Caffeine is definitely an alkaloid from the xanthine group of substances. Caffeine behaves as a activator of the central nervous system resulting in the heart to beat quicker as well as resulting in feelings of elevated energy, well-being as well as mental focus. Caffeine may be the only medicine which is found naturally or even included with broadly ingested foods. Additionally it is contained in the coffee. The amount of caffeine is decided coffee roast. More dark roasts have got greater proportion amounts of caffeine. Giving up coffee or even soda can make drawback signs and symptoms -just like headaches, tiredness, frustration, anxiousness – even though the brain's biochemistry readjusts. Coupled with extra work pressures, it may improve blood pressure level considerably, ultimately causing an elevated chance of long-term cardiovascular disease (Bashir and Ahsan 2019).

Caffeine plays a role in the decrease in production of hormones manufactured in large quantity in the course of youngsters – DHEA, melatonin, as well as others – which usually speeds aging. Simply because caffeine dehydrates the body, skin manages to lose water through the cells, leading to wrinkles as well as dropping. Caffeine will cause the urinary excretion of Ca, Mg, K, Fe, as well as trace minerals, be responsible for brittle bones. Several ladies are more inclined to be affected by caffeine than the others because of genetic factors; however, it seems like wise to suggest that all more mature ladies have sufficient nutritional calcium, as well as restrict caffeine consumption to a maximum of 300 mg on a daily basis. Mood disruptions just like anxiousness, depressive disorders, frustration, and a focus problem are irritated simply by long-term caffeine usage. As opposed to raising mental activity as numerous caffeine drinkers think, caffeine in fact reduces the flow of blood towards the brain as much as 30 per cent as well as adversely effects memory as well as mental performance. Mood disruptions might appear as soon as the catalyst effect of caffeine wears off – often around three hours later on forcing numerous to take more caffeine plus they might also seem throughout the healing period right after giving up caffeine.

Taking coffee is not recommended for the ladies who are expecting simply because they're more likely to have losing the unborn baby as well as stillbirth. These extra supplies, like cream, sugar along with other seasonings, can also add calories in your daily diet. Excessive caffeine within the body brings about anxiousness, trouble sleeping, insomnia, headaches, muscle tremors, gastrointestinal issues, diarrhea, nausea as well as irregular heartbeat. Coffee is really a powerful diuretic and having it excessively can result in lack of fluids. Lots of people who may have a lot more than 5-6 cups of coffee each day complain of dehydration and other connected health problems. Dehydration in addition dries out the skin, which makes it prone to facial lines as well as stretch-marks. Even though caffeine is recognized to help with short-term weight reduction, in the longer run heavy usage is a lot more prone to result in putting on weight.

31.8 Culinary Uses

Coffee berries, edible and slightly sweet, are eaten occasionally by children and field workers (Rogers 2009; Mejia and Ramirez-Mares 2014)). Coffee seeds have been chewed as a masticatory stimulant in its native Ethiopia since ancient times. Dried coffee beans (seeds) are roasted and ground and brewed to make the coffee, the most popular and widely consumed beverage in the world. Coffee is widely used as flavoring in ice cream, cakes, pastries, candies, and liqueurs. Fermented drink from the pulp is consumed in Arabia. Fruits and seeds have been used for vitamins, minerals. As a food additive, the seeds have been used for condiment/seasoning.

31.9 Conclusion

Results of many studies indicated that adequate consumption of coffee provides the prevention against diabetes, colorectal cancer and Parkinson's diseases. However, coffee may have the negative impact on body and increase the risk of developing hypertension and heart diseases, stroke, cardiac arrhythmias. The complex beverage, coffee contains thousands of compounds for flavor, aroma, bioactive components such as methylxanthines (*e.g.,* caffeine, theobromine, and theophylline). However, the amount of bioactive components mainly the caffeine depends on species, type, processing, storage as well as roasting process and temperature. A single serving of coffee provides between 20 and 675 mg of CGAs. CGAs constitute a family of esters made up of transcinnamic acids, mainly CA and ferulic acid, and quinic acid. Depending on the type of coffee and method of preparation, The caffeine content of a cup of coffee varies depending mainly on the brewing method, and also on the coffee variety. According to the USDA National Nutrient Database, a 240 ml cup of "coffee brewed from grounds" contains 95 mg caffeine, whereas an espresso (25 ml) contains 53 mg.

REFERENCES

Bae JH., Park JH., Im SS., Song DK. (2014). Coffee and health. Integrat, Med, Res.,3: 189- 191.

Bashir S., Ahsan F. (2019). Consumption Coffee: Health Perspectives and Drawbacks. Obes. Nutr J., 21(1): 1-4.

Brian J.,DistelbergBJ.,Staack A., K'dee D.,Elsen, MA., Sabate J. (2017).The Effect of Coffee and Caffeine on Mood, Sleep,and Health-Related Quality of Life*J. Caffei. Res.,* 7: 59-70.

Dam VRM., Feskens EJ. (2002). Coffee consumption and risk of type 2 diabetes mellitus. Lance.,360: 1477-1478.

Esquivel P., Jiménez VM. (2012). Functional properties of coffee and coffee by-products. Food Res. Int.,46: 488-95.

Fukushima Y., Tashiro T., Kumagai A., Ohyanagi H., Horiuchi T., Takizawa K., Sugihara N., Kishimoto Y., Taguchi C., Tani M., Kondo K. (2014).Coffee and beverages are the major contributors to polyphenol consumption from food and beverages in Japanese middle-aged women.J.Nutr Sci., 3: e48.2014 Oct 22. doi: 10.1017/jns.2014.19

Heckman Ma., Weil, J., de Mejia EG. (2010). Caffeine (1, 3, 7-trimethylxanthine) in foods: A comprehensive review on consumption,functionality, safety, and regulatory matters. J. Food Sci.,75: R77–R87.

Ludwig IA., Clifford MN., Lean MEJ., Ashihara H., Crozier A. (2014). Coffee: biochemistry and potential impact on health. *Food Funct.,* 5: 1695-1717. 10.1039/C4FO00042K pmid: 24671262

Matos LA., De Castro E Sousa JM., DeCarvalhoMelo-CavalcanteAA. (2016).Coffee: A health fuel-blot popular drinking. Int. J. Pharma. Pharmaceut. Sci.,8(5): 1-6.

McCuskerRR.,Fuehrlein B., Goldberger BA., Gold MS. (2006). Caffeine content of decaffeinated coffee. J. Anal.Toxicol., 30: 611-613.

Mejia EG., Ramirez-Mares MV. (2014). The impact of caffeine and coffee on our health.TrendsEndocrinol. Metabol.,25: 489-92.

Poole R., Kennedy OJ., Roderick P., Jonathan A Fallowfield JA., Peter C HayesPC., Julie Parkes J. (2017). Coffee consumption and health: umbrella review of meta-analyses of multiple health outcomes BMJ., 359: j5024. doi: 10.1136/bmj.j5024

Prediger RD. (2010) Effects of caffeine in Parkinson's disease: from neuroprotection to the management of motor and non-motor symptoms. J Alzhei. Dis., 20: S205-S220.

Reunanen A, Heliovaara M, Aho K (2003) Coffee consumption and risk of type 2 diabetes mellitus. Lancet 361: 702-3.

Rogers PJ. (2009). Caffeine, mood and mental performance ineveryday life. Nutr Bull. 32(Suppl. 1): 84–89.9.

Schwartz J., Weiss ST. (1992). Caffeine intake and asthma symptoms. Ann. Epidemiol.,2: 627-623.

Tavani A. (2004). Coffee, decaffeinated coffee, tea and cancer of the colon and rectum: A review of epidemiological studies. Cancer Caus. Cont., 15: 743-757.

Glossary of Medical Terms

A-   , an-: Lack of or without.

-ation: Indicates a process.

Abrasion: A cut or scrape that typically isn't serious.

Abscess: A tender, fluid-filled pocket that forms in tissue, usually due to infection.

Acute: A condition that comes on suddenly and is severe in nature, such as a broken bone or an asthma attack

AFR: Acute renal failure

a.c.: Before meals. As in taking medicine before meals.

a/g ratio: Albumin to globulin ratio.

ACL: Anterior cruciate ligament. ACL injuries are one of the most common ligament injuries to the knee. The ACL can be sprained or completely torn from trauma and/or degeneration.

Ad lib: At liberty. For example, a patient may be permitted to move out of bed freely and orders would, therefore, be for activities to be ad lib.

ADHD: Attention deficit hyperactivity disorder

ADR: Adverse drug reaction. If a patient is taking a prescription drug to treat high blood pressure disease

AIDS: Acquired immune deficiency syndrome

ALL: Acute lymphoblastic leukemia.

Anuric: Not producing urine. A person who is anuric is often critical and may require dialysis.

ANED: Alive no evidence of disease. The patient arrived in the ER alive with no evidence of disease.

ADH: Antidiuretic hormone

ARDS: Acute respiratory distress syndrome.

ARF: Acute renal (kidney) failure

ASCVD: Atherosclerotic cardiovascular disease. A form of heart disease.

Ageing Population: As people live longer, age-related diseases such as dementia and arthritis are becoming more prevalent.

ALS: Advanced life support. severe, but the duration is short.

Androgen replacement therapy (**ART**), often referred to as **testosterone replacement therapy** (**TRT**), is a form of hormone therapy in which androgens, often testosterone, are supplemented or replaced. It typically involves the administration of testosterone through injections, skin creams, patches, gels, pills, or subcutaneous pellets. ART is often prescribed to counter the effects of male hypogonadism.

Alkalinizing agentsare drugs used to manage disorders associated with low pH. For example, they may be used to treat acidosis due to kidney failure.

A**moisturizer**, or**emollient**, is a cosmetic preparation used for protecting, moisturizing, and lubricating the skin. These functions are normally performed by sebum produced by healthy skin. The word "emollient" is derived from the Latin verb*mollire*, to soften.

AKA: Above the knee amputation.

AMI: Acute myocardial infarction (heart attack)

Ambulatory: Refers to patients who are able to walk and are not bedridden.

Antipruritics,**abirritants**, or**anti-itch drugs**, are medications that inhibit the itching often associated with sunburns, allergic reactions, eczema, psoriasis, chickenpox, fungal infections, insect bites and stings like those from mosquitoes, fleas, and mites, and contact dermatitis and urticaria caused by

Antihistaminesare drugs which treat allergic rhinitis, common cold, influenza, and other allergies. Typically, people take antihistamines as an inexpensive, generic drug that can be bought without a prescription and provides relief from nasal congestion, sneezing, or hives caused by pollen, dust mites, or animal

Aromatase inhibitors(**AIs**) are a class of drugs used in the treatment of breast cancer in postmenopausal women and in men, and gynecomastia in men. They may also be used off-label to reduce estrogen conversion when supplementing testosterone exogenously. They may also be used for chemoprevention.

Anthropology: Study of how different cultures use plants.

b.i.d.: Twice daily. As in taking medicine twice daily.

B-ALL: B-cell acute lymphoblastic leukemia

Bibasilar: At the bases of both lungs. For example, someone with pneumonia in both lungs might have abnormal bibasilar breath sounds.

Biopiracy refers to biological resource theft by researchers or research groups without approval from the government. This frequently occurs when researchers take traditional knowledge without consent or take advantage of the cultures they are studying. As the people who found or developed the resources do not receive a portion of the earnings, biopiracy is regarded as unethical or comparable to theft.

BKA: Below the knee amputation.

BMP: Basic metabolic panel. Electrolytes (potassium, sodium, carbon dioxide, and chloride) and creatinine and glucose.

BP: Blood pressure. Blood pressure is recorded as part of the physical examination. It is one of the "vital signs."

BPD: Borderline personality disorder. A personality disorder.

BSO: Bilateral salpingo-oophorectomy. A BSO is the removal of both of the ovaries and adjacent Fallopian tubes and often is performed as part of a total abdominal hysterectomy.

Benign: Not cancerous.

Biopsy: A small sample of tissue that's taken for testing.

Biodiversity Conservation: By documenting the traditional uses of plants, ethnobotany contributes to the conservation of India's rich plant biodiversity.

Bioprospecting: The search for plant and animal species from which medicinal drugs and other commercially valuable compounds can be obtained.

BMI: Body mass index, a measure of body fat based on height and weight

C&S: Culture and sensitivity, performed to detect infection.

C/O: Complaint of. The patient's expressed concern.

cap: Capsule.

Ca: Cancer; carcinoma. For example, a patient who underwent treatment for cancer should assure that they are eating and drinking enough fluids daily, both during and after treatment.

CABG. Coronary artery bypass graft. A surgery involving the heart.

CBC: Complete blood count.

CC: Chief complaint. The patient's main concern.

cc: Cubic centimeters. For example, the amount of fluid removed from the body is recorded in ccs.

Chem panel: Chemistry panel. A comprehensive screening blood test that indicates the status of the liver, kidneys, and electrolytes.

CPAP: Continuous positive airway. Treatment for sleep apnea.

COPD: Chronic obstructive pulmonary disease.

CT: Chemotherapy. A type of treatment therapy for cancer.

CVA: Cerebrovascular accident (Stroke).

CDE: Complete dental (oral) evaluation.

Cancer: A group of diseases characterized by uncontrolled cell growth. Common types include breast, lung, and prostate cancer.

Cardi/o: Related to the heart.

Cardiovascular Diseases: These include heart attacks, strokes, and other heart-related conditions. They are the leading cause of death worldwide.

CBC (Complete Blood Count): A blood test that measures different components of the blood, including red and white blood cells.

Chronic: Signifies a recurring, persistent condition like heart disease.

Conservation: The protection and preservation of natural resources, including plants used in traditional medicine.

Contusion: A bruise.

Climate Change: Environmental changes are affecting health through increased heatwaves, natural disasters, and the spread of diseases

Clinical testing: The drug is used on people to confirm that it is safe to use.

Coughing: Can indicate respiratory infections or conditions like asthma. **Defibrillator**: A medical device that uses electric shocks to restore heart rythem.

Derm/a/o, dermat/o: Pertaining to the skin.

Diabetes: A chronic condition that affects how the body processes blood sugar.

Diarrhea: Often associated with gastrointestinal infections1.

Dementia: Growing public health issue as populations age, leading to significant care needs.

Discovery: The Drug Development process starts with Discovery, a process of identifying a new medicine.

Development: Chemicals extracted from natural products are used to make pills, capsules, or syrups for oral use. Injections for direct infusion into the blood drops for eyes or ears.

DM: Diabetes mellitus.

D/C or DC: Discontinue or discharge. For example, a doctor will D/C a drug. Alternatively, the doctor might DC a patient from the hospital.

DCIS: Ductal Carcinoma In Situ. A type of breast cancer. The patient is receiving treatment for Ductal Carcinoma.

DDX: Differential diagnosis. A variety of diagnostic possibilities are being considered to diagnose the type of cancer present in the patient.

DJD: Degenerative joint disease. Another term for osteoarthritis.

DNC, D&C, or D and C: Dilation and curettage. Widening the cervix and scrapping with a curette for the purpose of removing tissue lining the inner surface of the womb (uterus).

DNR: Do not resuscitate. This is a specific order not to revive a patient artificially if they succumb to illness. If a patient is given a DNR order, they are not resuscitated if they are near death and no code blue is called.

DOE: Dyspnea on exertion. Shortness of breath with activity.

DTR: Deep tendon reflexes. These are reflexes that the doctor tests by banging on the tendons with a rubber hammer.

DVT: Deep venous thrombosis (blood clot in a large vein).

Dys-: Abnormal, difficult, or painful.

ETOH: Alcohol. ETOH intake history is often recorded as part of patient history.

ECT: Electroconclusive therapy. A procedure used to control seizures (convulsions).

Edema: Swelling caused by fluid accumulation.

Embolism: An arterial blockage, often caused by a blood clot.

Epidermis: The outer layer of the skin.

EKG: Electrocardiogram, a way of monitoring the heart and testing for problems.

Encephal/o: Related to the brain.

Ethnobotany: The study of how people of a particular culture and region make use of indigenous plants.

Ethnology: Richard Evans Schultes is known as the **"Father of Ethnobotany."** The first person who defined ethnobotany was **John Harshberger** in the year **1895.** He defined it as "the study of plants used by primitive & aboriginal people." Ethnobotany is derived from "*Ethnology*" – study of culture and "*Botany*" – study of plants. Ethnobotany is a multidisciplinary field that examines the relationships between plants and human cultures.

Ethnopharmacology: The study of the medicinal use of plants by various ethnic groups.

Fatigue: Persistent tiredness not relieved by rest.

FDA Review: drug is sent to FDA before launching the drug into the market.

FDA post-Market Review: The drug is reviewed and monitored by FDA for the safety once it is available to the public.

Fever: Often a sign of infection.

Fracture: Broken bone or cartilage.

FX: Fracture.

FSH: Follicle-stimulating hormone. A blood test for follicle-stimulating hormone is used to evaluate fertility in women.

Gangreneis a type of tissue death caused by a lack of blood supply. Symptoms may include a change in skin color to red or black, numbness, swelling, pain, skin breakdown, and coolness. The feet and hands are most commonly affected.

Gonadotropinsare glycoprotein hormones secreted by gonadotropic cells of the anterior pituitary of vertebrates. This family includes the mammalian hormones follicle-stimulating hormone (FSH) and luteinizing hormone (LH), the placental/chorionic gonadotropins, human chorionic gonadotropin (hCG) and equine

Gastr/o: Related to the stomach.

Gland: An organ or tissue that produces and secretes fluids that serve a specific function.

HDL-C: High-density lipoprotein cholesterol, often called "good" cholesterol.

Headache: Can be a symptom of many conditions, including migraines and infections. **Hemat/o**: Pertaining to blood.

Horticulture: Study of management of useful plants i.e., fruits, vegetables or ornamental plants in orchards/home gardens.

Hypertension: High blood pressure, a condition where the force of the blood against the artery walls is too high.

Hypotension: Low blood pressure, where the blood pressure is lower than the normal range.

HPS: Hantavirus pulmonary syndrome. A type of contagious, infectious disease is transmitted by rats infected with the virus.

IBS: Irritable bowel syndrome (A medical disease that involves the gastrointestinal tract.)

IDDM: Insulin-dependent diabetes mellitus. Type 1 diabetes

Infectious Diseases: Diseases like HIV/AIDS, tuberculosis, and malaria continue to be major health challenges in many parts of the world.

Inpatient: A patient who requires hospitalization.

Intravenous: Indicates medication or fluid that's delivered by vein.

LDL-C: Low-density lipoprotein cholesterol, often called "bad" cholesterol.

Long COVID: Ongoing symptoms following COVID-19 infection, affecting daily life and requiring more research for effective treatments.

LyMacro-: Large in size.

Melan/o-: Black or dark in color.

tes: Electrolytes.

Malignant: Indicates the presence of cancerous cells.

Medicinal Plants: Plants used for medicinal purposes, often based on traditional practices

Mental Health Disorders: Conditions like depression, anxiety, and bipolar disorder are major contributors to disability and poor quality of life.

Micro-: Small in size.

-ology: The study of a particular concentration.

-osis: Indicates something that is abnormal.

-otomy: To cut into.

Muscle Aches: Common in viral infections like the flu.

My/o: Related to muscle.

Nosologyis the branch of medical science that deals with the classification of diseases. Fully classifying a medical condition requires knowing its cause, the effects it has on the body, the symptoms that are produced, and other factors. For example, influenza.

Obesity: A condition characterized by excessive body fat, which increases the risk of other health problems like heart disease, diabetes, and certain cancers. **Outpatient**: A patient who receives care without being admitted to a **- hospital**

Oste/o: Related to bone.

Paleobotany: It examines the fossil remains of plants, including algae, mosses, ferns, and higher plants like trees and flowering plants. It helps us to understand plant evolution, past climates, and interactions between plants and other organisms throughout that particular geological time.

Paleontology is the study of fossils. It is a scientific study of fossils or the prehistoric life that was present on the Earth billions of years ago. Paleontologists mainly study and analyze the fossils of those extinct species by using different physical, chemical, and biological processes or techniques.

Pathology: The study of diseases, their causes, and effects.

pathy: Disease or disease process.

-plasty: Surgical repair.

Pharmacognosy: The study of medicines derived from natural sources, including plants.

Phytochemistry: The branch of chemistry concerned with plants and plant

Potencyor**biological potency**is a measure of a drug's biological activity expressed in terms of the dose required to produce a pharmacological effect of given intensity. A highly potent drug evokes a given response at low concentrations, while a drug of lower potency evokes the same response only at higher concentrations. Higher potency does not necessarily mean greater effectiveness nor more side effects nor less side effects.

Poly-: Many.products.

Preclinical research: Drugs go under laboratory or animal testing, to ensure that they can be used on Humans.

Prognosis: The predicated outcome of disease progression and treatment.

Pseudo-: False or deceptive, usually in regard to appearance.

PT: Physical therapy, a type of treatment to help patients move and feel better

Pulmon/o: Refers to the lungs.

Rash: May indicate allergic reactions, infections, or autoimmune diseases. **Relapse**: Return of disease or symptoms after a patient has recovered.

Respiratory Infections: Lower respiratory tract infections, such as pneumonia, are significant causes of illness and death, particularly in children and the elderly.

Retro-: Behind or backward.

Rhin/o: Related to the nose.

Road Injuries: Traffic accidents are a leading cause of death and injury, particularly among young people.

Rx: Prescription, usually for medication but can also signify another treatment.

Shortness of Breath: Common in respiratory conditions like asthma or COVID-19.

Sclerosis: Hard or hardening.

Sustainable Use: Using plant resources in a way that meets current needs without compromising the ability of future generations to meet their needs. **Sutures**: Stitches, which are used to join tissues together as they heal.

Stasis: Slowing or stopping the flow of a bodily fluid.

Swelling: Often seen in infections, injuries, or chronic conditions like arthritis.

Traditional Knowledge: Knowledge, innovations, and practices of indigenous and local communities around the world.

Transplant: The removal of an organ or tissue from one body that is implanted into another.

TSH: Thyroid-stimulating hormone. A blood test for TSH is used to diagnose thyroid disease.

Vaccine: A substance that stimulates antibody production to provide immunity against disease.

Zoonotic disease: A disease that is transmissible from animals to humans.

Index